APPLIED CHEMISTRY AND CHEMICAL ENGINEERING

Volume 4

Experimental Techniques and
Methodical Developments

APPLIED CHEMISTRY AND CHEMICAL ENGINEERING

Volume 4

Experimental Techniques and
Methodical Developments

Edited by

A. K. Haghi, PhD
Lionello Pogliani, PhD
Eduardo A. Castro, PhD
Devrim Balköse, PhD
Omari V. Mukbaniani, PhD
Chin Hua Chia, PhD

Apple Academic Press Inc.
3333 Mistwell Crescent
Oakville, ON L6L 0A2 Canada

Apple Academic Press Inc.
9 Spinnaker Way
Waretown, NJ 08758 USA

Library and Archives Canada Cataloguing in Publication

Applied chemistry and chemical engineering / edited by A.K. Haghi, PhD, Devrim Balköse, PhD, Omari V. Mukbaniani, DSc, Andrew G. Mercader, PhD.
Includes bibliographical references and indexes.
Contents: Volume 1. Mathematical and analytical techniques --Volume 2. Principles, methodology, and evaluation methods --Volume 3. Interdisciplinary approaches to theory and modeling with applications --Volume 4. Experimental techniques and methodical developments --Volume 5. Research methodologies in modern chemistry and applied science.
Issued in print and electronic formats.
ISBN 978-1-77188-515-7 (v. 1 : hardcover).--ISBN 978-1-77188-558-4 (v. 2 : hardcover).--ISBN 978-1-77188-566-9 (v. 3 : hardcover).--ISBN 978-1-77188-587-4 (v. 4 : hardcover).--ISBN 978-1-77188-593-5 (v. 5 : hardcover).--ISBN 978-1-77188-594-2 (set : hardcover).
ISBN 978-1-315-36562-6 (v. 1 : PDF).--ISBN 978-1-315-20736-0 (v. 2 : PDF).-- ISBN 978-1-315-20734-6 (v. 3 : PDF).--ISBN 978-1-315-20763-6 (v. 4 : PDF).-- ISBN 978-1-315-19761-6 (v. 5 : PDF)
1. Chemistry, Technical. 2. Chemical engineering. I. Haghi, A. K., editor
TP145.A67 2017   660   C2017-906062-7   C2017-906063-5

Library of Congress Cataloging-in-Publication Data

Names: Haghi, A. K., editor.
Title: Applied chemistry and chemical engineering / editors, A.K. Haghi, PhD [and 3 others].
Description: Toronto ; New Jersey : Apple Academic Press, 2018- | Includes bibliographical references and index.
Identifiers: LCCN 2017041946 (print) | LCCN 2017042598 (ebook) | ISBN 9781315365626 (ebook) | ISBN 9781771885157 (hardcover : v. 1 : alk. paper)
Subjects: LCSH: Chemical engineering. | Chemistry, Technical.
Classification: LCC TP155 (ebook) | LCC TP155 .A67 2018 (print) | DDC 660--dc23
LC record available at https://lccn.loc.gov/2017041946

ABOUT THE EDITORS

A. K. Haghi, PhD

A. K. Haghi, PhD, holds a BSc in Urban and Environmental Engineering from the University of North Carolina (USA), an MSc in Mechanical Engineering from North Carolina A&T State University (USA), a DEA in applied mechanics, acoustics and materials from the Université de Technologie de Compiègne (France), and a PhD in engineering sciences from the Université de Franche-Comté (France). He is the author and editor of 165 books, as well as of 1000 published papers in various journals and conference proceedings. Dr. Haghi has received several grants, consulted for a number of major corporations, and is a frequent speaker to national and international audiences. Since 1983, he served as professor at several universities. He is currently Editor-in-Chief of the *International Journal of Chemoinformatics and Chemical Engineering* and the *Polymers Research Journal* and on the editorial boards of many international journals. He is also a member of the Canadian Research and Development Center of Sciences and Cultures (CRDCSC), Montreal, Quebec, Canada.

Lionello Pogliani, PhD

Lionello Pogliani, PhD, was Professor of Physical Chemistry at the University of Calabria, Italy. He studied Chemistry at Firenze University, Italy, and received his postdoctoral training at the Department of Molecular Biology of the C. E. A. (Centre d'Etudes Atomiques) of Saclay, France, the Physical Chemistry Institute of the Technical and Free University of Berlin, and the Pharmaceutical Department of the University of California, San Francisco, CA. Dr. Pogliani has coauthored an experimental work that was awarded the GM Neural Trauma Research Award. He spent his sabbatical years at the Centro de Química-Física Molecular of the Technical University of Lisbon, Portugal, and at the Department of Physical Chemistry of the Faculty of Pharmacy of the University of Valencia-Burjassot, Spain. He has contributed nearly 200 papers in the experimental, theoretical, and didactical fields of physical chemistry, including chapters in specialized books. He has also presented at more than 40 symposiums. He also published a book on the numbers 0, 1, 2, and 3. He is a member of the International Academy of

Mathematical Chemistry. He retired in 2011 and is part-time teammate at the University of Valencia-Burjassot, Spain.

Eduardo A. Castro, PhD

Eduardo A. Castro, PhD, is full professor in theoretical chemistry at the Universidad Nacional de La Plata and a career investigator with the Consejo Nacional de Investigaciones Cientificas y Tecnicas, both based in Buenos Aires, Argentina. He is the author of nearly 1000 academic papers in theoretical chemistry and other topics, and he has published several books. He serves on the editorial advisory boards of several chemistry journals and is often an invited speaker at international conferences in South America and elsewhere.

Devrim Balköse, PhD

Devrim Balköse, PhD, is currently a faculty member in the Chemical Engineering Department at the Izmir Institute of Technology, Izmir, Turkey. She graduated from the Middle East Technical University in Ankara, Turkey, with a degree in Chemical Engineering. She received her MS and PhD degrees from Ege University, Izmir, Turkey, in 1974 and 1977, respectively. She became Associate Professor in Macromolecular Chemistry in 1983 and Professor in process and reactor engineering in 1990. She worked as Research Assistant, Assistant Professor, Associate Professor, and Professor between 1970 and 2000 at Ege University. She was the Head of the Chemical Engineering Department at the Izmir Institute of Technology, Izmir, Turkey, between 2000 and 2009. Her research interests are in polymer reaction engineering, polymer foams and films, adsorbent development, and moisture sorption. Her research projects are on nanosized zinc borate production, ZnO polymer composites, zinc borate lubricants, antistatic additives, and metal soaps.

Omari V. Mukbaniani, DSc

Omari Vasilii Mukbaniani, DSc, is Professor and Head of the Macromolecular Chemistry Department of Iv. Javakhishvili Tbilisi State University, Tbilisi, Georgia. He is also the Director of the Institute of Macromolecular Chemistry and Polymeric Materials. He is a member of the Academy of Natural Sciences of the Georgian Republic. For several years he was a member of the advisory board of the *Journal Proceedings of Iv. Javakhishvili Tbilisi State*

University (Chemical Series) and contributing editor of the journal *Polymer News* and the *Polymers Research Journal*. He is a member of editorial board of the *Journal of Chemistry and Chemical Technology*. His research interests include polymer chemistry, polymeric materials, and chemistry of organosilicon compounds. He is an author more than 420 publications, 13 books, four monographs, and 10 inventions. He created in the 2007s the "International Caucasian Symposium on Polymers & Advanced Materials," ICSP, which takes place every other two years in Georgia.

Chin Hua Chia, PhD

Chin Hua Chia, PhD, is an Associate Professor at the School of Applied Physics, Faculty of Science and Technology at Universiti Kebangsaan Malaysia (National University of Malaysia), Malaysia. A recipient of the Young Scientist Award from the National University of Malaysia and the Malaysian Solid State Science and Technology (MASS) in 2012 and 2014, respectively, he is a member of several professional organizations and has published several book chapters and more than 100 articles in professional journals as well as has presented at many professional meetings. His core research interests include polymer, nanocomposites, chemistry of lignocellulosic materials, and magnetic nanomaterials.

Applied Chemistry and Chemical Engineering, 5 Volumes

Applied Chemistry and Chemical Engineering,
Volume 1: Mathematical and Analytical Techniques
Editors: A. K. Haghi, PhD, Devrim Balköse, PhD, Omari V. Mukbaniani, DSc, and
Andrew G. Mercader, PhD

Applied Chemistry and Chemical Engineering,
Volume 2: Principles, Methodology, and Evaluation Methods
Editors: A. K. Haghi, PhD, Lionello Pogliani, PhD, Devrim Balköse, PhD,
Omari V. Mukbaniani, DSc, and Andrew G. Mercader, PhD

Applied Chemistry and Chemical Engineering,
Volume 3: Interdisciplinary Approaches to Theory and Modeling with
Applications
Editors: A. K. Haghi, PhD, Lionello Pogliani, PhD, Francisco Torrens, PhD,
Devrim Balköse, PhD, Omari V. Mukbaniani, DSc, and Andrew G. Mercader, PhD

Applied Chemistry and Chemical Engineering,
Volume 4: Experimental Techniques and Methodical Developments
Editors: A. K. Haghi, PhD, Lionello Pogliani, PhD, Eduardo A. Castro, PhD,
Devrim Balköse, PhD, Omari V. Mukbaniani, PhD, and Chin Hua Chia, PhD

Applied Chemistry and Chemical Engineering,
Volume 5: Research Methodologies in Modern Chemistry and Applied Science
Editors: A. K. Haghi, PhD, Ana Cristina Faria Ribeiro, PhD,
Lionello Pogliani, PhD, Devrim Balköse, PhD, Francisco Torrens, PhD,
and Omari V. Mukbaniani, PhD

CONTENTS

LIST OF CONTRIBUTORS

A. M. M. Ali
Faculty of Applied Sciences, Universiti Teknologi MARA (UiTM), 40450 Shah Alam, Selangor, Malaysia

K. T. Archvadze
Food Industry Department, Georgian Technical University, 77 Kostava 0175, Tbilisi, Georgia
Department of Technology, Sukhumi State University, Ana Politkobskaia 9, 0186, Tbilisi, Georgia

Devrim Balköse
Department of Chemical Engineering, İzmir Institute of Technology, Gulbahçe Urla, Izmir, Turkey.
E-mail: devrimbalkose@gmail.com

Bogdana Bashta
Lviv Polytechnic National University, 12, St. Bandera Str., Lviv 79013, Ukraine

Emili Besalú
Departament de Química, Institut de Química Computacional i Catàlisi (IQCC), Universitat de Girona, 17003 Girona, Catalonia, Spain

Vladimir I. Binyukov
Emanuel Institute of Biochemical Physics, Russian Academy of Sciences, 4, Kosygin Str., Moscow 119334, Russia

Michael Bratychak
Lviv Polytechnic National University, 12, St. Bandera Str., Lviv 79013, Ukraine. E-mail: mbratych@polynet.lviv.ua

I. R. Chachava
Food Industry Department, Georgian Technical University, 77 Kostava 0175, Tbilisi, Georgia
Department of Technology, Sukhumi State University, Ana Politkobskaia 9, 0186, Tbilisi, Georgia

Tanmoy Chakraborty
Department of Chemistry, Manipal University Jaipur, Dehmi-Kalan, Jaipur 303007, India.
E-mail: tanmoy.chakraborty@jaipur.manipal.edu; tanmoychem@gmail.com

C. H. Chan
Faculty of Applied Sciences, Universiti Teknologi MARA (UiTM), 40450 Shah Alam, Selangor, Malaysia. E-mail: cchan_25@yahoo.com.sg

E. Chkhaidze
Faculty of Chemical Technology and Metallurgy, Georgian Technical University, Tbilisi, Georgia.
E-mail: ekachkhaidze@yahoo.com

Z. N. Chubinishvili
TSU Petre Melikishvili Institute of Physical and Organic Chemistry, 31 A. Politkovskaia Str., 0186 Tbilisi, Georgia

M. B. Gurgenishvili
TSU Petre Melikishvili Institute of Physical and Organic Chemistry, 31 A. Politkovskaia Str., 0186 Tbilisi, Georgia. E-mail: marina.gurgenishvili@yahoo.com

Jozef Haponiuk
Gdansk University of Technology, 11/12 G. Narutowicza Str., 80233 Gdansk, Poland

Ostap Ivashkiv
Lviv Polytechnic National University, 12, St. Bandera str., Lviv 79013, Ukraine

Ajith James Jose
Research Department of Chemistry, St. Berchmans College, Changanassery, Kerala, India

J. Vicente Julian-Ortiz
Departamento de Química Física, Unidad de Investigación de Diseño de Fármacos y Conectividad Molecular, Facultad de Farmacia, Universitat de València, Burjassot (València), Spain, and MOLware SL, Valencia, Spain. E-mail: jejuor@uv.es

D. Kharadze
Ivane Beritashvili Center of Experimental Biomedicine

N. Z. Khotenashvili
TSU Petre Melikishvili Institute of Physical and Organic Chemistry, 31 A. Politkovskaia Str., 0186 Tbilisi, Georgia

Ajay Kumar
Department of Mechatronics Engineering, Manipal University Jaipur, Dehmi-Kalan, Jaipur 303007, India

R. G. Liparteliani
TSU Petre Melikishvili Institute of Physical and Organic Chemistry, 31 A. Politkovskaia Str., 0186 Tbilisi, Georgia

Ludmila I. Matienko
Emanuel Institute of Biochemical Physics, Russian Academy of Sciences, 4, Kosygin Str., Moscow 119334, Russia. E-mail: matienko@sky.chph.ras.ru

T. I. Megrelidze
Food Industry Department, Georgian Technical University, 77 Kostava 0175, Tbilisi, Georgia
Department of Technology, Sukhumi State University, Ana Politkobskaia 9, 0186, Tbilisi, Georgia

Elena M. Mil
Emanuel Institute of Biochemical Physics, Russian Academy of Sciences, 4, Kosygin Str., Moscow 119334, Russia

Larisa A. Mosolova
Emanuel Institute of Biochemical Physics, Russian Academy of Sciences, 4, Kosygin Str., Moscow 119334, Russia

Güler Narin
Department of Chemical Engineering, Uşak University, Uşak, Turkey. E-mail: gulernarin@gmail.com

Filiz Özmıhçı Ömürlü
Department of Chemical Engineering, İzmir Institute of Technology, Gulbahçe Urla, Izmir, Turkey. E-mail: filizozmihci@gmail.com

Sukanchan Palit
Department of Chemical Engineering, University of Petroleum and Energy Studies, Energy Acres, Post Office Bidholi via Premnagar, Dehradun 248007, Uttarakhand, India. E-mail: sukanchan68@gmail.com, sukanchan92@gmail.com

G. Sh. Papava
TSU Petre Melikishvili Institute of Physical and Organic Chemistry, 31 A. Politkovskaia Str., 0186 Tbilisi, Georgia

Sh. R. Papava
TSU Petre Melikishvili Institute of Physical and Organic Chemistry, 31 A. Politkovskaia Str., 0186 Tbilisi, Georgia

Lionello Pogliani
Departamento de Química Física, Unidad de Investigación de Diseño de Fármacos y Conectividad Molecular, Facultad de Farmacia, Universitat de València, Burjassot (València), Spain, and MOLware SL, Valencia, Spain. E-mail: liopo@uv.es

S. Rafiei
Department of Textile Engineering, University of Guilan, Rasht, Iran

H. Ramli
Centre of Foundation Studies, Universiti Teknologi MARA, Cawangan Selangor, Kampus Dengkil, 43800 Dengkil, Selangor, Malaysia

Prabhat Ranjan
Department of Mechatronics Engineering, Manipal University Jaipur, Dehmi-Kalan, Jaipur 303007, India

Olena Shyshchak
Lviv Polytechnic National University, 12, St. Bandera Str., Lviv 79013, Ukraine

Heru Susanto
Department of Information Management, College of Management, Tunghai University, Taichung, Taiwan

L. V. Tabatadze
Food Industry Department, Georgian Technical University, 77 Kostava 0175, Tbilisi, Georgia
Department of Technology, Sukhumi State University, Ana Politkobskaia 9, 0186, Tbilisi, Georgia

Z. Sh. Tabukashvili
TSU Petre Melikishvili Institute of Physical and Organic Chemistry, 31 A. Politkovskaia Str., 0186 Tbilisi, Georgia

Paulose Thomas
Optoelectronic Lab, St. Berchmans College, Changanassery, Kerala, India

Aysun Topaloğlu
Department of Chemical Engineering, Izmir Institute of Technology, Gulbahce 35430, Urla Izmir, Turkey

Senem Yetgin
Food Engineering Department, Kastamonu University, Kastamonu, Turkey

Gennady E. Zaikov
Emanuel Institute of Biochemical Physics, Russian Academy of Sciences, 4, Kosygin Str., Moscow 119334, Russia

N. F. A. Zainal
Centre of Foundation Studies, Universiti Teknologi MARA Cawangan Selangor, Kampus Dengkil, 43800 Dengkil, Selangor, Malaysia

LIST OF ABBREVIATIONS

ACGH	array comparative genomic hybridization
ADC	azodicarbonamide
ADRs	adverse drug reactions
AFM	atomic force microscopy
AIDS	acquired immune deficiency syndrome
ANN	artificial neural network
ARDs	acireductone dioxygenases
ATR	attenuated total reflectance
AU-ROC	area under ROC
BET	Brunauer–Emmett–Teller
BL2O	balanced leave-two-out
CAR	chimeric antigen receptor
$CaSt_2$	calcium stearate
CBAs	chemical blowing agents
CDD	charged-couple devices
CFs	core functions
CO	carbon monoxide
COs	core orbitals
DDBJ	DNA database of Japan
DFT	density functional theory
DMF	dimethyl formamide
DNA	deoxyribose nucleic acid
DoEs	design of experiments
DOP	dioctyl phthalate
D–R	Dubinin–Radushkevich
DRIFT	diffuse reflectance infrared Fourier transform
DSC	differential scanning calorimeter
DTG	derivative TG
EHRs	electronic health records
EMBL	European Molecular Biology Laboratory
EMR	electronic medical records
ENR	epoxidized natural rubbers
FCC	fluid catalytic cracking
FCCU	fluidized catalytic cracking unit

FDR	false discovery rate
FE-SEM	field emission scanning electron microscopy
FIB	focused ion beam
FTIR	Fourier-transform infrared
FWO	Flynn–Wall–Ozawa
GPC	gel permeation chromatography
HAO	hydroxy-acrylic oligomer
HAP	hydroxyapatite
HIV	human immunodeficiency virus
HMPA	hexamethylphosphorotriamide
HNCO	cyanic acid
HO	heme oxygenase
HOMO	highest occupied molecular orbital
IT	information technology
K	kaolinite
KAS	Kissinger–Akahira–Sunose
KMTB	2-keto-4-(thiomethyl)butyrate
KoARD	Klebsiella oxytoca ARD
L1O	leave-one-out
LBHBs	low-barrier hydrogen bonds
LDPE	low-density polyethylene
Li	lithium
LP	linear programming
LPG	liquefied petroleum gas
LSDA	local spin density approximation
LUMO	lowest unoccupied molecular orbital
M	montmorillonite
MOEAs	multiobjective evolutionary algorithms
MOO	multiobjective optimization
MP	methylpyrrolidone
MPP	melamine polyphosphate
MSP	methionine salvage pathway
MTA	methylthioadenosine
N_2	nitrogen
NCBI	National Center for Biotechnology Information
NGS	next-generation sequencing
$NH_2CONHNHCONH_2$	hydrazodicarbinamide
NH_3	ammonia
NHCONHCONH	urazol

NMR	nuclear magnetic resonance
NOM	natural organic matter
NR	natural rubber
PAC	polyacrylate
PAN	polyacrylonitrile
PBAs	physical blowing agents
PCL	poly(ϵ-caprolactone)
PDB	protein data bank
PEO	poly(ethylene oxide)
pHEMA	poly(2-hydroxyethyl methacrylate)
PMA	poly(methyl acrylate)
PMMA	poly(methyl methacrylate)
PnBMA	poly(n-butyl methacrylate)
PPO	polyphenylene oxide
PPy	polypyrrole
PSPD	position-sensitive photo diode
PVC	polyvinyl chloride
RPM	revolution per minute
SaaS	Software as a Service
SAM	S-adenosyl methionine
SARS	severe acute respiratory syndrome
SCD	supercritical drying
SCF	supercritical fluid
SEM	scanning electron microscopy
SFOs	symmetrized fragment orbitals
SP	sodium perchlorate
SPEs	solid polymer electrolytes
SSIRs	superposing significant interaction rules
STEM	scanning transmission electron microscopy
SWNT	single-walled carbon nanotubes
TEM	transmission electron microscopy
TEOS	tetraethoxysilane
TG	thermogravimetric
TGA	thermogravimetric analysis
THF	tetrahydrofuran
TIBO	tetrahydroimidazo[4,5,1-jk][1,4]benzodiazepinone
TMOS	tetramethoxysilane
UHR-SEM	ultrahigh resolution-SEM
wpn	water molecules per nucleotide

XRD	X-ray diffraction
X-RD	X-ray diffractometry
XRF	X-ray fluorescence
ZnSt2	zinc stearate
USPEA	unsaturated/saturated poly(ester amide)s

PREFACE

Experimental techniques and methodical developments provide a detailed yet easy-to-follow treatment of various techniques useful for characterizing the structure and properties of engineering materials. With an emphasis on techniques most commonly used in laboratories, this book enables researchers and students to derive the maximum possible information from the experimental results obtained.

Not only does this book summarize the classical theories, but also it exhibits their engineering applications in response to the current key issues.

This timely volume provides an overview of new methods and presents experimental research in applied chemistry using modern approaches. Each chapter describes the principle of the respective method, as well as the detailed procedures of experiments with examples of actual applications and demonstrates the advantages and disadvantages of each physical technique. Thus, readers will be able to apply the concepts as described in the book to their own experiments.

This book introduces the current state-of-the-art technology in key engineering materials with an emphasis on the rapidly growing technologies. It takes a unique approach by presenting specific materials, then progresses into a discussion of the ways in which these novel materials and processes are integrated into today's functioning manufacturing industry.

It follows a more quantitative and design-oriented approach than other texts in the market, helping readers gain a better understanding of important concepts. They will also discover how engineering material properties relate to the process variables in a given process as well as how to perform quantitative engineering analysis of manufacturing processes.

This volume:

- Highlights some important areas of current interest in applied chemistry and chemical engineering
- Provides an up-to-date and thorough exposition of the present state of the art of applied chemistry and chemical engineering
- Describes the types of techniques now available to engineers and technicians, and discusses their capabilities, limitations, and applications

- Provides a balance between materials science and chemical aspects, basic and applied research, and high technology and high volume (low cost) composite development.
- Explains modification methods for changing of different materials properties
- Presents and reviews the necessary theoretical and background details
- Provides a detailed description of the experiment to be conducted and how the data could be tabulated and interpreted.

This volume presents research and reviews and information on implementing and sustaining interdisciplinary studies in science, technology, engineering, and mathematics.

Applied Chemistry and Chemical Engineering, 5-Volume Set includes the following volumes:

- Applied Chemistry and Chemical Engineering, Volume 1: Mathematical and Analytical Techniques
- Applied Chemistry and Chemical Engineering, Volume 2: Principles, Methodology, and Evaluation Methods
- Applied Chemistry and Chemical Engineering, Volume 3: Interdisciplinary Approaches to Theory and Modeling with Applications
- Applied Chemistry and Chemical Engineering, Volume 4: Experimental Techniques and Methodical Developments
- Applied Chemistry and Chemical Engineering, Volume 5: Research Methodologies in Modern Chemistry and Applied Science.

PART I
Polymer Chemistry and Technology

CHAPTER 1

SYNTHESIS OF UNSATURATED BIODEGRADABLE POLY (ESTER AMIDE)S AND STUDY OF THEIR THERMAL AND MECHANICAL PROPERTIES

E. CHKHAIDZE[1,*] and D. KHARADZE[2]

[1]Faculty of Chemical Technology and Metallurgy, Georgian Technical University, Tbilisi, Georgia

[2]Ivane Beritashvili Center of Experimental Biomedicine, Faculty of Chemical Technology and Metallurgy, Georgian Technical University, Tbilisi, Georgia

*Corresponding author. E-mail: ekachkhaidze@yahoo.com

CONTENTS

ABSTRACT

Biodegradable poly(ester amide)s of different structures containing unsaturated double bonds in main chain are received: homopoly(ester amide)s (UPEA) with 100% content of fumaric acid residues, unsaturated/saturated poly(ester amide)s (USPEA) with <100% content of fumaric acid residues, and saturated copoly(ester amide)s of L-leucine and L-phenylalanine with 100% content of fumaric acid residues. Their thermal properties are studied with the purpose of determination of application area and possibilities of cross-linking of unsaturated poly(ester amide)s using UV irradiation are shown.

1.1   INTRODUCTION

Polymers of biomedical designation are actively used today in different areas of medicine: as surgical, dental, orthopedical materials, in the form of medications for capsulation, for system of targeted delivery of medications, etc. At the same time, demand for polymer materials with complex of specific properties rapidly grows, since it is easily possible that these materials can possibly solve many significant problems of medicine. Among biomedical polymers of different classes, one of the key positions is held by AABB-structure poly(ester amide)s on the basis of natural amino acids,[1–6] which are characterized by interesting complex of properties: high skills of biodegradation and biocompatibility with tissues, hydrophylity, good material properties, etc. Getting of even more diverse poly(ester amide)s with predetermined properties is possible via their functionalization, with insertion of chemically active groups, long hydrophilic chain into polymers, etc.[7] One of the prospective ways of polymers functionalization is represented by synthesis of macromolecules containing unsaturated bonds both in main chain and in lateral chains.[8] Addition of desirable functional groups to unsaturated bonds, as well as reactions of multiple grafting, copolymerization, hybridization with other unsaturated polymers, for example, unsaturated polysaccharides (acryl-dextran, etc.), synthesis of multifunctional biodegradable hydrogels can be implemented, as it was made by Chu and colleagues,[9,10] getting of three-dimensional biodegradable systems with introduction of unsaturated bonds into biodegradable macromolecule of poly(ester amide)s is possible, etc.

Taking into account the abovementioned, we set a goal to insert unsaturated bonds into AABB-type poly(ester amide)s with the purpose of their functionalization and respective extension of application area. Introduction of unsaturated bonds into main chain of polymers was implemented by us using *fumaric acid*—unsaturated dicarboxylic acid that was main monomer of the work. It should be noted that due to significant place hold by fumaric acid in polymer industry and taking into account oil price increase in the last period, at this moment, the method of enzymatic synthesis of fumaric acid (*Rhizopus species*) with glucose is elaborated. Fumaric acid received using this method is three times cheaper compared with the same product obtained from oil. It is also noteworthy the circumstance that fermentation process proceeds with fixation of carbon dioxide.[11]

Unsaturated polyesters (polyester resins), which find extensive application in the last period, are the most widely used in practice next to conducting the cross-linking reaction of rubber containing unsaturated bonds. Biomedicine is one of the most interesting application areas of different classes of polymers containing unsaturated bonds. Aliphatic, unsaturated, biodegradable polyesters were tested in bone tissue repair surgery—in the form of hardening binding materials for receipt of bone prostheses. Structure of polymeric binding materials is selected such a way so that mutually opposite processes of polymer degradation/bone growth could continue until complete recovery of tissue. Application of unsaturated biodegradable polyesters and polyanhydrides is possible for construction of effective medication delivery system.[9,12–16]

As a result of conducted researches on the basis of natural amino acids, we received functional, biodegradable poly(ester amide)s containing different amount of unsaturated bonds in main chain. We studied their thermal and mechanical properties, and that allows us to identify the areas of their application.

1.2 SYNTHESIS OF POLYMERS

For receipt of functional polymers containing unsaturated bonds in main chain, we carried out a synthesis of di-*p*-nitrophenyl fumarate—*bis*-electrophyllic monomer—unsaturated activated diester of fumaric acid (Scheme 1.1) using two methods: acceptor-catalytic method in the solution and interphase method.

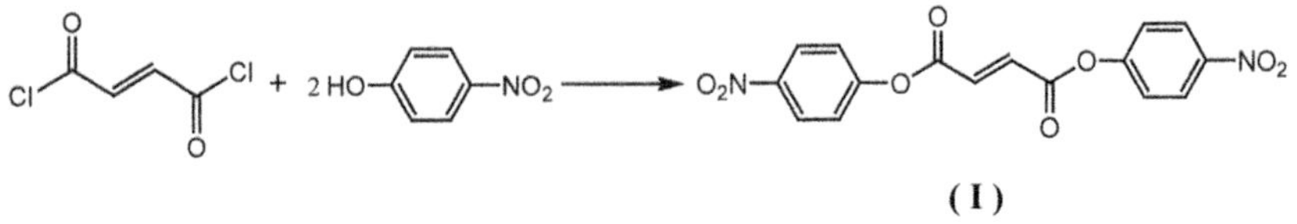

(I)

SCHEME 1.1

Synthesis of di-*p*-nitrophenyl ether of fumaric acid using interphase method gives us an opportunity to synthesize high-purity monomer with minimum losses. For purification of received product up to polycondensation purity, only single recrystallizing turned to be enough. Di-*p*-nitrophenyl ethers of saturated acids (e.g., sebacic acid) were also used by us as *bis*-electrophilic monomers. Synthesis of fumaric acid and di-*p*-nitrophenyl ether is described in details in our previous work.[17,18] Characteristics of this compound are given there, too.

Synthesis of di-*p*-toluene sulfonic acid salts of *bis*(α-amino acid) α,ω-alkylene diesters was carried out by us through direct condensation of free α-amino acids with α,ω-diols in the presence of monohydrate of *p*-toluene sulfonic acid (Scheme 1.2).

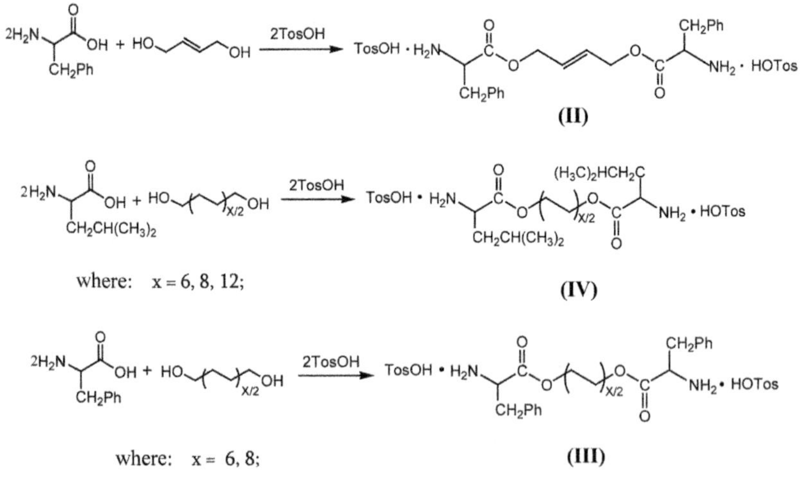

where: x = 6, 8, 12;   (IV)

where: x = 6, 8;   (III)

SCHEME 1.2

Salts were received with high yield (75–99%). Probable structures of obtained compounds were identified using FTIR and ^{1}H NMR spectral and elemental analyses.[17,18]

For synthesis of targeted polymers, we used method of so-called activated polycondensation.[19,20] According to this method, the polycondensation reaction implements under soft conditions that are necessary for getting rid of early polymerization of double bonds.

Using polycondensation of obtained monomers, we got saturated poly(ester amide)s of different structure (Scheme 1.3):

1. Unsaturated homopoly(ester amide)s on the basis of L-phenylalanine and unsaturated diol with 100% content of fumaric acid residues (UPEA) (**V**), via polycondensation of (**I**) and (**II**) compounds.
2. Unsaturated homopoly(ester amide)s with 100% content of fumaric acid residues (UPEA) (**VI a, b**) on the basis of L-phenylalanine (**III**) or L-leucine (**IV**) via polycondensation of obtained ethers with (**I**).
3. Unsaturated/saturated poly(ester amide)s on the basis of L-leucine with <100% content of fumaric acid residues (USPEA) (**IV**) via polycondensation of compound with (**I**) and ethers of di-*p*-nitrophenyl of sebacic acid.
4. Unsaturated copoly(ester amide)s on the basis of L-leucine and L-phenylalanine with 100% content of fumaric acid residues (co-UPEA) (**III**) via polycondensation of and (**IV**) compounds with (**I**).

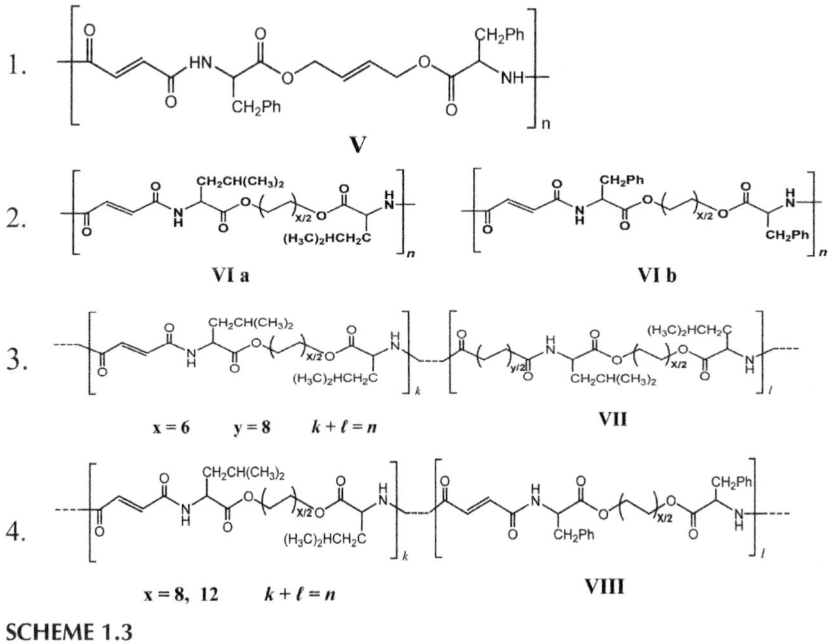

SCHEME 1.3

Synthesis and properties of each of them are described in details in Chkhaidze et al. (2011).[17]

Also, it is necessary to explain polymers designations. General formula of UPEAs, saturated homopoly(ester amide)s, is FA-[AA-x]. FA denotes polymers obtained on the basis of fumaric acid. Polymers contain fumaric acid residues –CO–CH=CH–CO– in main chain of polymers. AA denotes amino acids: L-phenylalanine or L-leucine, x is the number of methylene groups of diol and it is equal to 6, 8, and 12; for unsaturated diols, 2-butene-diol residue is denoted by 4en.

General formula of UPEAs, polymers received on the basis of fumaric acid and saturated acids, is FA/y$_{k/l}$-[AA-x]. FA denotes fumaric acid residues, y is the number of methylene groups in saturated acids –CO–(CH$_2$)$_y$–CO–, and y = 8. AA denotes L-phenylalanine or L-leucine, x is the number of methylene groups of diol and it is equal to 6. k and l are molar fractions of fumaric acid and saturated acid in initial reaction mixture, respectively.

General formula of co-UPEA, copolymers obtained on the basis of fumaric acid, is FA-[Leu-x]$_k$-[Phe-x]$_l$, FA denotes fumaric acid residues, Leu and Phe are amino acids, x is the number of methylene groups of diols, k and l is the ratio of monomers containing leucine and phenylalanines in terms of mol/mol. For instance, FA-Leu-12 denotes UPEA, which consists of fumaric acid and Leu-12, FA/8$_{40/60}$-Leu-6 denotes USPEA, which consists of 40 mol% of fumaric acid and 60 mol% of sebacic acid. FA-[Leu-6]$_{0.75}$-[Phe-6]$_{0.25}$ denotes co-UPEA, which consists of 100 mol% of fumaric acid, as well as 75 mol% of Leu-6 monomer (**IV**) and 25 mol% of Phe-6 monomer (**III**).

We studied properties of obtained polymers: solvability in organic solvents, biodegradation skills (enzyme hydrolysis) using gravity method in vitro experiments, under conditions most closely resembling to physiological conditions (pH 7.4; 37°C) with enzymes α-chymotrypsin and lipase. Reactions of polymers' chemical transformation were conducted by us using thioglycolic acid, 2-mercaptoethanole, and amino acid—β-alanine. All the abovementioned is described in the work.[17]

It should be noted that for polymers of biomedical designation, except the above-listed properties, it is necessary to study thermal and mechanical properties, since they have significant impact on other properties of polymers and also they determine their application area. For instance, biodegradation of crystalline polymers proceeds at far less rates than that of amorphous polymers. Polymers with high vitrification temperatures are prospective compounds in the form of extracted surgical and construction materials, while polymers with low vitrification temperatures are used in the form of artificial leather, stent coating, for getting of medication-controlled extraction systems, and polymers with good mechanical characteristics are prospective generally as surgical materials, reconstructive orthopedic surgery, for preparation of blood vessels stent, etc.

1.3 THERMAL PROPERTIES

Thermal properties of polymers were studied by us using differentiated scanning calorimetry (DSC). Measurements were conducted with the use of calorimeter "Perkin-Elmer DSC-7" in nitrogen medium, in the temperature range 20–200°C; heating rate 10K/min. Part of work we implemented by means of calorimeter "NETZSCH" DSC 200 PC" "PHOX," in the temperature range 20–150°C. Heating rate 10K/min. In general, assessment of thermal properties of polymers is possible by means of vitrification temperatures (T_g), change of specific heat capacity (Δc_p), melting temperature (T_m), and change of enthalpy (ΔH).[21,22] In case of limitation of intra- and extramolecular flexibility of polymeric chains, their T_g increases. In its turn, Δc_p shows us what conformational changes have taken place in amorphous part and gives us an opportunity to make certain conclusions on molecular interactions in polymers. Existence of T_m points at formation of ordered (probably crystalline) structure in polymers. Thermal characteristics of studied polymers are given in Table 1.1, while some DSC curves are presented in Figures 1.1–1.4. Thermal characteristics of polymers of all synthesized groups are given in the table. Thermal properties of analogues of saturated structures of some of them are also given as a comparison.

According to obtained results, we can conclude that unsaturated poly(ester amide)s are characterized by relatively high vitrification temperature T_g in comparison with corresponding saturated poly(ester amide)s that was anticipated and is related to increased rigidity of their chains (in other words, less mobility).

For instance, unsaturated polyester amide FA-Phe-6 compared with saturated 8-FA-Phe-6 of similar structure is characterized by ~70°C higher vitrification temperature (Table 1.1, samples N2, N3). FA-Phe-6 contains double unsaturated C=C bonds in diamide part of macromolecule, which precondition rigidity of main chain of macromolecule. In comparison with FA-Phe-6,

FA-Phe-4en, which contains C=C bonds both in diamide and in diester parts, has even higher vitrification temperature (Table 1.1, sample N1).

Poly(ester amide)s on the basis of amino acid leucine are characterized by generally more flexible chains compared with poly(ester amide)s of similar structure, which are received on the basis of phenylalanine.[17] In this case, insertion of unsaturated bonds in amide part of polymers has no significant impact on vitrification temperature of polymers (T_g) (Table 1.1, samples N4, N5), while it has bigger impact on melting temperature of polymers (increases by 50°C) (Figs. 1.1 and 1.2) and on degree of order. Though it should be noted that as a result of second scanning, that is, repeated measurement, this difference is leveled off.

TABLE 1.1 Thermal Characteristics of Homopoly(ester amide)s, Unsaturated/Saturated Poly(ester amide)s and Copoly(ester amide)s.

No.	Polymer	Scan I*			
		Tg (°C)	ΔCp (J/g*K)	Tm (°C)	ΔH (J/g)
1	FA-Phe-4en	109	–	~223	–
2	FA-Phe-6	92	–	~216	–
3	8-Phe-6	21.00	1.04	54.50	15.48
4	8-Leu-12	21.8	0.369	85.1	50.67
5	FA-Leu-12	22.2	0.573	137.9	30.38
6	FA-Leu-12 + phthalic anh	24.7	1.632	102.4	–
7	FA-Leu-12 + phthalic anh N6—after 3 months)	28.1	0.91	121.3	21.74
8	FA/8$_{25/75}$-Leu-6	32.9	1.189	–	–
9	FA/8$_{40/60}$-Leu-6	45.4	1.873	–	–
10	FA-[Leu-8]$_{0.75}$-[Phe-8]$_{0.25}$	27.2	2.831	–	–
11	FA-[Leu-12]$_{0.75}$-[Phe-8]$_{0.25}$	15.5	0.201	–	–

*Scan I—I measurement.

There is no significant difference between vitrification temperature of FA-Leu-12 blocked and unblocked with phthalimide (Table 1.1, samples N5, N6, and N7). It should be noted that blocking of side groups was implemented with the purpose of getting rid of cross-linking reaction. Certainly, though slight rise of T_g is observed after 3-month delay of FA-Leu-12 blocked by phthalimide (Samples N6, N7), presumably due to structural changes.

As was expected, increase of content of fumaric acid residues in unsaturated/saturated polymers FA/8$_{25/75}$-Leu-6 and FA/8$_{40/60}$-Leu-6 (samples N8 and N9) caused rise of vitrification temperature of polymers (T_g) that is expected result, since chain length increases, too. Change in specific heat capacity (Δc_p) of polymers points at the same process. Together with rise of vitrification temperature, the specific heat capacity and degree of order of polymers also increases.

Comparison of N10 and N11 samples of copolymers allows us to make conclusion that increase of length of methylene chain of diol reduces vitrification temperature of polymers (T_g), their specific heat capacity (Δc_p) and, respectively, increases amorphism degree.

Basically, only single value of vitrification temperature is received at the thermogram of polymers (Table 1.1, Figs. 1.1–1.4), that is, in the synthesis process amorphous polymers of nondomain structure are formed. Major part of poly(ester amide)s obtained on the basis of leucine is characterized by low vitrification temperature that is significant circumstance from the viewpoint of their processing.

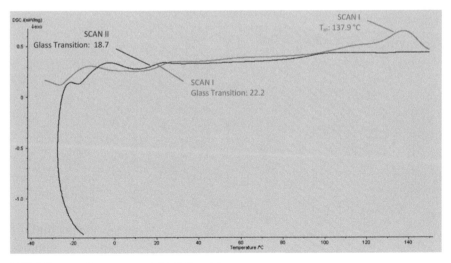

FIGURE 1.1   Thermogram of FA-L-Leu-12.

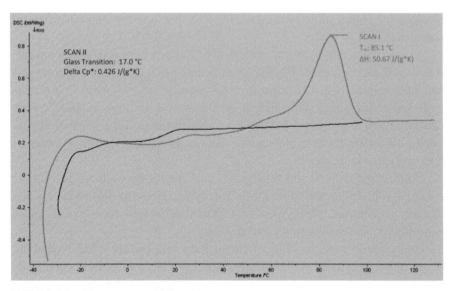

FIGURE 1.2   Thermogram of 8-Leu-12.

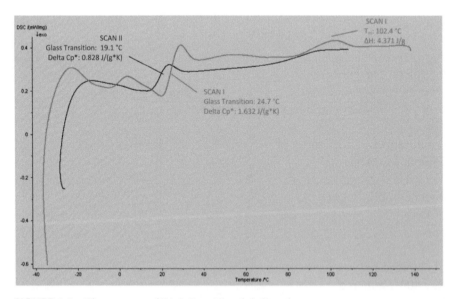

FIGURE 1.3   Thermogram of FA-L-Leu-12 + phthalic anh.

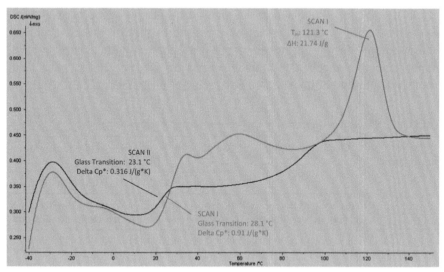

FIGURE 1.4 Thermogram of FA-L-Leu-12 + phthalic anh (after 3-month delay).

1.4 MECHANICAL PROPERTIES

As we mentioned earlier, for identification of application area of biomedical polymers, the knowledge of mechanical properties of polymers together with other abovementioned properties is important. At the same time, due to availability of double bonds, an improvement of mechanical properties of polymers is possible via photocross-linking reaction. We selected unsaturated/saturated polyester amide FA/8$_{25/75}$-Phe-6 (Mw = 69,700, Mn = 28,000, Mw/Mn = 2.49) for study of mechanical properties and determined mechanical properties of films of mentioned polymers with 0.3–0.5-mm thickness. Films were dried in vacuum at 50–70°C. Mechanical properties of films were studied using Mecmesin "Multitest 1-i." Samples in the form of so-called doggy bone were prepared.

With the purpose of establishment of possibilities of photocross-linking of obtained poly(ester amide)s, we conducted UV irradiation (using Intelli-Ray 400, Shuttered UV Flood light) of unsaturated/saturated polyester amide FA/8$_{25/75}$-Phe-6 without initiator and in the presence of 5% of photoinitiator Darocur-1173. Young modulus E (GPa), maximum force F_{max} (MPa), breaking strength σ (MPa), and breaking elongation ε (%) prior to irradiation and after 5-min irradiation were determined.

As the obtained results showed, significant change of mechanical characteristics (E, F_{max}, and σ) is not observed during film irradiation without photoinitiator. In the presence of 5% of photoinitiator Darocur-1173, improvement of mechanical characteristics of polymers takes place, and photocross-linking of unsaturated poly(ester amide)s occurs (Table 1.2).

TABLE 1.2   Mechanical Properties of Unsaturated/Saturated Polyester Amide FA/8$_{25/75}$-Phe-6.

Defined parameters	Original samples		Samples in the presence of photoinitiator Darocur-1173			
			Drying at 50°C		Drying at 70°C	
	Prior to irradiation	After irradiation	Prior to irradiation	After irradiation	Prior to Irradiation	After irradiation
Young modulus E (GPa)	1.64 ± 0.09	1.55 ± 0.03	1.08 ± 0.12	2.15 ± 0.20	1.75 ± 0.05	2.35 ± 0.02
Maximum force F_{max} (MPa)	30.5 ± 5.5	27.0 ± 3.0	27.0 ± 3.3	49.0 ± 3.0	40.0 ± 8.0	61.0 ± 0.3
Breaking strength σ (MPa)	19.0 ± 8.0	22.0 ± 2.0	36.7 ± 6.2	42.0 ± 7.0	35.0 ± 5.0	61.0 ± 0.3
Breaking elongation ε (%)	185.0 ± 25.0	205.0 ± 50.0	253.0 ± 32.0	185.0 ± 5.0	60.0 ± 20.0	4.5 ± 2.0

Thus, on the basis of preliminary data, we can suppose that photocross-linking of synthesized unsaturated polyester amide improves mechanical characteristics of material that make their use prospective as surgical materials, for example, in the form of photocross-linked, biodegradable blood vessel stent, etc.

As a result of carried out research, we received functional, biodegradable polymers on the basis of natural amino acids containing different number of unsaturated bonds in main chain, with good material properties, which are prospective for getting the biodegradable materials with high mechanical strength, for implanted artificial organs, biodegradable hydrogels, medication delivery systems, micro- and nanocapsulation, etc.

KEYWORDS

- **unsaturated/saturated poly(ester amide)s**
- **homopoly(ester amide)s**
- **copoly(ester amide)s**
- **unsaturated biodegradable polymers**
- **cross-linking**

REFERENCES

1. Katsarava, R.; Beridze, V.; Arabuli, N.; Kharadze, D.; Chu, C. C.; Won, C. Y. *J. Polym. Sci. Part A Polym. Chem.* **1999**, *37*, 391–407.
2. Gomurashvili, Z.; Katsarava, R.; Kricheldorf, H. R. *J. Macromol. Sci. Pure Appl. Chem.* **2000**, *37*, 215–227.
3. Okada, M.; Yamada, M.; Yokoe, M.; Aoi, K. *J. Appl. Polym. Sci.* **2001**, *81*, 2721–2734.
4. Jokhadze, G.; Machaidze, M.; Panosyan, H.; Chu, C. C.; Katsarava. R. *J. Biomater. Sci. Polym. Ed.* **2007**, *18*(4), 411–438.
5. Lee, S. H.; Szinai, I.; Karpenter, K.; Katsarava, R.; Jokhadze, G.; Chu, C. C.; Huang, Y.; Verbeken, E.; Bramwell, O.; De Scheerder, I.; Hong, M. K. *Coron. Artery Dis.* **2002**, *13*(4), 237–241.
6. Defife, K.; Grako, K.; Cruz-Aranda, G.; Price, S.; Chantung, R.; Macpherson, K.; Khshabeh, R.; Gopalan, S.; Turnell, W. G. *J. Biomater. Sci.* **2009**, *20*, 1495–1511.
7. Katsarava, R.; Kharadze, D.; Jokhadze, G.; Nepharidze, N. *Functional Polymers and Their Use in Scientific Research and Biomedicine*; Publishing House Technical University: Tbilisi, 2009; p 167.
8. Lou, X.; Detrembleur, C.; Lecomte, P.; Jerome, R. *e-Polymers,* **2002**, *034.*
9. Guo, K.; Chu, C. C. *J. Polym. Sci. Part A Polym. Chem.* **2005**, *3*, 3932–3944.
10. Guo, K.; Chu, C. C. *Biomaterials* **2007**, *28*, 3284–3294.
11. Engel, C. A. R.; Straathof, A. J. J.; Zijlmans, T. W.; Van Gulik, W. M.; Van der Wielen, L. A. M. *Appl. Microbiol. Biotechnol.* **2008**, *78*, 379–389.
12. Domb, A. J.; Laurencin, C. T.; Israeli, O.; Gerhart, T. N.; Langer, R. *J. Polym. Sci. Polym. Chem.* **1990**, *28*(5), 973–985.
13. Lacoudre, N.; Leborgne, A.; Sepulchre, M.; Spassky, N.; Djonlagic, J.; Jacovic, M. S. *Macromol. Chem.* **1986**, *187*(2), 341–350.
14. Guo, K.; Chu, C. C. *J. Biomed. Sci. Polym. Ed.* **2007**, *18*, 489–504.
15. Domb, A.; Ron, E.; Giannos, S.; Flores, S.; Kim, C.; Dow, R.; Langer, R. *14th Inter. Symposium on the Controlled Release of Bioactive Materials*; Lee, P. I., Leonhardt, B. A., Eds.; Elsevier: Toronto, 1987; p 138.
16. Domb, A. J.; Martinowitz, E.; Ron, E.; Giannos, S.; Langer, R. *Polym. Chem.* **1991**, *29*, 571–579.
17. Chkhaidze, E.; Tugushi, D.; Kharadze, D.; Gomurashvili, Z.; Chu, C. C.; Katsarava, R. *J. Macromol. Sci. A* **2011**, *48*(7), 455–555.

18. Katsarava, R.; Kharadze, D.; Japaridze, N.; Omiadze, T.; Avalishvili, L. *Makromol. Chem.* **1985,** *186,* 939–954.
19. Katsarava, R. D. Synthesis of Heterochain Polymers Using Chemical Active Monomers (Active Polycondensation) (Review). *Vysokomolec. Soed. A.* **1983,** *31,* 1555–1571.
20. Katsarava, R. D. Achievements and Problems Active Polycondensation. *Russ. Chem. Rev.* **1991,** *60,* 1419–1448.
21. Benzler, B. Dynamische Differenzkalorimetrie–Hohe Reproduzierbarkeit. *Plastverarbeiter* **1996,** *47,* 66.
22. Chaires, J. Calorimetry and Thermodynamics in Drug Design. *Annu. Rev. Biophys.* **2008,** *37,* 135–151.

CHAPTER 2

EXPERIMENTAL DESIGN AND PROPERTIES OF PVC FOAMS: AZODICARBON AMIDE AND ZINC OXIDE

FILIZ Ö. ÖMÜRLÜ and DEVRIM BALKOSE*

Department of Chemical Engineering, İzmir Institute of Technology, Gulbahçe, Urla, Izmir, Turkey

Corresponding author. E-mail: devrimbalkose@gmail.com

CONTENTS

ABSTRACT

In this work, flexible PVC foam films were prepared using azodicarbonamid (ADC) as chemical bowing agent and zinc oxide (ZnO) catalyst in polyvinyl chloride (PVC) plastisols. Zinc stearate and calcium stearate were added as thermal stabilizers for PVC. $ZnSt_2$ and $CaSt_2$were added to PVC as thermal stabilizers to plastisol. Foaming and gelation of the films occur simultaneously during heating the plastisols. The optimum ZnO and ADC concentration were analyzed by 2^3 experimental design by selecting density as the response. Design equation of density gives time as the main factor. On the other hand, design equation of induction time gives ZnO amount as the main factor.

2.1   INTRODUCTION

The use of polymer foams is extremely widespread in today's technology. Indeed, it is hard to think of going a day without coming across some sort of polymer foam. They have a part to play almost everywhere in our life such as disposable packaging of fast food, in sports and leisure products, the cushioning of furniture and insulation material, in military applications, in vehicles, in aircraft, in textiles, and in some other common applications. For industrial foam applications, selection of convenient polymers depends upon their properties, the economics of the foaming system and their ease of production.

Naturally occurring polymer foams have been known for a long time, such as sponges, cork. However, synthetic polymer foams have only been introduced to the market over the last 50 years or so. They continue to grow at a rapid pace because of their toughness, flexibility, resistance to chemicals and abrasion, lightweight, excellent strength/weight ratio, superior insulating abilities, energy-absorbing performance, and comfort features.[7] Thermal insulation, low density/lightness, and compressibility are all increased by foaming a polymer.

Polymer foams are composed of a solid and gas phase mixed together to form a foam, which happens too fast for the system to respond in a smooth form. The product foam has a polymer matrix with either air bubbles or air tunnels included in it, which is known as either closed-cell or open-cell structure. Closed-cell foams are generally more stiff, on the other hand open-cell foams are usually elastic.[24] The open-cell foams are most suitable for car seating, furniture, bedding, and acoustical insulation, among other uses. The closed-cell foams are best for thermal insulation.[7] There are many types of polymeric foams, for instance; polyurethane, polystyrene foam, biodegradable foam, polyolefin foams, and starch.

Polymer foams can be separated into either thermoplastics or thermosets, which are further divided into rigid or flexible foams. Although, the thermoplastics can usually be broken down and recycled, thermosets are harder to recycle since they are usually heavily cross-linked. Polymeric foams and foam technology were explained in detail by Klemper and Sendijarevic[12] and PVC foams were reviewed in detail by Wilkes et al.[27]

2.1.1 CHEMICAL AND PHYSICAL BLOWING AGENTS

The gas that is used in the foam is named as a blowing agent, and two types of blowing agents can be used to expand a polymer: physical and chemical. Chemical blowing agents (CBA) are chemicals that decompose to give gaseous products in the process.[24] CBA have two main advantages: they are easy to introduce into the polymer (mixing and handling) and they are easily processed with ordinary equipment that is to say no modification needed.

On the other hand, physical blowing agents (PBA) are gases that do not react chemically in the foaming process. Therefore they are inert to the polymer forming the matrix.[24] PBA provide the expansion gas by undergoing changes in their physical state. These changes may include volatilization of a liquid or release of a compressed gas to atmospheric pressure after its combination into a polymer while under pressure. Some examples of PBA can be given such as nitrogen, carbon dioxide, and light hydrocarbons; moreover, each one can be used alone or in cofoaming agent.

The most popular CBA for polyolefins is certainly azodicarbonamide (ADC) which is selected for its high gas yield 210–220 cm^3 of gas per gram of product) and the ability to match its decomposition temperature with the polymer's processing temperature using suitable activators (catalysts or kickers), such as transition metal salts (lead, cadmium, and zinc), polyols, urea, alcohol, amines, and organic acids. ADC decomposition is a function of particle size, heating rate, activator type and concentration, as well as degree of dispersion of the blowing agent and activator. ADC thermal decomposition is an exothermic process and several studies have reported its heat of reaction. The decomposition of ADC in polyethylene at high temperature (extent of reaction vs. time) produces an S-shaped curve typical of an autocatalytic reaction.[17] ADC has been used as cross-linking and blowing agent in plastics. Both rigid and flexible poly(vinyl chloride) (PVC) foams have been prepared by adding ADC as a blowing agent. Two types of end products are obtained by decomposition of ADC: solids such as urazol (NHCONHCONH) and hydrazodicarbon-amide (NH$_2$CONHNHCONH$_2$) serve as nucleating agents; and a gaseous mixture of cyanic acid (HNCO), ammonia (NH$_3$), carbon monoxide (CO), and nitrogen (N$_2$). ADC provides foam formation by decomposing at 195–216°C, according to the way of preparation and releasing a large volume of nitrogen and carbon monoxide.[17]

PVC is a unique thermoplastic which has the ability to be produced with wide range of properties by the choice of additive. The addition of different level of plasticizer to PVC produces materials that have different softnesses. This flexible material is preferred due to good chemical resistance, toughness, flexibility, their ease of production, and economic reasons.[27] PVC foams widely used in wind energy, marine, road, rail, aerospace, textile, military applications because of its excellent features, such as light weight, sound and thermal insulation, excellent strength/weight ratio.[7,22] PVC foams can be either be obtained by physical blowing agents or CBA. Microcellular PVC/thermoplastic foams were obtained by physical blowing with carbon dioxide.[21] The foaming parameters considerably influence the mechanical and structural and moisture absorption properties of renewable resin precursor and residual filler modified PVC foams obtained by high-pressure carbon dioxide.[23] Rigid PVC foams were also prepared by using CBA.[3] CBA are used in producing PVC foams by gelation of plastisols. The PVC plastisol composition and processing conditions effect the foam expansion and tear strength of foams.[17] Effects of plasticizers, such as phthalate esters and mixed phthalates on foaming of PVC plastisols, were investigated by Verdu et al.,[25] Zoller and Marcilla,[29–31] Yu et al. showed that the asphalt filled hydrophobic flexible PVC foam crumps had high oil sorption capacity.[28]

2.1.2 PVC THERMAL DEGRADATION AND STABILIZATION

Plastisols are dispersion of PVC particles in plasticizers. PVC undergoes thermal degradation when plastisols are heated.[11] While preparing PVC foam, PVC resin is dispersed with an appropriate plasticizer, blowing agent, and stabilizer, and then heated to 180–200°C for simultaneous gelation and foaming of the plastisol.[7] Moreover, degradation of PVC takes place during PVC foaming process starting at about 100°C by autocatalysis of evolved hydrogen chloride.[8] Dehydrochlorination is accompanied by polymer discoloration going from yellow to brown and black due to formation of conjugated double bonds as shown by eq 2.1. The dehydrochlorination of PVC gives hydrogen chloride gas and dehydrochlorinated PVC (dePVC) having conjugated double bonds.

$$PVC\ (s) \rightarrow dePVC\ (s) + HCL\ (g). \tag{2.1}$$

Different organic and inorganic compounds are used as heat stabilizers which are based on metal soaps. Zinc soaps are useful stabilizers, however, product zinc chloride which is a Lewis acid, accelerates dehydrochlorination after its concentration reached up to certain level.[4] Zinc soaps are used together with calcium soaps to eliminate the zinc chloride effect. $CaCl_2$ is formed which is not a Lewis acid. The reaction between $ZnCl_2$ and calcium stearate ($CaSt_2$) is shown in eq 2.2.

$$ZnCl_2 + CaSt_2 \rightarrow ZnSt_2 + CaCl_2. \tag{2.2}$$

The role of metal oxides on the thermal degradation of vinyl polymers suggest that the kinetics of the degradation is affected by metal oxides.[9] ZnO addition increased significantly decomposition rates and a modified mechanism was proposed to take into account this activation effect. ZnO added at a low level catalyzes the decomposition of ADC, without any detrimental effect on degradation of PVC.[18] Zinc maleate/ZnO mixtures has synergistic effect of thermal stability of PVC.[13]

Thermal degradation of PVC in the presence of metal oxides was investigated using a thermogravimetric (TG) method by Gupta and Visvanath.[9] It follows a two-step mechanism. In the first step chlorine free radical is formed as in the case of pure PVC, and in the second step chlorine free radical replaces oxygen from metal oxide to form metal chloride and oxygen free radical. Subsequently, the oxygen free radical abstracts hydrogen from PVC. Formation of metal chloride is the rate controlling step.

In the case of transition metal oxides coated with PVC, the thermal degradation can be correlated with the bulk properties such as the ability of electron transfer in the transition metals to form different valence states, and degradation may be described via an electron transfer (redox) mechanism which proceeds through formation of HCl.[9] The liberated HCl then reacts with metal oxide, and the rate is faster at elevated temperature. Thus, the activation energy is lower than that for the degradation of pure PVC. The statistical investigation of the thermal stability of PVC powder was investigated by Savrik et al.[19]

Response surface approach was used to optimize fusion properties of rigid foam PVC/clay composites.[15]

The objective of this study is to investigate the effects of time and zinc oxide on both ADC decomposition and foam formation. A 2^3 factorial design was done for optimum foaming conditions and the response was selected as density. Since the thermal stability was also affected, optimum conditions for highest induction time were also investigated.

2.2 MATERIALS AND METHOD

2.2.1 MATERIALS

PVC (Petvinil, P.38/74) was the product of Petkim Co. Dioctyl phthalate (DOP) and ADC were purchased from Merck Co. Zinc stearate $(ZnSt_2)$ and $CaSt_2$ taken from Baerlocher Co. and Viscobyk 5025 from BYK-Chemie and ZnO from Ege Chemical Co. was used in this study. The PVC powder was composed of spheres of a large particle size range from 10 nm to 20 μm.[21] The particle size and the aspect ratio of ZnO was 3.86 and 4.35 μm, respectively.[15]

2.2.2 PREPARATION OF THE PVC FOAM FILMS

In order to produce PVC foam films, PVC plastisol was prepared by mixing 100 parts PVC, 80 parts DOP, 5 parts Viscobyk 5025, 1 part $CaSt_2$, and 1 part $ZnSt_2$. Heat stabilizers and PVC were dried at 50°C for about 1 h before mixing. This mixture was stirred until it became a homogeneous suspension. This sample was separated into 20 g aliquots. Four samples were prepared having different amounts of ADC and ZnO. About 0.2 g ADC and 0.25 g ZnO were added into the first 20 g. A total of 0.2 g ADC and 0.25 g ZnO, 0.2 g ADC and 0.1 g ZnO, 0.4 g ADC and 0.25 g ZnO, 0.4 g ADC and 0.1 g ZnO powders were added to the second, third, fourth 20 g, respectively. The plastisol samples were cast onto glass substrate with the aid of an automatic film applicator (Sheen 1133N) with 90 μm gap size at a speed of 50 mm/s. Films were allowed to gel and foam in oven (Venticell) at 190°C for 5 and 10 min to see the effect of heating time on the foaming. The experiments were repeated one more time to optimize the data for analysis of variance.

2.2.3 EXPERIMENTAL DESIGN

2^3 experiments were made to minimize the density of the films. The variables affecting the density are ZnO amount, ADC amount, and the time of heating at 190°C. High and low values of each variable were chosen as shown in Table 2.1. The designs were carried out by using Design of Expert (Trial version 8.0). The same variables were also used in optimizing the induction time of the prepared composites.

TABLE 2.1   The Selected High and Low Values of the Variables for Minimizing the Density of the Films.

Variable	High	Low
ZnO content (g/20 g plastisol)	0.25	0.10
ADC content (g/20 g plastisol)	0.4	0.2
Time (min)	10	5

2.2.4 DENSITY AND PORE VOLUME

Density of PVC foams was measured by using Sartorius density measurement kit.[26] Water was used as the liquid causing buoyancy. To determine the sample's buoyancy, floating sample was immersed by a metal sieve. The negative weight displayed by the balance corresponded to the buoyancy acting on the sample in the liquid. Density of foam samples was calculated by using eq 2.3.

$$\rho = \frac{W_a \rho_{fl}}{0.99983 G} + 0.0012, \tag{2.3}$$

where G is the buoyancy force, ρ_{fl} the fluid density, ρ the foam density, and W_a is the dry weight of foam. The percentage of the volume of pores ($V_{totalpore}$) of the foam samples was calculated by using eq 2.4.

$$V_{totalpore} = (1 - \frac{\rho}{\rho_{Theoretical}}) \times 100. \tag{2.4}$$

Theoretical density values ($\rho_{Theoretical}$) were calculated from the mass of the sample and the geometric volume of the samples.

2.2.4.1 CHARACTERIZATION OF THE FILMS

Scanning electron microscopy (SEM) was used to obtain PVC plastisol foam films cell structure and to view the internal configuration of the cells. Films were broken after immersing in liquid nitrogen (77K) and fracture surfaces were coated with a thin gold film and examined by using SEM (Philips XL-30S FEG).

The density of the films was measured by using "Archimedes" apparatus. The average of four measurements was used in density determination by taking the water density as 0.99983 g/cm^3 at 25°C.

Tensile testing of the films was made according to ASTM D882. Stress-strain diagrams for samples with ZnO and without ZnO were obtained by tensile testing using Testometric tensile testing machine. Five samples from each set were cut to thin strips 10 ×100 mm in size and stretched uniaxially with a cross-head speed of 10 mm/min.

2.2.5 THERMAL GRAVIMETRIC ANALYSIS

The effect of ZnO on ADC decomposition was investigated by TG analysis using Shimadzu TGA-51. Samples with ADC/ZnO mass ratios of 0.2:0.04, 0.2:0.08, 0.2:0.08 were prepared by through mixing of the powders Small amount (10 mg) of powders was placed in an alumina pan for each run and analysis were carried out with a heating rate of 5, 10, 20°C/min under nitrogen gas flow up to 600°C. The TG analysis was also made for films prepared by 10 min heating at 190°C. The films were heated at 10°C/min rate from room temperature up to 600°C.

2.2.6 THERMAL STABILITY OF PVC

Thermal stability of PVC can be determined by 763 PVC Thermomat instrument. It is equipped with two heating blocks each with four measuring positions. Each block can be individually heated, therefore two sets of four samples can be measured at two different temperatures or eight samples can be measured at the same temperature. Also the measurements at the individual measuring positions can be started individually. The films were cut into 0.5 cm × 0.5 cm small squares and 0.5 g of samples was put into reaction vessels. The heating blocks of PVC Thermomat were heated to 190°C. The reaction vessels that contained PVC plastigel films were placed into the heating blocks. HCl formed was taken up by a nitrogen gas stream and trans-ferred into the measuring vessels.[2] The deionized water absorbed HCl gas and the change in conductivity of the solution was determined with respect to time by the conductimeter. Therefore, the decomposition process was monitored by measuring the conductivity of aqueous HCl solution.[2]

When PVC plastigels are heated in PVC Thermomat in the presence of nitrogen gas, the conductivity of water, which nitrogen gas is passed, changes with respect to time. The period when the conductivity starts to increase is called as induction time, and the period when the conductivity value reaches to 50 mS/cm is called as stability time. This value is the maximum accept-able level of degradation.

The photographs of the films as made and after the PVC Thermomat test were recorded by a digital camera.

2.3 RESULTS AND DISCUSSION

2.3.1 CHARACTERIZATION OF PVC FOAMS

Adhesion and cross-linking of the swollen PVC particles forming a gel and release of the gases from ADC occur simultaneously while heating PVC plastisols. When the temperature and heating periods were increased PVC particles were swollen with the plasticizer and fused to each other. The gases released by ADC were entrapped in the fused plastigel forming the porous structure. Porous structured PVC matrix composites are obtained by this method. The microstructures of the composites were investigated by SEM. SEM micrographs of the fracture surfaces of the composites obtained by immersing in liquid nitrogen are seen in Figures 2.1 and 2.2. The plastigel film without ZnO and ADC is expected to be nonporous when heated for 10 min at 190°C. However, there were unfused PVC particles or closed air bubbles covered with fused PVC in the film as seen in Figure 2.1. This film had a lower density (1.139 g/cm^3) than expected for a nonporous film (1.165 g/cm^3) and thus it has around 2% void volume.

The pores were elliptical in shape and they had a broad size distribution as seen in SEM micrographs of PVC foams in Figure 2.2. Their average dimensions in width and length and their standard deviation and their aspect ratio are reported in Table 2.2.

The maximum sized pores were obtained for the plastigel containing 0.4 g ADC and 0.25 g ZnO per 20 g of plastisol.

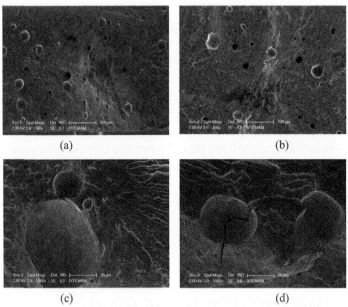

FIGURE 2.1 SEM micrographs of PVC plastisol film heated for 10 min at 190°C. (a) and (b) at 100× magnification, and (c) and (d) at 1000× magnification.

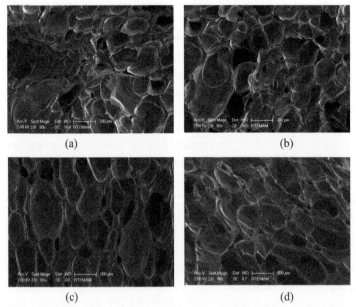

FIGURE 2.2 SEM micrographs of PVC plastisol foam films fracture surfaces (for 10 min at 190°C) consist of these ratios: (a) 0.2 g ADC:0.25 g ZnO, (b) 0.2 g ADC:0.1 g ZnO, (c) 0.4 g ADC:0.25 g ZnO, and (d) 0.4 ADCg:0.1g ZnO per 20 g plastisol.

TABLE 2.2   The Dimensions of the Elliptical Foams and Their Aspect Ratio.

ADC mass (g/20 g)	ZnO mass (g/20 g)	Width (µm)		Length (µm)		Aspect ratio
		Average	Standard deviation	Average	Standard deviation	
0.2	0.25	155.2	67.8	166.3	94.0	0.9
0.2	0.1	168.2	67.4	267.2	98.2	0.6
0.4	0.25	188.6	76.9	396.4	164.0	0.5
0.4	0.1	143.2	63.6	306.3	43.1	0.5

2.3.1.1   DENSITY

Plastisols and samples containing ZnO and ADC are tested by using density Archimedes' kit. Density of the samples was calculated using eq 2.3. Table 2.3 gives the densities and pore volume % of the films prepared by 5 and 10 min heating at 190°C. The maximum pore volume (77.42%) obtained was for the film with 1.9% ADC and 1.2% ZnO (0.4 g ADC and 0.25 g ZnO). This was close to the pore volume obtained by Şahin et al.[18] for the plastisol with 2% ADC and 2.5% ZnO that was heated for 12 min at 190°C.

TABLE 2.3   Density of the Composites After Heating at 190°C for 5 and 10 min.

ADC (g/20 g)	ZnO (g/20 g)	Density (g/cm³)		Pore volume (%)	
		5 min heated	10 min heated	5 min heated	10 min heated
0	0	1.139	1.158	2.23	0.60
0.2	0.25	1.161	0.315	0.34	72.96
0.2	0.10	1.160	0.386	0.43	66.87
0.4	0.25	1.164	0.263	0.09	77.42
0.4	0.10	1.156	0.467	0.77	59.91

ADC is known to promote foam formation by decomposing at 195–216°C depending on the mode of preparation and releasing a large volume of nitrogen and carbon monoxide gas.[17] Finely dispersed filler additives are known to reduce this decomposition temperature by activating ADC and providing nucleation sites for gas evolution. The foam formation for PVC composites were related with the decomposition of ADC. The density of the composites was lowered, since ADC gave gaseous products that were entrapped in plastigel. However, as seen in Table 2.3, for 5 min heating period, the foam formation did occur because there was not enough time for gelation of PVC plastisol to entrap the gaseous products from the decomposition of ADC. However, after 10 min heating, the density of the composites drops very sharply and bubble formation within the composites was obtained due to decomposition of ADC and confinement of the gases in gelled plastisol.

2.3.1.2   MECHANICAL BEHAVIOR OF PVC FOAMS

Samples with and without ADC and ZnO are tested for mechanical properties using Uniaxial Tensile Testing Machine. Representative stress strain diagrams of the samples without and with ADC and ZnO are shown in Figure 2.3.

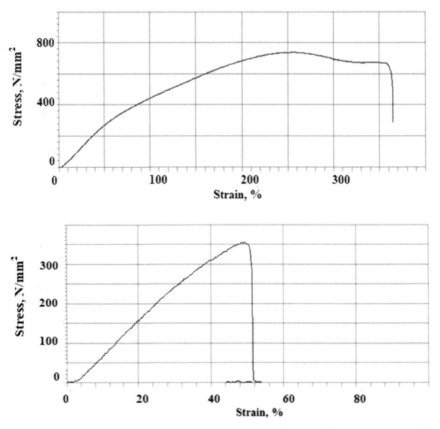

FIGURE 2.3   Stress–strain curves for films (a) without and (b) with 0.2g ADC and 0.25g ZnO.

Elastic modulus of the films is found from the slope of the stress strain curve at the origin using eq 2.5.

$$Elastic\ modulus = Tensile\ Strength/Strain. \tag{2.5}$$

According to equal assumption of the slab model, the tensile strength (σ_c) of a composite equals to

$$\sigma_c = \sigma_f + \sigma_p\ (1 - \varphi_f), \tag{2.6}$$

where φ_f and σ_p are volume fraction of filler and tensile strength of the plastic phase, respectively.[6] The filler in the present case is air in the foams and its tensile strength is zero.

Elastic modulus of the foams can be predicted using elastic modulus of the plastic phase and ratio of the densities of the foam (ρ) to the plastic (ρ_p).[1]

$$E_c = E_p\ (\rho/\ \rho_p)^2. \tag{2.7}$$

The tensile strength and elastic modulus were predicted in eqs 2.6 and 2.7 and reported in Table 2.4.

TABLE 2.4   Mechanical Properties of Samples PVC and PVC Film for 10 min at 190°C.

Composition		Tensile strength (Mpa)		Strain at break (%)	Elastic modulus (Mpa)	
ADC (g/20 g)	ZnO (g/20 g)	Experimental	Theoretical		Experimental	Theoretical
0	0	600	600	355	436	436
0.2	0.25	345	162	50	690	32
0.2	0.1	333	114	55	605	48
0.4	0.25	330	75	51	647	22
0.4	0.1	238	132	36	661	70

The film without ZnO and ADC had higher tensile strength than the foamed films as reported in Table 2.4. However, the films had higher tensile strength values than predicted by eq 2.6. This could be attributed to the simultaneous cross-linking of PVC by ADC during foaming process. The modulus of elasticity values of the foamed films was also much higher than predicted by eq 2.7. The tensile strength of the films in the present study was nearly 10 fold higher than the films prepared in a previous study made by Şahin et al.[18] The only difference in the two studies is that using the PVC dispersant Viscobyk 5025. The agglomeration of PVC particles during plastisol preparation was prevented and the area of contact between plasticizer swollen particles was higher, that made better fusion of the particles.

2.3.2 THERMAL CHARACTERIZATION OF AZODICARBONAMIDE

The effect of ZnO and ZnO concentration was studied by TGA. TG curves of ZnO and ADC mixtures at 5 and 10°C/min rates are seen in Figures 2.4 and 2.5.

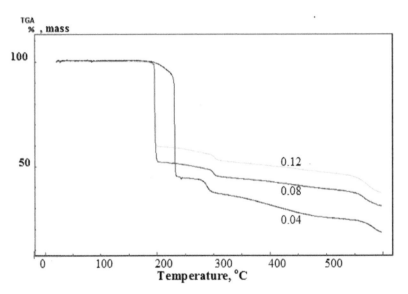

FIGURE 2.4   TG curves for mixtures with different ZnO content for 10°C/min heating rate. The parameters on the curves are g ZnO per 0.2 g ADC.

As depicted in Table 2.5 the decomposition temperature of ADC decreases as ZnO% increases. The mixtures with 0.04/0.2 and 0.12/0.2 ZnO/ADC ratio started to evolve gases at 219.04°C and 188.83°C, respectively.

While N_2 gas and CO are the gaseous products of ADC solid residue consists of urazol and hydro-ADC. The solid residue also decomposes to give gaseous products at higher temperature. ZnO has two effects in ADC decomposition: lowering the decomposition onset temperature and leading to complete gasification of ADC. When ZnO ratio is increased, ADC decomposition onset temperature decreased nearly 30°C from 220°C to 190°C. Indeed, the addition of ZnO was found to accelerate the thermal decomposition of ADC by increasing the reaction rate of the first step of ADC decomposition by an order of magnitude using only 0.1%.[17]

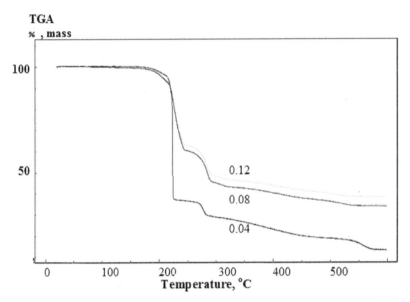

FIGURE 2.5 TG curves for mixtures with different ZnO content for 10°C/min heating rate. The parameters on the curves are g ZnO per 0.2 g ADC.

TABLE 2.5 Onset Temperature and the Mass Loss % at 300°C and 600°C for ZnO and ADC Mixtures at 20°C/min Heating Rate.

ADC concentration (wt%)	ZnO concentration (wt%)	ADC mass (g)	ZnO mass (g)	Onset of mass loss/°C	Mass loss (%) At 300°C	At 600°C
83.33	16.67	0.2	0.04	219.04	60.27	78.27
71.43	28.57	0.2	0.08	200.01	42.33	58.84
62.50	37.50	0.2	0.12	188.83	43.68	59.49

The kinetics of the decomposition of ADC was investigated by Kissinger method[11] in the present study.

$$\ln(\beta/T_{max}^{2}) = \left\{ \ln AR / E + \ln\left[(1-\alpha_{max})n-1\right]\right\} - E / RT_{max},\qquad(2.8)$$

where β is the rate of heating, T_{max} is temperature at which the rate is maximum, α_{max} is the extent of reaction at T_{max}, n is the order of reaction, A is pre-exponential factor, and E is the activation energy.

TABLE 2.6 TGA Results Obtained from ADC and ZnO Powders Decomposition at 5°C/ min, 10°C/min, and 20°C/min Heating Rates.

β (°C/min)	ZnO (%)	ADC (%)	T_{max} (K)	Mass loss at T_{max} (%)	$\ln(\beta/T_{max}^2)$
5	16.66	83.33	496.84	63.20	−10.80
	28.57	71.42	499.91	38.85	−10.81
	37.50	62.50	500.26	37.19	−10.82
10	16.66	83.33	504.61	56.17	−10.14
	28.57	71.42	469.96	48.46	−10.00
	37.50	62.50	468.72	40.26	−9.99
20	16.66	83.33	502.76	60.27	−9.44
	28.57	71.42	480.93	42.33	−9.35
	37.50	62.50	469.64	43.67	−9.30

However, no fit of the data given in Table 2.6 to Kissinge plot (eq 2.8) shown in Figure 2.6 was obtained. This was attributed to the heterogeneous nature of the reaction. Two solid powders were mixed and the contact surface of the particles of ADC and ZnO were not the same at each point. The degree of dispersion of the ADC and ZnO strongly effected the decomposition of ADC as suggested by Robledo-Ortiz et al.[17]

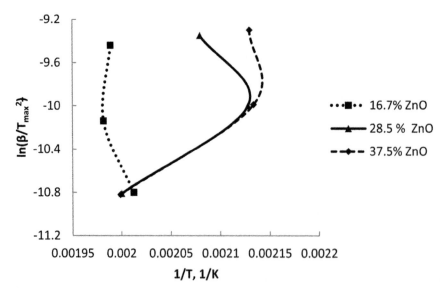

FIGURE 2.6 Kissinger plot of ADC and ZnO powder mixtures.

2.3.3 THERMAL CHARACTERIZATION OF THE FILMS

The degradation behavior of plastisol and composites after foaming process was investigated by TG analysis and TG curves are given in Figure 2.7a.

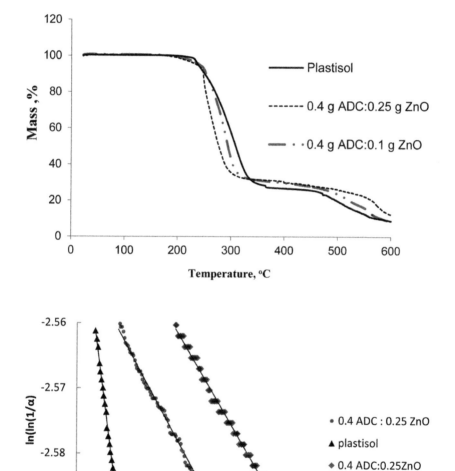

FIGURE 2.7 (a) TG curves and (b) Broido's plot of plastisol 0.4 g ADC:0.25 g ZnO polymer composite and 0.4 g ADC:0.1 g ZnO polymer composite.

As depicted in Figure 2.7a, ZnO addition has also a catalytic effect on PVC thermal degradation. The onset of mass loss for dehydrochlorination was lower for the films with ADC and ZnO than that of the plastisol. It was 225°C, 206°C, and 198°C for the plastisol, for the films with 0.4 g ADC and 0.1 g ZnO per 20 g plastisol, and 0.4 g ADC and 0.25 g ZnO for 20 g plastisol, respectively.

The activation energy of dehydrochlorination was found by using Broido's method shown by eq 2.9.[5]

$$\ln (\ln 1 / \alpha) = - E / RT + C,$$

(2.9)

where α is the extent of reaction, E is the activation energy of the decomposition, T is the absolute temperature, R is the gas constant, and C is a constant. The activation energy of the mass loss was found from the slopes of the lines in Figure 2.7b as 16, 4.2, and 3.8 kJ/mol for the plastisol, for the films with 0.4 g ADC and 0.1 g ZnO per 20 g plastisol and 0.4 g ADC and 0.25 g ZnO for 20 g plastisol, respectively, for the mass range of 98% and 96%. Presence of ADC and ZnO accelerated the rate of dehydrochlorination by lowering the activation energy of the dehydrochlorination reaction.

2.3.4 THERMAL STABILITY OF THE COMPOSITES

The films prepared by heating at 190°C for 10 min had light yellow color as seen in Figure 2.8. After PVC Thermomat test (190°C, approximately after 25 min) they were all blackened. At long heating periods at 190°C in PVC Thermomat, PVC undergoes severe dehydrochlorination reaction and they appeared as black due to absorption of visible light by the long sequences of conjugated double bonds.

Representative conductivity versus time curves of PVC foam films measured by PVC Thermomat are given in Figure 2.9. Each experiment was made two times to observe reproducibility. The average values of the induction and stability times of the films at 190°C are reported in Table 2.7 for films initially heated at 190°C for 5 and 10 min.

Addition of ZnO and ADC into plastisol decreased the induction time and stability time for both 5 min heating and 10 min heating periods. PVC was not severely degraded during foaming process since the induction times were around 2–4 min for the foams.

PVC dehydrochlorination at high temperature has been reduced by addition of the $ZnSt_2$ and $CaSt_2$ heat stabilizers. Reaction of these stabilizers with

HCl greatly reduces autoaccelerating effect of HCl released by dehydrochlorination. According to Balkose et al., metal soap added to PVC interacts with HCl and decreases its degradation effect as seen in eq 2.10 and carboxylate groups of metal soaps are exchanged with labile chlorine atoms of PVC and form thermally stable bonds.[4]

| (a) | (b) | (c) | (d) |

| (e) | (f) | (g) | (h) |

FIGURE 2.8 The picture of the films as prepared and after PVC thermomat tests. As prepared (a) 0.2 g ADC:0.25 g ZnO, (b) 0.2 g ADC:0.1 g ZnO, (c) 0.4 g ADC:0.25 g ZnO, and (d) 0.4 ADCg:0.1g ZnO. After PVC thermomat test (e) 0.2 g ADC:0.25 g ZnO, (f) 0.2 g ADC:0.1 g ZnO, (g) 0.4 g ADC:0.25 g ZnO, and (h) 0.4 ADCg:0.1g ZnO.

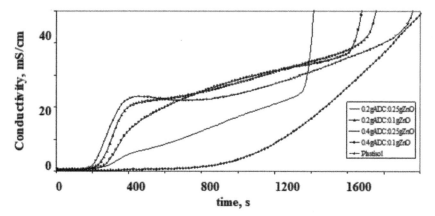

FIGURE 2.9 The change of conductivity of the aqueous solution with respect to time for PVC films at 190°C for 5 min (first trial).

TABLE 2.7   Average Values of Induction and Stabilization Time at 190°C for Films Initially Heated at 190°C for 5 and 10 min.

Sample composition		Induction time (min)		Stability time (min)	
ADC (g/20 g)	ZnO (g/20 g)	5 min heated	10 min heated	5 min heated	10 min heated
0	0	18.25	17.20	33.48	30.53
0.2	0.25	4.07	2.94	24.88	19.35
0.2	0.1	3.81	2.54	29.72	23.62
0.4	0.25	3.41	3.07	32.61	23.08
0.4	0.1	3.54	2.08	33.40	23.88

$$ZnSt_2 + 2HCl \rightarrow ZnCl_2 + HSt. \qquad (2.10)$$

$ZnSt_2$ and $CaSt_2$ are used together because of their synergetic effect. The $ZnCl_2$ formed by the reaction of $ZnSt_2$ with HCl had also accelerating effect on dehydrochlorination reaction. Thus, it was converted to inactive products by reacting with $CaSt_2$ as shown by eq 2.2.

ZnO is capable of neutralizing HCl forming $ZnCl_2$ as shown in eq 2.11. It is a fact that, when $ZnCl_2$ concentration exceeds a certain level it also accelerates dehydrochlorination reaction extensively.

$$ZnO + 2HCl \rightarrow ZnCl_2 + H_2O. \qquad (2.11)$$

During the decomposition of ADC, gas formation was obtained and it acts as blowing agent. Porous PVC matrix composites are obtained. As seen in Table 2.2, after 5 min heating, the foam formation did not occur since there was not enough time to decompose ADC and give gaseous products. However, after 10 min heating, the density of the composites drops very sharply and bubble formation within the composites was obtained. The 0.4 g ADC and 0.25 g ZnO per 20 g plastisol (1.9 wt% ADC–1.2 wt% ZnO) and 10 min heating time should be selected for the homogenously distributed and maximum pore formation. The induction and stability time of composites with ADC and ZnO were lower than that of PVC without them. This indicated that ADC and ZnO had accelerating effect on PVC dehydrochlorination. In the heating period of 10 min for preparation of PVC foams, ADC decomposed and even if PVC was dehydrochlorinated, it was not to an extent that will cause a severe discoloration.

2.3.5 DESIGN OF EXPERIMENT

To understand the optimum ADC and ZnO concentration for the foam formation 2^3 experimental design was applied using the response as density as seen in Table 2.8. Equation 2.12 gives the design equation for main factors and interaction. Figures 2.10 and 2.11 give the design cube and 3D plot of the desirability of ADC and ZnO concentration.

TABLE 2.8 Design Table for Response Density and Variables ADC Concentration, ZnO Concentration, and Time.

Experiment number	ADC (g/20g)	ZnO (g/20g)	Time (min)	Density (g/cm³)
1	0.2	0.10	10	0.324
2	0.2	0.25	5	1.159
3	0.2	0.10	5	1.145
4	0.4	0.25	5	1.165
5	0.4	0.10	5	1.158
6	0.4	0.10	10	0.326
7	0.4	0.25	5	1.163
8	0.2	0.25	10	0.382
9	0.4	0.10	5	1.155
10	0.4	0.25	10	0.297
11	0.2	0.25	5	1.163
12	0.2	0.25	10	0.506
13	0.4	0.10	10	0.608
14	0.2	0.10	5	1.175
15	0.2	0.10	10	0.448
16	0.4	0.25	10	0.402

$$\text{Density} = 2.18 - 1.07 \times \text{ADC} - 1.63 \times \text{ZnO} - 0.20 \times \text{time} + 6.28 \times \text{ADC} \times \text{ZnO}$$
$$+ 0.21 \times \text{ADC} \times \text{time} + 0.32 \times \text{ZnO} \times \text{time} - 1.21 \times \text{ADC} \times \text{ZnO} \times \text{time}. \tag{2.12}$$

According to the density design equation the most important parameter for the foam formation is the time. The interaction between ADC–ZnO and time mostly affected the foam formation for the system.

The desirability approach is one of the most widely used methods for the optimization of multiple response processes. It is based on the idea that the "quality" of a product that has multiple quality characteristics, with one of them outside of some "desired" limits, is completely unacceptable. The general approach is desirability between 0 and 1. First convert the response density is at its goal, then desirability is 1, if the response is outside of an acceptable region desirability is 0. According to the desirability graph the 0.4 g ADC: 0.25 ZnO per 20 g plastisol (1.9 wt% ADC–1.2 wt% ZnO) and 10 min should be selected for the homogenously distributed foam formation.

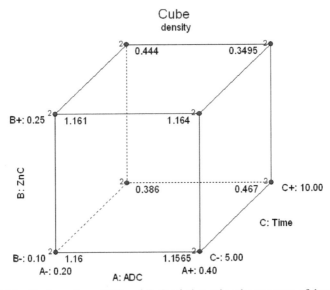

FIGURE 2.10   Design cube—ADC and ZnO relation using the response of density.

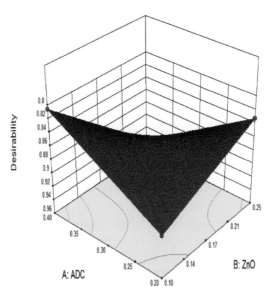

FIGURE 2.11   3D plot of desirability—ADC and ZnO relation using the response of density.

To find optimum ADC and ZnO concentration for the thermal stability 2^3 experimental design was applied using the response as induction time. Equation 2.13 gives the design equation for main factors and interaction. Figures 2.12 and 2.13 give the design cube and 3D plot of the desirability on ADC and ZnO concentration.

$$\text{Induction time} = 5.37 - 1.825 \times \text{ADC} + 5.2 \times \text{ZnO}$$
$$- 0.34533 \times \text{time} - 22 \times \text{ADC} \times \text{ZnO}$$
$$+ 0.35833 \times \text{ADC} \times \text{time} - 0.16667 \times \text{ZnO} \times \text{time} \qquad (2.13)$$
$$+ 1.76667 \times \text{ADC} \times \text{ZnO} \times \text{time}.$$

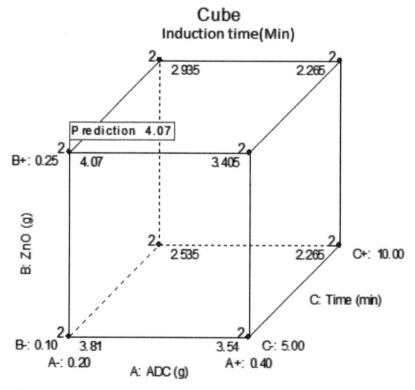

FIGURE 2.12 Design cube—ADC and ZnO relation using the response of induction time.

According to the induction time design equation the most important parameter for the foam formation is the ZnO amount. The interaction between ADC–ZnO and time mostly affected the induction time for the system. The desirability graph show that the 0.2 g ADC and 0.25 g ZnO per

20 g plastisol (0.98 wt% ADC–1.2 wt% ZnO) and 5 min should be selected for longest induction time. However, since no foaming occurs in 5 min the foams should be prepared by 10 min heating, sacrificing to a small extent in thermal stability of PVC.

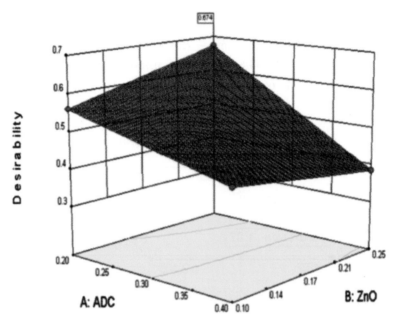

FIGURE 2.13   3D plot of desirability—ADC and ZnO relation using the response of induction time.

2.4   CONCLUSIONS

In this work, flexible PVC foam films with elastic modulus in the range of 550 and 690 MPa were prepared using CBA ADC and ZnO additive from PVC plastisols. ZnO was added to lower the onset of ADC decomposition temperature nearly 20°C as shown by TG analysis. In order to prevent dehydrochlorination of PVC, $ZnSt_2$ and $CaSt_2$ were added to PVC as thermal stabilizers to plastisol. The plastisol with ADC and ZnO should be heated at least for 10 min at 190°C to obtain the foamed films. For 5 min heating, foam formation was not obtained. There was not enough time for decomposition of ADC to give gaseous products. The other possibility is the incomplete gelation of the plastisol to entrap the evolved gases. However, after 10 min heating, the density of the composites was lowered very sharply and bubble formation within the composites was obtained. PVC plastigel foam films with elliptical cells with 143–396 μm dimensions were formed.

The optimum ZnO and ADC concentration were analyzed by 2^3 experimental design by selecting density as the response. Design equation of density gives time as the main factor. On the other hand, design equation of induction time gives ZnO amount as the main factor.

Addition of ZnO and ADC into plastisol decreases the thermal stability of the films as shown by TG analysis and PVC Thermomat tests. The onset of the mass loss, induction time, and the stability time was lower for the films with ADC and ZnO than that of PVC plastisol without them. Even if dehydrochlorination was accelerated by ADC and ZnO, it was not to a high extent to cause a severe discoloration in the 10 min heating period. Thus, it can be concluded that flexible foam films could be obtained by using the chemical blowing agent ADC in PVC plastisols. Foaming and gelation of the films occur simultaneously during heating the plastisols. The foamed films could be extensively used in packing and textile industry.

ACKNOWLEDGMENT

The authors thank Yagmur Kardeslik and Rukiye Turkay, chemical engineers who graduated from Izmir Institute of Technology, for their contribution to experimental work.

KEYWORDS

- blowing agent
- azodicarbonamide
- stabilizers
- polyvinyl chloride
- dehydrochlorination

REFERENCES

1. Alian, A. M.; Abu-Zahra, N. Mechanical Properties of Rigid PCV-clay Composites. *Polym. Plast. Technol. Eng.* **2009**, *48*, 1014–1019.

2. Atakul, S.; Balkose, D.; Ulku, S. Synergistic Effect of Metal Soaps and Natural Zeolite on Poly(Vinyl Chloride) Thermal Stability. *J. Vinyl. Addit. Technol.* **2005**, *11*, 47–56.

3. Azimipour, B.; Marchant, F. Effect of Calcium Carbonate Particle Size on PVC Foam. *J. Vinyl. Addit. Technol.* **2006**, *12*, 55–57.

4. Balkose, D.; Gokcel, H. I.; Göktepe, E. Synergism of Ca/Zn Soaps in Poly(Vinyl Chloride) Thermal Stability. *Eur. Polym. J.* **2001**, *37*, 1191–1197.

5. Broido, A. A. Simple, Sensitive Graphical Method of Treating Thermogravimetric Analysis Data. *J. Polym. Sci. Part B: Polym. Phys.* **1969**, *7*, 1761–1773.

6. Crawford, R. J. *Plastic Engineering;* Pergamon Press: New York, 1981.

7. Demir, H.; Sipahioglu, M.; Balkose, D.; Ulku, S. Effect of Additives on Flexible PVC Foam Formation. *J. Mater. Process. Technol.* **2008**, *195*, 144–153.

8. Garcia, J. C.; Marcilla, A. Rheological Study of the Influence of the Plasticizer Concentration in the Gelation and Fusion Processes of PVC Plastisols. *Polymer* **1998**, *39*, 3507–3514.

9. Gupta, M. C.; Viswanath, S. G. Role of Metal Oxides in the Thermal Degradation of Poly(Vinyl Chloride). *Ind. Eng. Chem. Res.* **1998**, *37*, 2707–2712.

10. Jiménez, A.; Torre, L.; Kenny, J. M. Thermal Degradation of Poly(Vinyl Chloride) Plastisols Based on Low-migration Polymeric Plasticizers. *Polym. Degrad. Stab.* **2001**, *73*, 447–453.

11. Kissinger, H. E. Variation of Peak Temperature with Heating Rate in Differential Thermal Analysis. *J. Res. Nat. Bur. Stand.* **1956**, *57*, 217–221.

12. Klemper, D.; Sendijarevic, V. *Handbook of Polymeric Foams and Foam Technology*, 2nd ed.; Hanser Publishers: Munich, 2004.

13. Li, G. X.; Wang, M.; Huang, X.; Li, H.; He, H. Effect of Zinc Maleate/Zinc Oxide Complex on Thermal Stability of Poly(Vinyl Chloride). *J. Appl. Polym. Sci.* **2015**, *132*(7) (Article Number 41464).

14. Moghri, M.; Khakpour, M.; Akbarian, M.; Saeb, M. R. Employing Response Surface Approach for Optimization of Fusion Characteristics in Rigid Foam PVC/Clay Nanocomposites. *J. Vinyl. Addit. Technol.* **2015**, *21*, 51–59.

15. Ozmihci, F. O.; Balkose, D. Effects of Particle Size and Electrical Resistivity of Filler on Mechanical, Electrical, and Thermal Properties of Linear Low-Density Polyethylene–Zinc Oxide Composites. *J. Appl. Polym. Sci.* **2013**, *130*, 2734–2743.

16. Radovanovic, R.; Jaso, V.; Pilić, B.; Stoiljković, D. Effect of PVC Plastisol Composition and Processing Conditions on Foam Expansion and Tear Strength. *Hem. Indus.* **2014**, *68*, 701–707.

17. Robledo-Ortiz, R. J.; Zepeda, C.; Gomez, C.; Rodrigue, D.; Gonzalez-Nunez, R. Non-isothermal Decomposition Kinetics of Azodicarbonamide in High-Density Polyethylene Using a Capillary Rheometer. *Polym. Test.* **2008**, *27*, 730–735.

18. Sahin, E.; Mahlicli, F. Y.; Yetgin, S.; Balkose, D. Preparation and Characterization of Flexible Poly(Vinyl Chloride) Foam Films. *J. Appl. Polym. Sci.* **2012**, *125*, 1448–1455.

19. Savrik, S. A.; Erdogan, B. C.; Balkose, D.; Ulku, S. Statistical Thermal Stability of PVC. *J. Appl. Polym. Sci.* **2010**, *116*, 1811–1822.

20. Savrik, S. A.; Balkose, D.; Ulutan, S.; Ulku, S. Characterization of Poly(Vinyl Chloride) Powder Produced by Emulsion Polymerization. *J. Therm. Anal. Calorim.* **2010**, *101*, 801–806.

21. Shakarami, K.; Doniavi, A.; Azdast, T.; Aghdam, K. M. Microcellular Foaming of PVC/NBR Thermoplastic Elastomer. *Mater. Manuf. Process.* **2013**, *28*, 872–878.

22. Shi, A.; Zhang, G.; Zhao, C. Study of Rigid Cross-Linked PVC Foams with Heat Resistance. *Molecules* **2012**, *17*, 14858–14869.
23. Singh, V. V.; Rangasai, M. C.; Singh, V. V.; Rangasai, M. C.; Bhosale, S. H.; Ghoshet, S. B. Renewable Resin Precursor and Residual Filler Modified Poly(Vinyl Chloride) Foam, Investigation of Microcellular Development and Moisture Absorption Behavior. *Polym. Plast. Technol. Eng.* **2015**, *54*, 1263–1269.
24. Sivertsen, K. Polymer Foams. *3.063, Polym. Phys.* 2007. http://ocw.mit.edu/courses/materials-science-and-engineering/3-063-polymer-physics-spring-2007/assignments/polymer_foams.pdf (accessed June 12, 2016).
25. Verdu, J.; Zoller, A.; Marcilla, A. Plastisol Foaming Process: Decomposition of the Foaming Agent, Polymer Behavior in the Corresponding Temperature Range and Resulting Foam Properties. *Polym. Eng. Sci.* **2013**, *53*, 1712–1718.
26. WEB 1: *Sartorius Density Determination Kit User's Manual*; Sartorius AG: Goettingen, 1992. https://scaleman.com/index.php/aitdownloadablefiles/download/aitfile/aitfile_id/1414/ (accessed June 12, 2016).
27. Wilkes, C. E.; Daniels, C. A.; Summers, J. W. *PVC Handbook*; 2005 (Chapter 5). http://www.hanser.de/3-446-22714-8 (accesed June 12, 2016).
28. Yu, N.; Kakhramanl, Y. N.; Gadzhieva, R. S. Sorption Properties of Sorbents Based on Polyvinyl Chloride Foam. *Chem. Technol. Fuels Oils* **2012**, *47*, 464–469.
29. Zoller, A.; Marcilla, A. Soft PVC Foams: Study of the Gelation, Fusion and Foaming Processes. III. Mixed Phthalate Ester Plasticizers. *J. Appl. Polym. Sci.* **2012**, *124*, 2691–2701.
30. Zoller, A.; Marcilla, A. Soft PVC Foams: Study of the Gelation, Fusion, and Foaming Processes. II. Adipate, Citrate and Other Types of Plasticizers. *J. Appl. Polym. Sci.* **2011**, *122*, 2981–2991.
31. Zoller, A.; Marcilla, A. Soft PVC Foams: Study of the Gelation, Fusion, and Foaming Processes. I. Phthalate Ester Plasticizers. *J. Appl. Polym. Sci.* **2011**, *121*, 1495–1505.

CHAPTER 3

THERMAL AND TRIBOCHEMICAL PROCESSES IN POLYPHENYLENE OXIDE IN THE PROCESS OF FRICTION

M. B. GURGENISHVILI*, G. SH. PAPAVA, R. G. LIPARTELIANI, SH. R. PAPAVA, Z. N. CHUBINISHVILI, N. Z. KHOTENASHVILI, and Z. SH. TABUKASHVILI

TSU Petre Melikishvili Institute of Physical and Organic Chemistry, Tbilisi State University, 31 A. Politkovskaia Str., 0186 Tbilisi, Georgia

Corresponding author. E-mail: marina.gurgenishvili@yahoo.com

CONTENTS

ABSTRACT

Structural changes of polyphenylene oxide in the process of thermal treatment and friction were studied by mass spectrometric method. It was shown that under the terms of treatment at 300°C, destructive structuring processes take place in a polymer. Besides, Friss regrouping takes place that leads to branching of macromolecules and forms gel fraction. Investigation of tribochemical processes during friction showed that character of tribochemical processes in a polymer is conditioned by the formation of complex structure of pressed specimens containing branching, cross-linked polymer, and low molecular fraction as a result of friction at the border of friction surface.

3.1 INTRODUCTION

Antifriction self-lubricating polymer materials, which are obtained by traditional (polyamide. polyethylene, polycarbonate, polytetrafluorine ethylene, polyimide, polyarylate, formaldehyde resin et al) as well as comparatively new thermally stable polymers (polyimide, polyarylate),[1] are widely used in various spheres of technology. But alongside with the progress of technology, the demand on materials characterized by high exploitation indices and exploited on especially high charges (P) and velocities (V), and correspondingly on high heat conduction tear and wear resistant increase.[2] Therefore, creation of a new, filled multicomponent system in which these properties are well fused with good physical, chemical, and technological properties still remains as actual theoretical and practical problem of physics, chemistry, tribochemistry, and tribotechnology of modern polymer materials.

Creation of multicomponent, self-lubricating antifriction material is a complex scientific problem, which needs thorough study of the processes going on during their processing/treatment and friction. It needs study of the role of its separate components in order to reveal inadequate behavior of the whole system.

As is known, at the exploitation of aggregated at high rates and high loading of friction aggregates (even at low friction coefficient), we observe isolation of abundant heat. If antifriction system is characterized by low heat conduction and thermal stability, aggregate starts deformation and breaks down because of rapid heating of friction surface. Owing to this, the sphere of application of polymer antifriction materials is limited.

In special literature, there are references about polyethylene oxide—a heterochain polymer, as one of the most perspective materials. It was developed in the sixties of the last century by the American Company "General Electric."[3] It belongs to thermoresistant polymers and possesses a series of unique exploitation properties.

3.2 EXPERIMENTAL PART

Study of thermal stability of initial and pressed polyphenylene oxide (PPO) was carried out by mass spectrometry. Specters were taken at the temperature interval 25–500°C, after every 50°C. Energy of ionization electrons equaled to 70 e.v.

Investigation of tribochemical processes of specimens of PPO was performed on a device, installed in a chamber of mass spectrometer, in immediate closeness of ionization zone,[4] at the drum rotation velocity $V = 1$ m/s and pressure $P = 0.1$ Mpa.

3.3 RESULTS AND DISCUSSION

Figure 3.1 shows that thermal destruction of the initial PPO proceeded in two stages, reaching maximum emission of initial products at temperatures 200°C and 450°C. At 200°C, emission of xylene (m/z 106) and toluene (m/z 91) is observed. Toluene is used as a solvent for the synthesis of PPO and is preserved in trace quantity in ready product, while xylene is present in toluene in the form of premixes.

The main thermal gradation of PPO is observed at 450°C. It is accompanied with emission of products with c m/z 122, 107, 91, 135, 242.

Comparison of the obtained results with thermal destruction of model substances (dimethyl benzol, 2,6- and 3,5-dimethyl phenol, diphenyl methane, diphenyl ether) showed that at such conditions, destruction of the main chain of a polymer takes place by the formation of[5]:

m/z 122   m/z 107   m/z 91   m/z 77

A

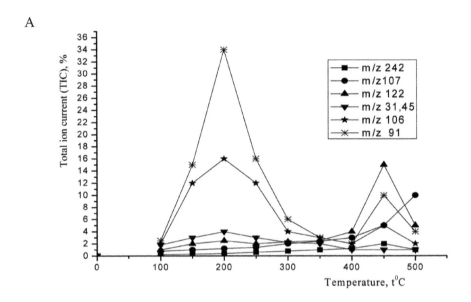

B

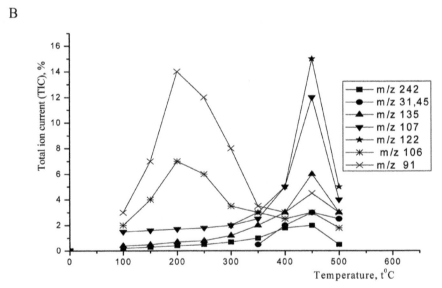

FIGURE 3.1   Effect of temperature on products of destruction in initial (A) and pressed at 300°C (B) olygophenylene oxide.

Appearance of a product with m/z 135 in the specter is probably conditioned by the Friss regrouping process at high temperature (450–500°C) (5) according to a scheme:

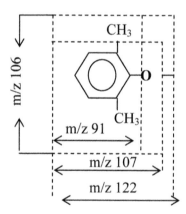

At the same time, at thermal destruction of purified PPO in mass specter ion m/z 135 is not observed, which refers to high thermal stability of a polymer.

Thus, results of the study of properties of industrial specimen of PPO polymer showed that at thermal impact (450°C), polymer destruction takes place in the main chain (bond phenyl-o-phenyl is destructed).

Alongside with it, in such conditions, Friss regrouping can take place (emission of m/z 135). Low molecular admixes present in a polymer, toluene, xylol as well as traces of dimmers, trimmers reduce thermal stability of PPO to 200°C and can lead to changes in its structure at its treatment.

Investigation of molecular mass characteristics of industrial PPO after its treatment by pressing at temperature 300°C showed that in this case, definite decrease of specific viscosity of a polymer from 0.47 to 0.45 dl/g and slight increase in its molecular mass from 56,000 to 65,000 are observed which might be conditioned by the formation of branched structure at a polymer. After pressing, formation of gel-fraction is observed in a polymer that amounts to approximately 4%.[6]

Formation of crossed structures in PPO in the process of treatment might be conditioned at the expense of oxidation of a side methyl group as well as recombination of free macro radicals created at thermal destruction.

Investigation of tribochemical processes of specimens of PPO was performed on a device, installed in a chamber of mass spectrometer, in immediate closeness of ionization zone,[4,7] at the drum rotation velocity $V =$ 1 m/s and pressure $P = 0.1$ Mpa (Fig. 3.2).

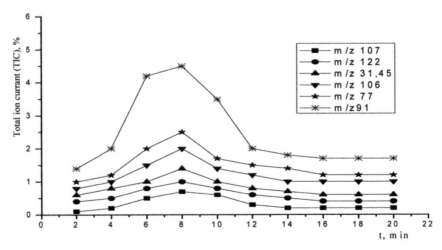

FIGURE 3.2 Influence of friction duration on the emission of the main products of tribodestruction from industrial specimen of polyphenylene oxide (PPO).

Investigation has shown that maximum emission of volatile products is observed in the process of running-in the first 5–10 min. At the increase of duration of friction run-in, the quantity of emitted products decreases.

The main products of tribodestruction are benzyl radical—$(C_6H_5CH_2,$ m/z 91), phenyl radical (C_6H_5, m/z 77), xylol ($C_6H_4(CH_3)_2$, m/z 106), traces of 2,6-xylenol ($C_6H_3(CH_3)_2OH$, m/z 122), as well as the products with mass number m/z 31 (CH_2OH) and m/z 45 (COOH), which probably appear as a result of oxidation of methyl side groups.

At the comparison of tribo- and thermal destruction of PPO,[8] we can see that in the process of friction, in the principle the products are emitted, which are inherent to low-temperature thermal destruction of initial PPO (m/z 106 and 91) taking place, as it was shown earlier, at 200°C.

We can suppose that initially thermally less stable products of low molecular fractions can undergo tribodestruction and be emitted as volatile products. At the same time, at the friction of PPO, we observe appearance of definite amount of ions with m/z 122, m/z 107, and m/z 45 and m/z 31 in the specter, which can prove that together with tribodestruction of low molecular fractions, at the friction cleavage of the side methyl groups (–CH$_3$) takes place together with destruction in the main chain.[9]

While investigating the process of friction of PPO, purified by their repeated precipitation from the solution in chloroform by acetone (Fig. 3.3), it was proved that period of run-in, in this case, increases from 6–7 to 12–13 min.

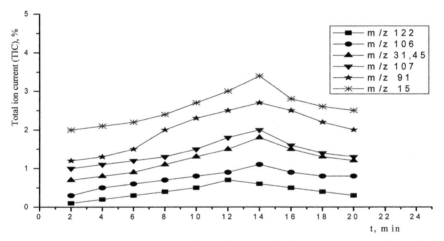

FIGURE 3.3 Influence of duration of friction on emission of the main products from specimens of repeatedly precipitated polyphenylene oxide.

In distinct from thermal destruction (Fig. 3.3), the main products of tribodestruction are m/z 15 (CH$_3$– group), m/z 91 (benzyl radical), m/z 107 (cresol), as well as small quantity of products with m/z 106 and 122. This time, phenyl radical is absent in the specter.

3.4 CONCLUSIONS

The obtained results show that the process of friction of purified PPO is accompanied mainly by detachment of methyl groups. The main chain at this moment is destructed at a less degree compared with that of industrial polymer.

Difference in tribodestruction and thermal destruction of purified PPO is that in the process of friction, cresol radical is formed at the expense of detachment of methyl group (m/z 15), definite amount of benzyl radical is emitted; while at thermal destruction, the main destruction product is xylenol radical (m/z 122). Apparently, this is conditioned by formation of "secondary" thermally stable structure in PPO in the process of friction on the surface of friction at the expense of recombination of free radicals which are formed as a result of thermo- and mechanical effect. Absence of low molecular fractions in a polymer, which actively participates in tribome-chanical reactions, slows down a process of tribodestruction of PPO, which finally leads to decrease of wear.

On the basis of the obtained results, we can suppose that character of tribomechanical processes in industrial PPO is conditioned by complex structure of pressed specimens which contain branched, crossed polymer and low molecular fraction.

This structure of industrial PPO is characterized, compared with other polyheteroarylenes,[10] with lower friction coefficient, up to 120–150°C, which makes expedient its application as a binding agent in antifriction plastic masses.

KEYWORDS

- polymer
- Friss regrouping
- tribomechanical reactions
- mass spectrometry
- polyphenylene oxide

REFERENCES

1. Hai, A. S. Aromatic Polyethers. *Adv. Polym. Sci.* **1966**, *4*, 496–527.
2. Korshak, V. V.; Gribova, I. A.; Krasnov, A. P.; Pavlov, S. V.; Slonimsky, G. A.; Askadsky, A. A. Chemical Structure of Polyheteroarylene Rings and Their Wear Resistance. *DAN USSR* **1985**, *282*(8), 654–659.
3. Conley, R. J. Studies of the Stability of Condensation Polymers in Oxygen-containing Atmospheres. *Macromolecules* **1967**, *A1*(1), 81–106.
4. Elyasherg, M. E. Expert Systems for Defining Structure of Organic Molecules by Spectral Methods. *Achiev. Chem.* **1999**, *68*, 579.
5. Iachoweiz, J.; Krjszewsk, K. P. Thermal Degradation of Poly-(2,6-Dimethyl-1,4-Phenylene Oxide). *J. Appl. Polym. Sci.* **1978**, *10*, 2891–2901.

6. Kelleher, P. G.; Jassie, L. B.; Gesner, B. D. Thermal Oxidation and Photo-oxidation of Poly-(2,6-Dimethyl-1,4-Phenylene Oxide). *J. Appl. Polym. Sci.* **1967,** *11*(1), 137–144.

7. Nekrasov, Y. S.; Zhokhov, V. E.; Aderikha, V. N.; Sorokin, D. S.; Gribova, I. A.; Krasnov, A. P.; Korshak, V. V. A Device for Investigation of Tribochemical Processes and Rigid Chain Polymers by Mass-spectrometry. *Frict. Wear* **1983,** *4*(1), 33–43.

8. Gurgenishvili, M. B. Development and Investigation of Antifriction Plastic Masses on the Base of Polyphenylene Oxide. Candidate's Thesis, Moscow, 1988, pp 33–43.

9. Popov, V. A.; Kolubaev, A. V. Analysis of Mechanisms of Formation of Surface Layers at Friction. *Frict. Wear* **1997,** *18*(6), 818–827.

10. Korshak, V. V.; Gribova, I. A.; Krasnov, A. P.; Pavlova, S. A.; Slonimsky, G. A.; Askadsky, A. A. Chemical Structure of Polyheteroarylene Rings and Their Wear Resistance. *DAN USSR* **1985,** *282*(8), 654.

CHAPTER 4

COMPATIBILITY AND THERMAL PROPERTIES OF POLY(ETHYLENE OXIDE) AND NATURAL RUBBER-GRAFT-POLY(METHYL METHACRYLATE) BLENDS

N. F. A. ZAINAL[1], C. H. CHAN[2,*], and A. M. M. ALI[2]

[1]Centre of Foundation Studies, Universiti Teknologi MARA Cawangan Selangor, Kampus Dengkil, 43800 Dengkil, Selangor, Malaysia

[2]Faculty of Applied Sciences, Universiti Teknologi MARA (UiTM), 40450 Shah Alam, Selangor, Malaysia

*Corresponding author. E-mail: cchan_25@yahoo.com.sg

CONTENTS

ABSTRACT

Natural rubber (NR) grafted with 40 %mol poly(methyl methacrylate) (PMMA) as side chains designated as NR-*g*-PMMA and poly(ethylene oxide) (PEO) were used to prepare polymer blends by solution casting method. Thermal stability of the PEO/NR-*g*-PMMA blends was investigated by thermogravimetric analysis (TGA). The decomposition temperatures of the blends reduce slightly at elevating amount of NR-*g*-PMMA. It can be deduced that PEO has a better thermal stability as compared with NR-*g*-PMMA. Thermal properties of PEO/NR-*g*-PMMA blends were examined using differential scanning calorimeter (DSC). The PEO/NR-*g*-PMMA blends exhibit compatibility as indicated by the depression of crystallinity (X^*) and melting temperature (T_m) of PEO at certain mass fraction of PEO in the blends (W_{PEO}). As for the intermolecular interactions between PEO and NR-*g*-PMMA in the blends, Fourier transform infrared (FTIR) spectroscopy was used. PEO and NR blends are immiscible, but PEO and PMMA blends are miscible. Hence, it is interesting to elucidate the phase behavior of PEO/NR-*g*-PMMA, where there may be the existence of miscible interfacial region of PEO and PMMA between the PEO and NR phases.

4.1   INTRODUCTION

Natural rubber (NR), a renewable resource, is an unsaturated elastomer and become an important material in the rubber industry, especially for the manufacturing of rubber tyres.[1] However, due to the existence of its unsaturated hydrocarbon and nonpolar characteristic in the chain structure, the uncured NR suffers from limitations for practical applications. Those limitations include low flame resistance, sensitive to heat and oxidation, limited resistance to chemicals and solvents, and poor ozone and weathering performance. Therefore, various methods have been developed to alter NR properties through physical and chemical modifications. Improvement in certain properties through chemical modification such as oil and flame resistance, hardness and tensile strength, and gas permeability has been reported.[2-4] Graft polymerization of vinyl,[5-7] methyl methacrylate,[8,9] and styrene[8,10,11] monomers onto NR backbone was attempted for desired properties of chemically modified NR. Besides, hydrohalogenation, halogenation, oxidation, epoxidation, ozonolysis, hydrogenation, carbine addition, cyclization, etc. are some of the common approaches for chemical modifications of NR in order to regulate the physical and mechanical properties of chemically modified NR.[12,13]

Binary polymer blending of amorphous and semicrystalline polymers is commonly in practice. Apart from acquiring synergistic effects of parent polymers, blending of polymers may lead to preparation of new materials with tunable properties for specific applications, for example, for usage as solid polymer electrolytes (SPEs) [i.e., salt is added to polymer(s)]. SPEs are used for ion batteries or other electrochromic devices.[14-17] For polymer blend, the term miscibility used to describe a molecularly dispersed system, and immiscibility, on the contrary, is coexistence of phases of the individual constituents. Compatible polymer blend is immiscible polymer blends that may have strong intermolecular interaction or miscible phase at the interfacial region of the two distinct phases of the parent polymers.

Poly(ethylene oxide) (PEO) and epoxidized natural rubbers (ENR) are immiscible blends.[18-22] PEO and NR blends are most likely immiscible as well. Blends of PEO and poly(methyl methacrylate) (PMMA) are miscible.[23-29] When PEO is blended with natural rubber-*graft*-PMMA (NR-*g*-PMMA), compatible blend may be prepared because good interfacial adhesion could be achieved if PEO likes PMMA-*graft*. Enhancement of conductivity while retaining acceptable mechanical properties of PEO/NR-*g*-PMMA-based SPEs as compared with binary system of PEO and salt may be obtained.

SPEs with lithium salt dissolved in PEO display low ionic conductivity ($\sigma_{DC} \sim 10^{-5}$ S cm^{-1}) at room temperature and prohibit them from practical applications.[30] Blending of PEO with ENR with addition of inorganic salt enhances conductivity as compared with system with PEO and inorganic salt at same salt content. This may be due to the preferential percolation of inorganic salt in PEO phase in relative to ENR phase.[21,31-34] However, these studies are silent on the mechanical properties and it is expected that the PEO/ENR-based SPEs suffer from poor mechanical properties due to immiscibility of the systems. LiClO$_4$-added PEO/PMMA 75/25 blend (m/m) (a miscible system) has lower ion conductivity than that of the PEO + LiClO$_4$ system but recorded enhanced conductivity as compared with PMMA + LiClO$_4$ systems at mass fraction of salt (W_S) at 0.091. Li salt is more soluble in PEO phase as compare with PMMA phase.[28]

The deficiencies encountered for PEO/ENR-based and PEO/PMMA-based SPEs may be approached by blending PEO and NR-*g*-PMMA with addition of inorganic salt, where PMMA-*graft* likes PEO. In this case, improved interfacial adhesion between PEO phase and NR phase can be targeted *via* miscible or compatible PEO/PMMA interfacial region. Percolation of salt can be facilitated in PEO phase without moving to NR phase when PEO is in excess.

Hence, the thermal properties and phase behavior of PEO/NR-*g*-PMMA are of interest in this study as there may be the existence of interfacial region of PEO and PMMA between the PEO and NR phases. The phase behavior–performance relationship of ternary systems (PEO/NR-*g*-PMMA SPEs, i.e., with addition of inorganic salt) will be revealed in forthcoming publication.

This work will focus on binary blends of PEO/NR-*g*-PMMA, where PMMA-*graft* accounts of 40 %mol of the graft copolymer. Thermal stability of the blends was studied using thermogravimetric analysis (TGA) by estimation of thermal decomposition temperature. Compatibility and thermal properties of the blends were elucidated by using differential scanning calorimeter (DSC). Fourier transform infrared (FTIR) spectroscopy was used to study the intermolecular interaction between the polymers in the blends.

4.2 EXPERIMENTAL

4.2.1 MATERIALS AND SAMPLE PREPARATION

PEO (Thermo Fisher, Pittsburgh, USA) and NR-*g*-PMMA [Green HPSP (M) Sdn Bhd, Petaling Jaya, Malaysia] were purified before the blend preparation. For purification, PEO was dissolved in chloroform ($CHCl_3$) (Merck, Darmstadt, Germany) and precipitated in *n*-hexane (Merck, Darmstadt, Germany). NR-*g*-PMMA was purified before further used through removal of macrogel and removal of homopolymers using selective extraction of homopolymers from the graft copolymer. The free NR homopolymer (ungrafted NR) was extracted with light petroleum ether at 40–50°C for 24 h and the remaining product was dried at 80°C in the oven for 24 h. Subsequently, the residue was further extracted in acetone at 40–50°C for 24 h to remove free PMMA homopolymer. The remaining product was dried to a constant weight at 80°C. The characteristics of PEO and NR-*g*-PMMA used in this work are tabulated in Table 4.1.

For the preparation of the blends, solution casting technique was used to prepare free-standing PEO/NR-*g*-PMMA films. These films were prepared from 2% (m/m) (m denotes mass) solutions of two polymers in tetrahydrofuran (THF) (Merck, Darmstadt, Germany). The solution was stirred for 24 h at 50°C until the polymers were completely dissolved. Next, the homogenous solution was poured into Teflon dish and dried under the fume hood for overnight at room temperature before dried at 50°C for 24 h in an oven. Then, the film was dried again under nitrogen atmosphere at 80°C for 30 min and lastly vacuum dried at 50°C for 48 h. All the films were kept in desiccators before characterization. PEO/NR-g-PMMA blends at compositions (m/m) of 0/100, 20/80, 40/60, 50/50, 60/40, 80/20, and 100/0 were prepared.

TABLE 4.1 Characteristic of PEO and NR-g-PMMA Used in This Work.

Constituents	PEO	NR-g-PMMA
M_η^a (g mol^{-1})	300,000	
M_w^b (g mol^{-1})		1,000,000
M_n^b (g mol^{-1})		450,000
T_m^c (°C)	65	–
ΔH_{ref} (J g^{-1})	188.3[f]	
T_g^d (°C)	−58	−63
		(−66, −2)[g]
T_d^e (°C)	374	350
Molecular structure		

[a]Viscosity average molar mass estimated by supplier.
[b]Mass average or number average molar mass as estimated in this work by GPC. Poly(styrene) with low polydispersity was used as standard.
[c]Melting temperature by DSC as determined in this work.
[d]Glass transition temperature after quench cooling by DSC as determined in this work.
[e]Decomposition temperature by TGA as determined in this work.
[f]Melting enthalpy adopted from Lide.[35]
[g]T_gs values of NR-g-PMMA with 57 %mol PMMA.[36]
[h]x = mol fraction of PMMA-graft = 0.4.
y = mol fraction of NR backbone = 0.6.

4.2.2 CHARACTERIZATION

4.2.2.1 GEL PERMEATION CHROMATOGRAPHY

The mass average (M_w) and number average (M_n) molar mass of neat polymers of NR-g-PMMA was studied using a Waters gel permeation chromatography (GPC) (Milford, US) coupled with Waters Styragel columns HR 3 or HR 5E CHCl$_3$ (Merck, Darmstadt, Germany). High-performance liquid chromatography was used as the mobile phase throughout the experiment with flow rate 1.0 mL/min and solvent with sample concentration of 10.0 mg mL^{-1}. Polystyrene standards (Easical, Darmstadt, Germany) (M_w = 665–6,989,000 g mol^{-1} with narrow polydispersity) with low polydispersity were used to prepare a calibration curve.

4.2.2.2 ¹H NUCLEAR MAGNETIC RESONANCE SPECTROSCOPY

Neat NR-g-PMMA was analyzed by using ¹H nuclear magnetic resonance (NMR). ¹H NMR spectrum of NR-g-PMMA was measured at 25°C on a Bruker Avance 300 spectrometer (Switzerland). It was operated at a resonance frequency of 300 MHz at point 60,000 with pulse width 10 μs and pulse interval of 18 s at 2000–3000 scan number. Sample of NMR was performed by using 1% m/v deuterated chloroform ($CDCl_3$) solutions of the rubber. The polymer solution was heated in the oven at 50°C for overnight or until the polymer dissolved in the solvent completely where a clear solution was observed. The integrated peak areas of the NMR spectra were used to calculate the %mol of grafted PMMA content on NR backbones using eq 4.1. The %mol of grafted PMMA content on NR backbones is calculated using the integrated peak area assigned to methoxy protons ($-OCH_3$) of grafted PMMA and olefinic proton (=CH–) of NR.

$$\% \text{Mol P MMA of NR-g-PMMA} = \frac{(A_{3.5}/3)}{(A_{3.5}/3) + A_{5.1}} \times 100\% \tag{4.1}$$

Quantities $A_{3.5}$ and $A_{5.1}$ represent integrated peak areas of the methoxy proton of PMMA at 3.5 ppm and olefinic proton of isoprene repeating unit at 5.1 ppm, respectively. From Figure 4.1, $A_{3.5}$ represents the total peak area value of three protons in methoxy of PMMA. Therefore, $A_{3.5}$ is divided by 3 to obtain the peak area value for one proton of methoxy as shown in eq 4.1.

4.2.2.3 THERMOGRAVIMETRIC ANALYSIS

Thermal stability of the PEO/NR-g-PMMA blends was studied using TGA Q500 (TA instrument, Delware, USA). About 8 mg of each sample were used for each experiment. All samples were heated from 30°C to 600°C in nitrogen atmosphere at the heating rate of 10°C min^{-1}. The activation energy of the thermal decomposition of the sample was determined through the semilogarithm of first derivative mass retained as the function of reciprocal of temperature.

4.2.2.4   DIFFERENTIAL SCANNING CALORIMETER

TA Q2000 (TA Instruments, Delware, USA) equipped with RCS90 refrigerator cooling system (TA Instruments, Delware, USA) and nitrogen gas purging was used to determine the values of glass transition temperature (T_g), change of heat capacity (ΔC_p), melting temperature (T_m), melting enthalpy (ΔH_m), and crystallinity (X^*) of the blends in the heating cycle. The DSC was calibrated with indium and sapphire standard. The sample was quenched cooled to −90°C for 5 min and heated to 80°C at a rate of 10°C min⁻¹. Similar thermal procedure was applied in sapphire standard run and baseline run. Quantity of T_g is taken at half ΔC_p value. To determine the C_p value of sample, the heat flow signal from the sample was compared with the baseline of sapphire standard which is of known C_p. Both curves were corrected by zero or baseline correction experiment whereby an empty reference and an empty sample pan were placed in the furnace. Quantity ΔC_p was estimated from the glass transition of the onset and endset of the heat flow curve. The melting peak maximum was taken as T_m value. Quantity ΔH_m is estimated from the area corresponding to the melting endotherm.

4.2.2.5   FTIR SPECTROSCOPY

The physical intermolecular interactions of PEO and NR-g-PMMA were examined by FTIR spectroscopy. FTIR spectrum collection was carried out using the attenuated total reflectance (ATR) accessory with diamond crystal window on Nicolet 6700 ATR–FTIR (Thermo Scientific, Madison, USA). The spectra were recorded in absorbance mode in the range of 600–4000 cm⁻¹ with 16 scans at a resolution of 2 cm⁻¹ at room temperature (~25°C). The sample was analyzed at three spots to ensure that the FTIR analysis was based on a representation region of sample.

4.3   RESULTS AND DISCUSSION

4.3.1   CHARACTERIZATION OF NR-G-PMMA

Grafting of PMMA onto NR backbone can be easily verified by ¹H NMR by characterizing the extracted NR-g-PMMA after petroleum ether and acetone extractions. A typical ¹H NMR spectrum of extracted NR-g-PMMA is shown in Figure 4.1. The absorption bands at the chemical shifts of 3.6 and 5.1

ppm, which indicate the signals of (−OCH₃) proton of the PMMA and the (=CH−) proton of the poly(isoprene), respectively, can be clearly observed. The %mol of PMMA grafted onto NR obtained from eq 4.1 is 40 %mol.

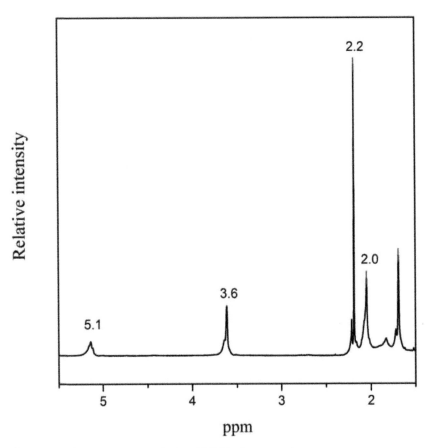

FIGURE 4.1 ¹H NMR spectra of NR-*g*-PMMA after extraction.

4.3.2 THERMAL STABILITY OF THE BLENDS

The TGA curve of %mass retained as a function of temperature is shown in Figure 4.2. The temperature onsets of %mass retained curve as a function of temperature ($T_{d'onset}$), and the temperature of maximum decomposition rate of the %mass retained curve ($T_{d'inflection}$) of PEO/NR-*g*-PMMA blends are shown in Table 4.2. Quantity T denotes temperature.

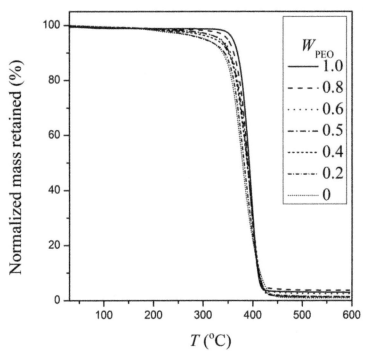

FIGURE 4.2 TGA thermograms of %mass retained curves as a function of temperature for PEO/NR-*g*-PMMA blends.

TABLE 4.2 $T_{d,onset}$, $T_{d,inflection}$, and %mass Loss of PEO/NR-*g*-PMMA Blends up to 600°C.

Polymer Blends, W_{PEO}	$T_{d,onset}$ (°C)	$T_{d,inflection}$ (°C)	%mass Loss
1.0	374	392	96
0.8	369	389	96
0.6	368	390	97
0.5	365	389	98
0.4	362	389	99
0.2	355	383	98
0.0	350	380	99

The thermal degradation of NR-*g*-PMMA shows that the materials are stable up to 165°C. Beyond that, a first minor degradation step takes place with a low mass loss of about 2–3%, and then a second (major) degradation step occurs between 320°C and 450°C. Both quantities of $T_{d,onset}$ and $T_{d,inflection}$ of the samples increase when the content of PEO increases. This may indicate that the blends are more thermally stable with higher content of PEO. All samples are thermally degraded at temperature 400–500°C.

4.3.2.1 TEMPERATURE DEPENDENCE OF THE RATE OF THERMAL DEGRADATION

The temperature dependence of the rate of thermal degradation in terms of Hoffman's Arrhenius-like relationship is discussed. The relationship for first derivatives of the mass retained (%) versus reciprocal of temperature is shown in eq 4.2.

$$\left|\text{deriv. mass retained}(\%)\right| \; \alpha \exp\left(-\frac{E_A}{RT}\right), \tag{4.2}$$

where R is the gas constant and E_A is the activation energy of the thermal degradation process of the sample. The estimation of quantities of E_A can be extracted from the slope of the semilogarithmic plots of |deriv. mass retained (%)| against $1/T$ as shown in the linear region as illustrated by the solid linear regression curve for $W_{PEO} = 0.8$ in Figure 4.3.

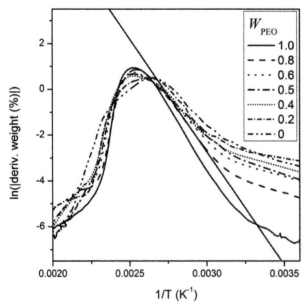

FIGURE 4.3 Plot of ln(|deriv. mass retained|) against $1/T$. Solid regression curve is after eq 4.2 for $W_{PEO} = 0.8$ for the linear region.

In Table 4.3, the E_A values for PEO are 59 kJ mol^{-1}, while NR-g-PMMA is 39 kJ mol^{-1}. This implies that neat PEO is more thermally stable than NR-g-PMMA. The E_A values for the blends show decreasing trend with higher content of NR-g-PMMA. Quantities of E_A are in agreement with the T_d values. This suggests that blends with higher content of PEO are thermally more stable.

TABLE 4.3 Activation Energy of the Rate of Thermal Degradation of PEO/NR-g-PMMA Blends.

Polymer Blends, W_{PEO}	E_A/R (K)	Activation Energy, E_A (kJ mol^{-1})
1.0	7122	59
0.8	6966	58
0.6	5744	48
0.5	5405	45
0.4	5096	42
0.2	4851	40
0	4750	39

4.3.3 COMPATIBILITY AND MELTING BEHAVIOR OF PEO/NR-G-PMMA BLENDS

4.3.3.1 GLASS TRANSITION

One of the common experimental approaches in determining the miscibility or compatibility of blend systems is estimation of T_g value(s) of the blends *via* thermal analysis, let us say by DSC.[37] However, this approach is applicable only for polymer pairs which have difference in T_gs for more than 20°C.[38] From DSC analysis shown in Figure 4.4, the T_g values for both PEO and NR-g-PMMA are −58°C and −63°C, respectively. The T_gs values of NR-g-PMMA with 57 %mol PMMA content[36] are −66°C (for NR backbone) and −2°C (for PMMA-*graft*) using DSC. The T_g of PMMA-*graft* is weak and sometimes it is rather challenging to be estimated precisely by linear heating rate of the heating cycle as for this study, where the T_g value of the PMMA-*graft* for the NR-g-PMMA in this study cannot be revealed clearly.

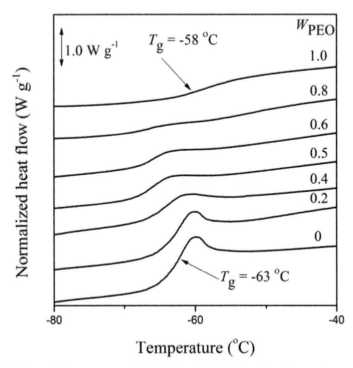

FIGURE 4.4 DSC thermograms of PEO/NR-g-PMMA blends heating cycle.

TABLE 4.4 Quantities T_g and ΔC_p of PEO/NR-g-PMMA.

W_{PEO}	T_g (°C)	ΔC_p [J/(K·°C)]
1	−58	0.131
0.8	−67	0.140
0.6	−66	0.135
0.5	−66	0.174
0.4	−64	0.205
0.2	−63	0.332
0	−63	0.387

Quantities ΔC_p of the superimposed glass transition of PEO and NR-g-PMMA in the blends are also evaluated with elevating amount of NR-g-PMMA as illustrated in Table 4.4. If PEO and NR-g-PMMA is immiscible, quantity ΔC_p of this superimposed glass transition may be an additive quantity from the quantities ΔC_p of the parent polymers. ΔC_p value of this superimposed glass transition of PEO and NR-g-PMMA in the blends is a measure of mass fraction of the amorphous microphase of PEO and NR backbone in the mixtures.

In molten state, PEO and NR-*g*-PMMA is immiscible and compatibility of PEO and PMMA-*graft* most likely can be noted. Upon cooling, liquid–solid phase separation transpires, where PEO starts to crystallize. PMMA-*graft* of NR-*g*-PMMA may not be able to be fully rejected out from the spherulitic growth front of PEO since PEO chain likes PMMA-*graft*. On the other hand, PEO has no special affinity to NR backbone, and hence, rather complete segregation of NR phase from PEO spherulites can normally be observed when PEO is in excess. According to this hypothesis, the amorphous phase of the blends may comprise a mixture of compatible interfacial region of PEO and PMMA-*graft*, which is neighboring of PEO and NR backbone phases. The amorphous phase of the blends is in equilibrium with PEO crystalline phase.

Figure 4.5 demonstrates negative deviation of additive quantity of ΔC_p for this glass transition from the quantities ΔC_p of the parent polymers for

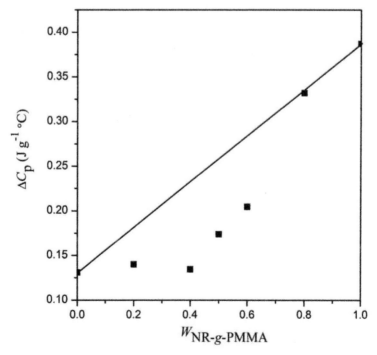

FIGURE 4.5 Change of heat capacity of the superimposed glass transition of PEO and NR backbone in the blends. Solid curve marks the additive quantity of the superimposed glass transition from the quantities ΔC_p of the parent polymers.

$0.2 \leq W_{\text{NR-g-PMMA}} < 0.8$. If the quantities ΔC_{p} is of close approximation to the solid curve, PEO and NR-g-PMMA is immiscible and PEO is not compatible with PMMA-*graft*. The negative deviation from the solid curve for $0.2 \leq W_{\text{NR-g-PMMA}} < 0.8$ may point that the amorphous phase of the blends may comprise amixture of compatible interfacial region of PEO and PMMA-*graft*, which is neighboring of PEO and NR backbone phases. Values of T_{g} and ΔC_{p} cannot be noted for mixture of compatible interfacial region of PEO and PMMA-*graft* under the experimental condition. Hence, lower content of PEO neighboring with NR backbone may lead to the depression of ΔC_{p} of this superimposed glass transition.

4.3.3.2   MELTING BEHAVIOR

DSC thermograms of PEO in blends for the heating cycle are shown in Figure 4.6. Both of the T_{m} and ΔH_{m} were obtained from the maximum peak and the area under the peak of the thermograms, respectively. As indicated in Figure 4.6, a single melting of semicrystalline PEO peak in the blends is observed. No observable melting peak for NR-g-PMMA is reported as it is an amorphous polymer.

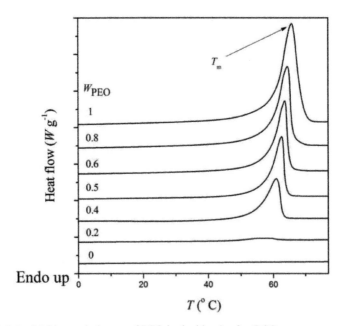

FIGURE 4.6   Melting endotherms of PEO in the blends after DSC.

As shown in Figure 4.7, the T_m values of PEO in PEO/NR-g-PMMA systems show a decreasing trend when the NR-g-PMMA content increases. As observed, there is only slight depression of the T_m when $W_{PEO} \geq 0.6$. However, depression of T_m of PEO can be seen clearly when $W_{PEO} < 0.6$. This indicates that the system may have a mixture of compatible interfacial region of PEO and PMMA-*graft* which is in agreement with the results of ΔC_p of the blends. Values of T_m of PEO obtained for PEO/NR-g-PMMA systems depress from 66°C ($W_{PEO} = 1$) to 55°C ($W_{PEO} = 0.2$). Elucidation of compatibility of PEO and PMMA-*graft* at interfacial region of the blends requires further investigation which will be reported in forthcoming publication.

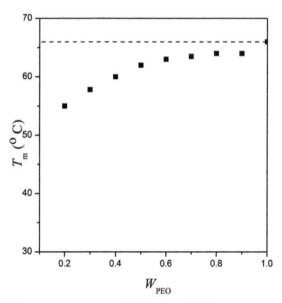

FIGURE 4.7   Melting temperatures of PEO/NR-g-PMMA blends against W_{PEO}. Dashed curve represents the constancy of melting temperature of PEO in the blends.

4.3.4   CRYSTALLINITY OF PEO/NR-G-PMMA BLENDS

The crystallinity (X^*) of the PEO phase in the blends was determined by the ΔH_m of PEO as shown in eq 4.3;

$$X^* = \left(\frac{\Delta H_m}{\Delta H^o_{ref} \times W_{PEO}} \right), \tag{4.3}$$

where $\Delta H°_{ref} = 188.3 \text{ J g}^{-1}$ is the enthalpy of 100% crystallinity of PEO. As indicated in Figure 4.8, the X^* of PEO in the blends decreases with increasing NR-g-PMMA content. The decrease in X^* is noted with elevating content of NR-g-PMMA. This implies that the blends be compatible at the interfacial region. This result supports the results of ΔC_p and T_m which point toward the compatible interfacial region of PEO and PMMA-$graft$ in the PEO/NR-g-PMMA blends.

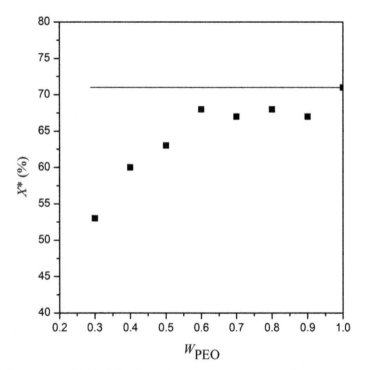

FIGURE 4.8 Crystallinity (X^*) of PEO/NR-g-PMMA blends against W_{PEO}. Solid curve marks the constancy of crystallinity of the blends.

4.3.5 FTIR SPECTROSCOPY

Figure 4.9 shows the FTIR spectra of neat PEO and NR-g-PMMA. The assignment of characteristic bands of both neat PEO and NR-g-PMMA is done based on references cited. The characteristic bands of PEO and NR-g-PMMA and the references are tabulated in Tables 4.5 and 4.6, respectively. It is observed that the characteristic bands of both neat PEO and NR-*g*-PMMA are relatively close to the characteristic bands extracted from the literature.

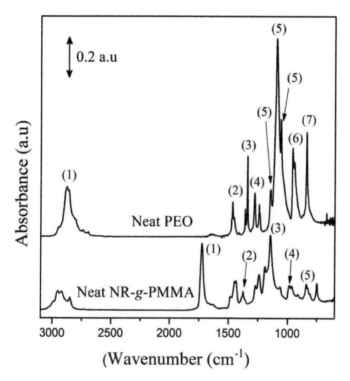

FIGURE 4.9   FTIR spectra of neat PEO and NR-*g*-PMMA.

TABLE 4.5   Characteristic Bands of PEO.

No.	Wavenumber(s) (cm⁻¹)	Reference Wavenumber(s) from Literature (cm⁻¹)	Assignment	Reference
1	2882	2885	$V_s(CH_2)$	39
2	1466	1466	$\delta_{as}(CH_2)$	39,40
3	1359, 1341	1360, 1343	$\omega(CH_2)$	39–41
4	1277	1278	$\omega(CH_2)$	39, 42
5	1145, 1093, 1059	1145, 1093, 1060	$\nu(C–O–C)$	39, 41
6	957	958	$\rho_{as}(CH_2)$	39, 42
7	841	844	$\rho_{as}(CH_2)$	43

TABLE 4.6 Characteristic Bands of NR-*g*-PMMA.

No.	Wavenumber (cm⁻¹)	Reference Wavenumber from Literature (cm⁻¹)	Assignment	Reference
1	1729	1732	ν(C=O)	44
2	1376	1390	δ(O–CH$_3$)	45
3	1146	1140	(C–O)	44
4	837	835	(=C–H)	44
5	984	990	ν_s(C–O–C)	45

In Figure 4.10(a), two sharp absorption bands at 1359 and 1341 cm⁻¹ representing the crystalline phase of PEO are observed. With the addition of NR-*g*-PMMA into the blends, there is a slight shifting of absorption bands at 1341 cm⁻¹, but no shifting of the wavenumber of absorption band at 1359 cm⁻¹. The absorbances of the band reduce as the content of NR-*g*-PMMA increases. In Figure 4.10(b), the middle absorption bands of ν(C–O–C) mode at 1093 cm⁻¹ corresponds to the amorphous phase of PEO, while the two shoulders represent the crystalline phase of PEO at 1059 and 1145 cm⁻¹. As observed, there is a slight shifting of absorption bands of ν(C–O–C) mode of amorphous phase of PEO to higher frequency from 1093 to 1101 cm⁻¹ and become broader in shape. As for the shoulders, only a slight shifting of the absorption bands observed at 1145 cm⁻¹ band and no clear shifting at 1059 cm⁻¹. If PMMA-*graft* is similar to PEO, the PEO spherulites will not be perfect and hence the crystalline structure of PEO will be somehow affected. The shifting of absorption bands at 1341, 1145 (crystalline phase of PEO), and 1093 cm⁻¹ (amorphous phase of PEO) may imply the compatibility of interfacial region of PEO and PMMA-*graft*. As shown in Figure 4.10(c), the decrease in the absorbance of the C=O stretching is qualitatively proportionate to the composition of the NR-*g*-PMMA in the blends with slight shifting of the wavenumber of absorption from 1726 to 1729 cm⁻¹. As W_{PEO} increases, the absorption band becomes broader in shape. The FTIR analysis shows that there is no specific molecular interaction between PEO and NR-*g*-PMMA. The compatibility of interfacial region of PEO and PMMA-*graft* in the blends may be due to the entropy effects as reported in most studies of PEO/PMMA blends.[46,47]

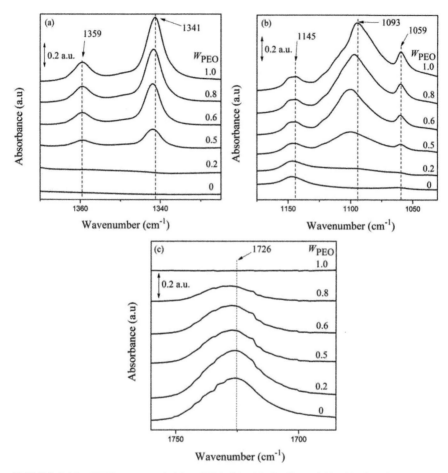

FIGURE 4.10   FTIR spectra of: (a) $\omega(CH_2)$, (b) $v(C-O-C)$, and (c) $v(C=O)$ of PEO/NR-g-PMMA blends.

4.4   CONCLUSIONS

PEO/NR-g-PMMA thin films were prepared using solution casting method. From thermal stability and the activation energy of rate of thermal decomposition studies, the blends show higher thermal stability with higher content of PEO. These blends are immiscible; however, compatibility of PEO and PMMA-$graft$ can be noted. Based on ΔC_p studies, the negative deviation at $0.2 \leq W_{NR-g-PMMA} < 0.8$ suggests the existence of a mixture of compatible interfacial region of PEO and PMMA-$graft$ in the blends. The depression of the T_m and X^* values further supports the compatibility at interfacial region of PEO and PMMA-*graft*. Based on FTIR studies, there is no specific intermolecular interaction between the parent polymers; however, it point toward only weak interaction as discussed in FTIR analysis.

ACKNOWLEDGMENT

The authors would like to express gratitude toward the Ministry of Education Malaysia for financing the project under Research Acculturation Grant Scheme (RAGS) [600-RMI/RAGS 5/3 (145/2014)] grant. Special thanks to Green HPSP (M) Sdn Bhd, Petaling Jaya, Malaysia for complimentary samples of NR-*g*-PMMA used in this study.

KEYWORDS

- poly(ethylene oxide)
- natural rubber-*graft*-poly(methyl methacrylate)
- thermogravimetric analysis
- differential scanning calorimeter
- Fourier transform infrared spectroscopy

REFERENCES

1. Kohjiya, S.; Ikeda, Y., Eds. *Chemistry, Manufacture and Applications of Natural Rubber*; Elsevier: Netherlands, 2014.
2. Brosse, J. C.; Campistron, I.; Derouet, D.; El Hamdaoui, A.; Houdayer, S.; Reyx, D.; Ritoit-Gillier, S. *J. Appl. Polym. Sci.* **2000**, *78*, 1461.
3. Derouet, D.; Radhakrishnan, N.; Brosse, J. C.; Boccaccio, G. *J. Appl. Polym. Sci.* **1994**, *52*, 1309.
4. Jams, L. W.; Sasaki, A. *Polym. Plast. Technol. Eng.* **2003**, *42*, 711.
5. Thiraphattaraphun, L.; Kaitkamjornwong, S.; Prasassarakich, P.; Damronglerd, S. *J. Appl. Polym. Sci.* **2001**, *81*, 439.
6. Prasassarakich, P.; Sintoorahat, P.; Wongwisetsirikul, N. *J. Chem. Eng. Jpn.* **2001**, *34*, 253.
7. Arayapranee, W.; Prasassarakich, P.; Rempel, G. L. *J. Appl. Polym. Sci.* **2002**, *83*, 3001.
8. Arayapranee, W.; Prasassarakich, P.; Rempel, G. L. *J. Appl. Polym. Sci.* **2003**, *89*, 74.
9. Kalkornsurapranee, E.; Sahakaro, K.; Kaesaman, A.; Nakason, C. *J. Appl. Polym. Sci.* **2009**, *114*(1), 587–597.
10. Suriyachi, P.; Kaitkamjornwong, S. *Rubber Chem. Tech.* **2004**, *77*, 930.

11. Suksawad, P.; Yamamoto, Y.; Kawahara, S. *Eur. Polym. J.* **2011**, *47*, 330–337.

12. Senyek, M. L. *Encyclopedia of Polymer Science and Engineering,* 2nd Edition; Mark, H. F., et al., Eds.; John Wiley and Sons: New York, 1989; Vol. 8.

13. Onchoy, N.; Phinyocheep, P. *Rubber Chem. Technol.* **2016**, *89*(3), 406–418.

14. Chandra, S. *Laboratory Experiments Manual*, 6th Asian Conference Cum International Research Training Workshop on Solid State Ionics, Varanasi, India 1998.

15. Gnanaraj, J. S.; Karekar, R. N. *Polymer* **1997**, *38*, 3709–3712.

16. Reddy, V. S. C.; Wu, G. P.; Zhao, C. X.; Jin, W.; Zhu, Q. Y.; Chen, W.; Sunil, M. *Curr. Appl. Phys.* **2007**, *7*, 655–661.

17. Scrosati, B.; Croce, F.; Panero, S. *J. Power Sources* **2001**, *100*, 93–100.

18. Chan, C. H.; Kammer, H. W.; Sim, L. H.; Harun, M. K. *Rubber: Types, Properties and Uses*; Popa, G. A., Ed.; Nova Science Publishers, Inc.: New York, 2010.

19. Sim, L. H.; Chan, C. H.; Kammer, H. W. 2011; pp 499–503. doi:10.1109/CSSR.2010.5773829.

20. Nawawi, M. A.; Sim, L. H.; Chan, C. H. *Int. J. Chem. Eng. Appl.* **2012**, *3*(6), 410–412.

21. Chan, C. H.; Kammer, H. W. *J. Appl. Polym. Sci.* **2008**, *110*(1), 424–432.

22. Chan, C. H.; Sulaiman, S. F.; Kammer, H. W.; Sim, L. H.; Harun, M. K. *J. Appl. Polym. Sci.* **2011**, *120*(3), 1774–1781.

23. Karim, S. R. A.; Sim, L. H.; Chan, C. H. *Adv. Mater. Res.* **2013**, *812*, 267.

24. Parizel, N.; Lauprete, F.; Monnerie, L. *Polymer* **1997**, *38*, 3719.

25. Martuscelli, E.; Pracella, M.; Yue, W. P. *Polymer* **1984**, *25*, 1097.

26. Cimmino, S.; Martuscelli, E.; Silvestre, C. *Makromol. Chem.* **1990**, *191*, 2447.

27. Elberanchi, A.; Daro, A.; David, C. *Eur. Polym. J.* **1999**, *35*, 1217.

28. Karim, S. R. A.; Sim, L. H.; Chan, C. H.; Ramli, H. *Macromol. Symp.* **2015**, *354*(1), 374–383.

29. Aram, E.; Ehsani, M.; Khonakdar, H. A.; Jafari, S. H.; Nouri, N. R. *Fibers Polym.* **2016**, *17*(2), 174–180.

30. Bruce, P. G. *Electrochim. Acta* **1995**, *40*, 2077–2086.

31. Noor, S. A. M.; Ahmad, A.; Talib, I. A.; Rahman, M. Y. A. *Ionics* **2011**, *17*(5), 451–456.

32. Noor, S. A. M.; Ahmad, A.; Talib, I. A.; Rahman, M. Y. A. *Ionics* **2010**, *16*(2), 161–170.

33. Noor, S. A. M.; Ahmad, A.; Rahman, M. Y. A.; Talib, I. A. *J. Appl. Polym. Sci.* **2009**, *113*(2), 855–859.

34. Chan, C. H.; Sim, L. H.; Kammer, H. W.; Tan, W. *Am. Inst. Phys. Conf. Proc.* **2012**, *1455*(1), 197–207.

35. Lide, D. R. *CRC Handbook of Chemistry and Physics*, 87th Edition; Taylor & Francis: New York, 2006; pp 6–112.

36. Yusoff, S. N. H. M. *Modified Natural Rubber Solid Polymer Electrolytes;* Universiti Teknologi MARA: Malaysia, 2013.

37. Anand, K. K.; Vasile, C. *Handbook of Polymer Blends and Composites*; Smithers Rapra Publishing, 2003.

38. Flory, P. J. *Principles of Polymer Chemistry*; Cornell University Press: Ithaca, New York, 1953.

39. Yoshihara, T.; Tadokoro, H.; Murahashi, S. *J. Chem. Phys.* **1964**, *41*, 2902–2911.

40. Gondaliya, N.; Kanchan, D. K.; Sharma, P.; Joge, P. *Mater. Sci. Appl.* **2011**, *2*, 1639–1643.

41. Manoratne, C. H.; Rajapakse, R. M. G.; Dissanayake, M. A. K. L. *Int. J. Electrochem. Sci.* **2006**, *1*, 32–46.

42. Sim, L. H.; Gan, S. N.; Chan, C. H.; Yahya, R. *Spectrochem. Acta* **2010**, *76*, 287–292.
43. Matsuura, H.; Miyazawa, T. *J. Polym. Sci. A-2* **1969**, *7*, 1735–1744.
44. Nakason, C.; Kaesaman, A.; Yimwan, N. *J. Appl. Polym. Sci.* **2003,** *87*, 68–75.
45. Alias, Y.; Ling, I.; Kumutha, K. *Ionics* **2005**, *11*(5–6), 414–417.
46. Fernandes, A. C.; Barlow, J. W.; Paul, D. R. *J. Appl. Polym. Sci.* **1986**, *32*, 5481.
47. Lodge, T. P.; Wood, E. R.; Haley, J. C. *J. Polym. Sci.* **2006**, *44*, 756.

CHAPTER 5

THERMAL PROPERTIES AND INTERMOLECULAR INTERACTION OF BINARY POLYMER BLENDS OF POLY(ETHYLENE OXIDE) AND POLY(N-BUTYL METHACRYLATE)

H. RAMLI[1], C. H. CHAN[2,*], and A. M. M. ALI[2]

[1]*Centre of Foundation Studies, Universiti Teknologi MARA, Cawangan Selangor, Kampus Dengkil, 43800 Dengkil, Selangor, Malaysia*

[2]*Faculty of Applied Sciences, Universiti Teknologi MARA, 40450 Shah Alam, Selangor, Malaysia*

Corresponding author. E-mail: cchan_25@yahoo.com.sg

CONTENTS

Applied Chemistry and Chemical Engineering: Volume 4

ABSTRACT

Poly(ethylene oxide) (PEO) and poly(n-butyl methacrylate) (PnBMA) polymer blends were prepared via solution casting method. Thermogravimetric analysis (TGA) was used to determine the decomposition temperature and the thermal stability of the binary polymer blends. Decomposition temperatures (T_d) of the polymer blends were estimated from the onset of the mass loss curves of TGA thermograms. Quantities T_d of the blends increase when the content of PEO increases indicated that the blends were more thermally stable with higher content PEO. Thermal properties of PEO/PnBMA polymer blends were investigated by differential scanning calorimeter. The results of glass transition temperature (T_g) and change of heat capacity (ΔC_p) suggest that PEO and PnBMA are immiscible for entire blend composition. The crystallinity (X^*) and melting temperature (T_m) of the polymer blends were estimated. There is insignificant variation of T_m of PEO in PEO/PnBMA blends, except for high content of PnBMA in the blends. This implied that these blends are immiscible under the experimental conditions agreed with results of T_g and ΔC_p. The values of X^* of PEO in the PEO/PnBMA blends do not deviate from the solid curve that marks the constancy of the crystallinity of PEO in blends. Fourier transform infrared spectroscopy was used to study the intermolecular interaction between PEO and PnBMA and suggests that the crystalline region of PEO is not interrupted by the addition of PnBMA at high content of PEO, only dipole–dipole interaction between PEO and PnBMA.

5.1 INTRODUCTION

Besides the acquiring synergistic effects of parent polymers, one of the prime motivating factors for blending of polymers is to widen and/or modify the range of properties of existing polymers and to develop new materials with the desired combination of properties for specific applications, for example, as the solid polymer electrolytes (SPEs) for ion batteries or other electrochemical devices.[1-6] Miscible blends are molecularly dispersed systems. Characteristic for immiscible blends is coexistence of phases of the parent constituents.

From the point of view of phase behavior, one has to distinguish miscible and immiscible blends. Polymer pairs, especially those with high molar masses, are generally immiscible. Therefore, most of the commercial polymer blend systems are multiphase systems. The performance (e.g., mechanical, transport, optical properties, etc.) of polymer blends depends mainly on the properties of the polymeric components, phase behavior, and the blend morphology.

The classic SPEs with lithium (Li) salt dissolved in poly(ethylene oxide) (PEO) display low conductivity ($\sigma_{DC} \sim 10^{-5}$ S cm^{-1}). Most of the binary systems of polymer and salt do not exhibit sufficient conductivity ($\sim 10^{-4}$ S cm^{-1}) for applications at ambient temperature. Hence, not only homopolymers have been used as the polymer host but also polymer blends,[2,3,7,8] block copolymers,[9–13] polymer composites,[14–19] etc. for the efforts to enhance the conductivity of SPEs. We focus on poly(ethylene oxide)-blend-based SPEs.

The conductivity of some of these miscible PEO-blend-based SPEs such as PEO/poly(methyl methacrylate) (PMMA)/Li,[20] PEO/poly(methyl acrylate) (PMA)/Li,[21] PEO/polyacrylate (PAC)/Li,[8,22] PEO/poly(propylene oxide) (PPO)/Li,[23] and PEO/poly(ϵ-caprolactone) (PCL)/Li[24] is lower than that of the binary PEO–salt systems.[25,26] PMA, poly(n-butyl methacrylate) (PnBMA), and PMMA belong to the acrylate family. PEO/PMA blends[21,27,28] and PEO/PMMA systems[3,29,30] are miscible, phase separation occurs for these systems when added with relatively high content of salt. Conductivity of these SPEs is influenced by the phase behavior of the systems.

Is PnBMA miscible with PEO with or without the addition of salt? Miscibility of PEO/PnBMA blends will have an effect on glass transition temperature (T_g). Quantity T_g may play an important role in determining the overall physical properties of polymer blends (e.g., conductivity in the case of SPEs). Quantity T_g of PMMA ($\sim 105°C$) is higher than PnBMA ($\sim 32°C$) and lowest for PMA ($\sim 10°C$). This makes PnBMA to be another potential candidate to be used as polymer host for SPEs. The T_g of PnBMA is lower than PMMA, which may be important for ion transport and on the other hand, it is higher than PMA, which may be important for appreciable mechanical strength for PEO/PnBMA SPEs. The phase behavior–performance relationship of ternary systems (PEO/PnBMA SPEs) will be revealed in forthcoming publications.

In this study, salt-free PEO/PnBMA blends are discussed. For high molar mass PEO and PnBMA, miscibility of the polymer blends on the molecular scale can be assessed by thermal characterization on the thermal properties such as quantities of T_g and melting temperature (T_m) and intermolecular interaction using Fourier transform infrared (FTIR) spectroscopic methods. Besides, thermal stability of the blends is studied as well.

5.2 EXPERIMENTAL PART

5.2.1 MATERIALS AND PREPARATION OF THE BLENDS

The characteristic of the polymers used in this chapter is listed in Table 5.1. All polymers used were purified before the blending process. For purification, the polymers were dissolved in chloroform (Merck, New Jersey, USA), followed by concentration of the polymer solutions using rotary evaporator. The pure PEO was precipitated in *n*-hexane (Merck, New Jersey, USA) while PnBMA was precipitated in methanol (Merck, New Jersey, USA).

TABLE 5.1   Characteristics of PEO and PnBMA.

Characteristics	PEO	PnBMA
M_w^a (g mol^{-1})	300,000	337,000
T_m^b (°C)	66	–
ΔH_{ref} (J g^{-1})	188[e]	–
T_g^c (°C)	−53	32
T_d^d (°C)	373	259
Source	Thermo Fisher, Pittsburgh, USA	Sigma-Aldrich, St. Louis, USA
Molecular structure		

[a] Weight-average molecular weight as determined by the suppliers.
[b] Melting temperature of PEO estimated in this study.
[c] Glass transition temperature estimated in this study.
[d] Decomposition temperature estimated in this study.
[e] Melting enthalpy of 100% crystallinity of PEO.[31]

PEO/PnBMA blends at compositions (w/w) of 0/100, 20/80, 40/60, 60/40, 80/20, and 100/0 were prepared via solution casting method. W_{PEO} and W_{PnBMA} represent mass fraction of PEO and PnBMA in the blends, respectively. Thin films of the blends were prepared from 2% (w/w) solutions of the polymers in tetrahydrofuran (THF) (Merck, New Jersey, USA). The solution was stirred for 24 h at 50°C until all the polymers were completely dissolved and then was casted into a Teflon dish. The solution was air-dried in a fume hood overnight at room temperature. The dried film was further dried at 50°C for 24 h in an oven. Then, the film was heated for 2 h at 80°C under nitrogen atmosphere before it was vacuum dried for 24 h at 50°C. All the films were kept in desiccators (with relative humidity around 40%) before further characterization.

5.2.2 MATERIAL CHARACTERIZATION

5.2.2.1 THERMOGRAVIMETRIC ANALYSIS

Thermal stability of the blends was studied using TGA Q500 (TA Instrument, Delware, USA). About 10 mg of each sample of the blends was heated in nitrogen atmosphere from 30°C to 600°C at the heating rate of 10°C min^{-1}. The onset temperature of the mass loss curve as a function of temperature was taken as the decomposition temperature (T_d) of the sample.

5.2.2.2 DIFFERENTIAL SCANNING CALORIMETER

Differential scanning calorimeter (DSC) Q2000 (TA Instrument, Delware, USA) was used to estimate the glass transition temperature (T_g), change of heat capacity (ΔC_p), melting temperature (T_m), melting enthalpy (ΔH_m), and crystallinity (X^*) of the systems during the heating run. Before sample analysis, the DSC was calibrated using an indium and sapphire standards. In order to minimize the thermo-oxidative degradation, nitrogen gas was purged throughout the analysis. About 10 mg of each sample of the blends was annealed at 80°C and quenched cool to −90°C at the heating rate of 10°C min^{-1}. Similar thermal procedure was applied in sapphire standard run and baseline run. Quantity of T_g was taken at half ΔC_p value. To determine the C_p value of sample, the heat flow signal from the sample was compared with the baseline of sapphire standard which is of known C_p. Both curves were corrected by zero or baseline correction experiment whereby an empty reference and an empty sample pan were placed in the furnace. Quantity of ΔC_p was estimated from the glass transition of the onset and endset of the heat flow curve. The melting peak maximum was taken as T_m value. Quantity ΔH_m was estimated from the area corresponding to the melting endotherm.

The X^* value of PEO was estimated from the ΔH_m in DSC trace, $X^* = \dfrac{\Delta H_m}{\Delta H_{ref}}$

5.2.2.3 *FOURIER TRANSFORM INFRARED*

The intermolecular interactions of PEO and PnBMA were examined using sampling method of attenuated total reflectance (ATR) with diamond crystal window on Nicolet 6700 ATR–FTIR (Thermo Scientific, Madison, USA). The sample was scanned at frequency range of 600–4000 cm^{-1} at a resolution of 2 cm^{-1} with 32 scans. The sample was scanned at several spots to ensure that the analysis was based on a representation region of sample.

5.3 RESULTS AND DISCUSSIONS

5.3.1 *THERMAL STABILITY OF NEAT POLYMERS AND BLENDS OF PEO/PNBMA*

Figure 5.1 shows the TGA thermal decomposition profile of PEO/PnBMA blends from 50°C to 600°C. Both neat polymers show one-step decomposition process, with T_d at 373°C (for PEO) and 259°C (for PnBMA) and approximate 100% mass loss at temperature more than 400°C. The thermal decomposition of blends proceeds in two steps, which corresponds to each component.

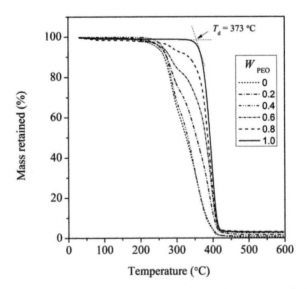

FIGURE 5.1 Percent mass retained curves as a function of temperature for PEO/PnBMA blends.

When the normalized mass loss fraction of PEO in the blends is plotted against its mass fraction of PEO (W_{PEO}) (*c.f.* Fig. 5.2), it correlates well to the blend composition. Besides, the percent mass loss of the blends increases drastically when temperature exceeds 250°C. This indicates that the blends are thermally stable up to roughly 200°C. Quantities T_d of the blends increase when the content of PEO increases indicates that the blends are more thermally stable with higher content PEO.

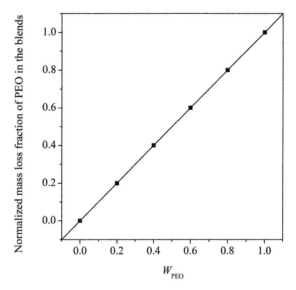

FIGURE 5.2 Plot of normalized mass loss fraction of PEO in the blends against mass fraction PEO. Solid curve represents normalized mass fraction of PEO = W_{PEO}.

5.3.2 MISCIBILITY OF PEO/PNBMA BLENDS

5.3.2.1 GLASS TRANSITION TEMPERATURE

Figure 5.3 shows the DSC traces of the PEO/PnBMA blends during the heating run. PEO has a T_g located at −53°C. When content of PnBMA is elevating in the blends, the T_g values of PEO in the blends remain almost constant. Quantity T_g of neat PnbMA is at 32°C. We note here, T_g of PnBMA in the blends, especially at high content of PEO, is difficult to be estimated as the glass transition superimposes with the start of the melting endotherm of PEO. Quantities ΔC_p of PEO in the blends are also evaluated after addition of PnBMA. Quantity ΔC_p of PEO increases monotonously with increasing W_{PEO}. Referring to Figure 5.4, constancy of quantity ΔC_p of PEO corresponding to the content of PEO in the blends is recorded. The results of T_g and ΔC_p suggest that PEO and PnBMA are immiscible for entire blend composition.[4,6]

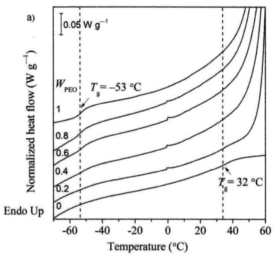

FIGURE 5.3 DSC traces of different compositions of PEO/PnBMA blends. Dashed curves represent T_g estimated for PEO and PnBMA.

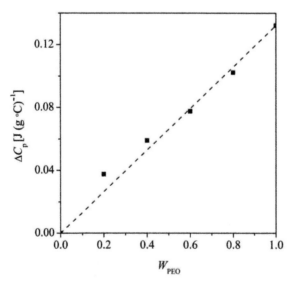

FIGURE 5.4 Changes in heat capacity of PEO in the blends. Dashed curve marks the constancy of change in heat capacity of PEO in the blends.

5.3.2.2   MELTING BEHAVIOR OF PEO/PNBMA BLENDS

Quantities T_m and ΔH_m of PEO in the blends are obtained from the peak and the area under the melting endotherm, respectively, during the heating cycle. A single melting peak of PEO is observed in the blends (see Fig. 5.5). PnBMA is an amorphous polymer, thus it does not have melting peak. Figure 5.6 shows the variations of T_m of PEO in PEO/PnBMA blends as a function of mass fraction of PEO. There is insignificant variation of T_m of PEO in PEO/PnBMA blends, except for high content of PnBMA in the blends, where a slight depression of apparent T_m can be noticed. This implies that these blends are immiscible under the experimental conditions. This result is in agreement with results of T_g and ΔC_p.

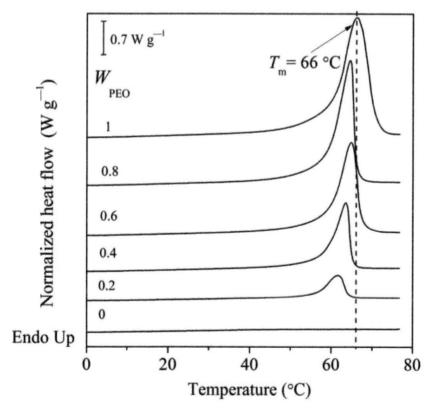

FIGURE 5.5   DSC traces for melting endotherms of PEO in the blends.

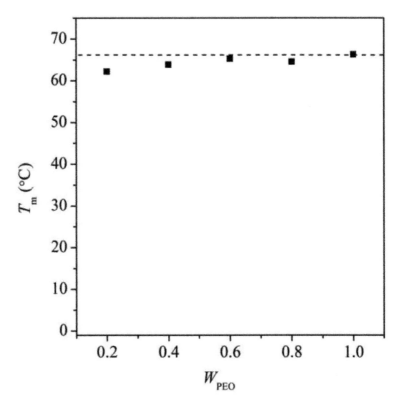

FIGURE 5.6 Apparent melting temperatures of PEO in PEO/PnBMA blends. Dashed curve represents the constancy of apparent melting temperature of PEO in the blends.

5.3.3 CRYSTALLINITY OF PEO IN PEO/PNBMA BLENDS

The relative degree of crystallinity (X^*) of PEO phase for the PEO/PnBMA blends is calculated from the ΔH_m values in the heating cycle using $X^* = \dfrac{\Delta H_m}{\Delta H_{ref}}$. Quantity ΔH_m is the enthalpy of fusion of the blend and $\Delta H_{ref} = 188.3 \ J \ g^{-1}$ is the enthalpy of fusion of 100% crystalline PEO.[31] For immiscible PEO/PnBMA (semicrystalline/amorphous) blends, the crystallization behavior of the PEO in the blends is expected to have close approximation as in the neat PEO. This is due to the crystallization of PEO in the blends which takes place within the domains of nearly neat PEO and is largely unaffected by the presence of the amorphous PnBMA. This in principle is true for the matrix-droplet or cocontinuous microstructures of the blends where values of T_m and X^* of PEO in the blends remain constant when the content of PEO is not too low. Figure 5.7 presents quantity X^* of PEO in the PEO/PnBMA blends as a function of W_{PEO}. The values of X^* of PEO in the PEO/PnBMA blends do not deviate from the solid curve inserted in Figure 5.7. The solid curve marks the constancy of the crystallinity of PEO in blends.

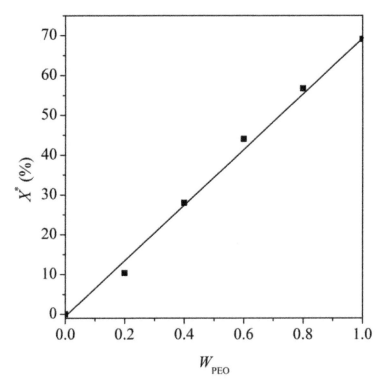

FIGURE 5.7 Crystallinity of PEO in PEO/PnBMA blends versus W_{PEO}. Solid curve represents constancy of crystallinity for PEO in the blends.

5.3.4 FTIR SPECTROSCOPY

Figure 5.8 illustrates the FTIR spectra of neat PEO and PnBMA. The characteristic bands of both PEO and PnBMA are tabulated in Table 5.2. The assignment of characteristic absorption bands of neat PEO and PnBMA is done based on the references cited. The characteristic absorption bands of the spectra of both neat PEO and PnBMA show good approximation to the characteristic absorption bands extracted from the literature.

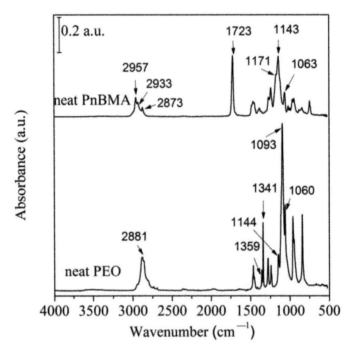

FIGURE 5.8   FTIR spectra of neat PEO and PnBMA.

TABLE 5.2   Characteristic Bands for PEO and PnBMA.

Polymer	Assignment	Wave number (cm⁻¹)	Reference
PEO	CH$_2$ stretching	2881	32
	CH$_2$ wagging	1359 and 1341	32
	C–O–C stretching	1144, 1093, 1060	33
PnBMA	CH$_3$ stretching	2957	34
	CH$_2$ stretching	2933 and 2873	34
	C=O stretching	1723	35
	C–O–C stretching	1171, 1143, 1063	35

The decrease in the absorbance of the C=O stretching is qualitatively proportionate to the composition of the PnBMA in the blends as shown in Figure 5.9(a). However, there is no shifting of the wave number of absorption band of C=O stretching implying that the parent polymers exhibit only dipole–dipole interaction. The center of the triplet absorption bands of C–O–C stretching appears as a triplet at 1093 cm⁻¹ that corresponds to the amorphous part of PEO while the two shoulders representing the crystalline phase of PEO at 1144 and 1060 cm^{-1} [see Fig. 5.9(b)].[36,37] Overlapping of the vibrational mode of the C–O–C shoulders of PEO at 1144 and 1060 cm^{-1} with the two C–O–C stretching absorption bands of PnBMA at 1143 and 1063 cm^{-1} predominates. The characteristic absorption bands of CH$_2$ wagging at 1359 and 1341 cm^{-1} represent the crystalline phase of PEO.[4,8,36,37] It is clearly seen that there is no significant shifting of the wave numbers of C–O–C stretching (1093 cm^{-1}) and CH$_2$ wagging mode (1359 and 1341 cm^{-1}) of PEO in the blends at high content of PEO. This suggests that the crystalline region of PEO is not interrupted by the addition of PnBMA at high content of PEO.

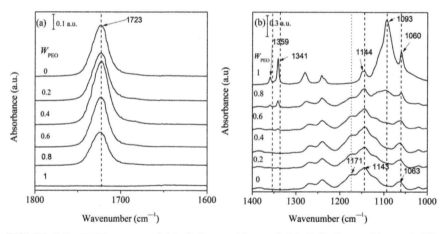

FIGURE 5.9   FTIR spectra of (a) C=O stretching and (b) C–O–C stretching and CH$_2$ wagging of the PEO/PnBMA blends.

5.4   CONCLUSIONS

Thin films of PEO/PnBMA were successfully prepared via solution casting method. Quantities T_d of the blends increase when the content of PEO increases indicating that the blends are more thermally stable with higher content of PEO. The results of T_g and ΔC_p suggest that PEO and PnBMA are immiscible for entire blend composition. There is insignificant variation of T_m of PEO in PEO/PnBMA blends, except for high content of PnBMA in the blends, where a slight depression of apparent T_m can be noticed. This implies that these blends are immiscible under the experimental conditions agreed with results of T_g and ΔC_p. The values of X^* of PEO in the PEO/

PnBMA blends do not deviate from the solid curve that marks the constancy of the crystallinity of PEO in blends. FTIR analysis suggests that the crystalline region of PEO is not interrupted by the addition of PnBMA at high content of PEO, only dipole–dipole interaction takes place between PEO and PnBMA.

ACKNOWLEDGMENT

The authors would like to thank the Ministry of Education Malaysia for the "Research Acculturation Grant Scheme" (RAGS) (600-RMI/RAGS 5/3 (145/2014)) supporting this study.

KEYWORDS

- poly(ethylene oxide)
- poly(*n*-butyl methacrylate)
- polymer blends
- thermogravimetric analysis
- differential scanning calorimeter
- Fourier transform infrared

REFERENCES

1. Tian Khoon, L.; Ataollahi, N.; Hassan, N. H.; Ahmad, A. *J. Solid State Chem.* **2016**, *20*, 203–213.
2. Sim, L. H.; Chan, C. H.; Kammer, H. W. *Mater. Res. Innov.* **2011**, *15*, 71–74.
3. Abd Karim, S. R.; Sim, L. H.; Chan, C. H.; Zainal, N. F. A.; Kassim, M. A. *Adv. Mater. Res.* **2013**, *812*, 267–270.
4. Chan, C. H.; Sim, L. H.; Kammer, H. W.; Tan, W. *Am. Inst. Phys. Conf. Proc.* **2012**, *1455*, 197–207.
5. Hashifudin, A.; Sim, L. H.; Chan, C. H.; Kammer, H. W.; Yusoff, S. N. H. M. *Polym. Res. J.* **2013**, *7.2*, 195–204.
6. Sim, L. H.; Gan, S. N.; Chan, C. H.; Kammer, H. W.; Yahya, R. *Mater. Res. Innov.* **2009**, *13*, 278–281.
7. Chan, C. H.; Kammer, H. W. *J. Appl. Polym. Sci.* **2008**, *110*, 424–432.
8. Sim, L. H.; Gan, S. N.; Chan, C. H.; Yahya, R. *Spectrochim. Acta A* **2010**, *76*, 287–292.

9. Sarapas, J. M.; Saijo, K.; Zhao, Y.; Takenaka, M.; Tew, G. N. *Polym. Adv. Technol.* **2016**, *27*, 946–954.

10. Xue, C.; Meador, M. A. B.; Zhu, L.; Ge, J. J.; Cheng, S. Z. D.; Puttharanat, S.; Eby, R. K.; Khalfan, A.; Bennett, G. D.; Greenbaum, S. G. *Polymer* **2006**, *47*, 6149.

11. Young, W. S.; Kuan, W. F.; Epps, T. H. *J. Polym. Sci.* **2014**, *52*, 1–16.

12. Sadoway, D. R. *J. Power Sources* **2004**, *129*, 1–3.

13. Niitani, T.; Shimada, M.; Kawamura, K.; Dokko, K.; Rho, Y. H.; Kanamura, K. *Electrochem. Solid-state Lett.* **2005**, *8*, 385–388.

14. Mogurampelly, S.; Ganesan, V. *Macromolecules* **2015**, *48*, 2773–2786.

15. Ganapatibhotla, L. V. N. R.; Maranas, J. K. *Macromolecules* **2014**, *47*, 3625–3634.

16. Alloin, F.; D'Aprea, A.; Kissi, N. E.; Dufresne, A.; Bossard, F. *Electrochim. Acta* **2010**, *55*, 5186.

17. Stephan, A. M.; Nahm, K. S. *Polymer* **2006**, *47*, 5952–6964.

18. Bloise, A. C.; Tambell, C. C.; Franco, R. W. A.; Donso, J. P.; Magon, C. J.; Souza, M. F.; Rosario, A. V.; Pereira, E. C. *Electrochim. Acta* **2001**, *46*, 1571.

19. Croce, F.; Appetechi, G. B.; Persi, L.; Scrosati, B. *Nature* **1998**, *394*, 456–458.

20. Osman, Z.; Ansor, N. M.; Chew, K. W.; Kamarulzaman, N. *Ionics* **2005**, *11*, 431–435.

21. Halim, S. I. A.; Chan, C. H.; Winie, T. *Macromol. Symp.* **2017**, *371*, 114–124.

22. Hashifudin, A.; Sim, L. H.; Chan, C. H.; Ramli, H. *Compos. Interface* **2014**, *21*, 797–805.

23. Watanabe, M.; Nagano, S.; Sanui, K.; Ogata, N. *Solid State Ion.* **1986**, *18–19*, 338–342.

24. Chiu, C. C.; Chen, H. W.; Kuo, S. W.; Huang, C. F.; Chang, F. C. *Macromolecules* **2004**, *37*, 8424–8430.

25. Berthier, C.; Gorecki, W.; Minier, M.; Armand, M. B.; Chabagno, J. M.; Rigaud, P. *Solid State Ion.* **1983**, *11*, 91–95.

26. Reddy, M. J.; Chu, P. P. *Electrochim. Acta* **2002**, *47*, 1189–1196.

27. Pedemonte, E.; Burgisi, G. *Polymer* **1994**, *35*, 3719–3721.

28. Wieczorek, W.; Such, K.; Florianczyk, Z.; Przyluski, J. *Electrochim. Acta* **1992**, *9*, 1565–1567.

29. Osman, Z.; Ansor, N. M.; Chew, K. W.; Kamarulzaman, N. *Ionics* **2005**, *11*, 431–435.

30. Jeddi, K.; Qazvini, N. T.; Jafari, S. H.; Khonakdar, H. A. *J. Polym. Sci. Pol. Phys.* **2010**, *48*, 2065–2071.

31. Cimmino, S.; Pace, E. D.; Martuscelli, E.; Silvestre, C. *Makromol. Chem.* **1990**, *191*, 2447–2454.

32. Li, X.; Hsu, S. L. *J. Polym. Sci. Pol. Phys.* **1984**, *22*, 1331.

33. Tadokoro, H.; Chatani, Y.; Yoshihara, T.; Tahara, S.; Murahashi, S. *Makromol. Chem.* **1964**, *73*, 109–127.

34. Thomas, E. A.; Zupp, T. A.; Fulghum, J. E.; Fredley, D. S.; West, J. L. *Mol. Cryst. Liq. Crys. A.* **1994**, *250*, 193–208.

35. Pekel, N.; Güven, O. *J. Appl. Polym. Sci.* **1998**, *69*, 1669–1674.

36. Halim, S. I. A.; Chan, C. H.; Sim, L. H. *Macromol. Symp.* **2016**, *365*, 95–103.

37. Abd Karim, S. R.; Sim, L. H.; Chan, C. H.; Ramli, H. *Macromol. Symp.* **2015**, *354*, 374–383.

CHAPTER 6

CROSS-LINKING OF EPOXY– ISOCYANATE MIXTURES IN THE PRESENCE OF HYDROXYL-ACRYLIC OLIGOMER

OSTAP IVASHKIV[1], BOGDANA BASHTA[1], OLENA SHYSHCHAK[1], JOZEF HAPONIUK[2], and MICHAEL BRATYCHAK[1,*]

[1]*Department of Chemical Technology of Oil and Gas Processing, Lviv Polytechnic National University, 12, St. Bandera Str., Lviv 79013, Ukraine*

[2]*Department of Polymers Technology, Gdansk University of Technology, 11/12 G. Narutowicza Str., 80233 Gdansk, Poland*

Corresponding author. E-mail: mbratych@polynet.lviv.ua

CONTENTS

ABSTRACT

The possibility of preparing epoxy–urethane coatings based on epoxy–isocyanate mixtures using epoxy resin, hydroxyl-acrylic oligomer, hardener, and isocyanate has been considered. Stepwise method of epoxy–urethane films formation has been developed. The effect of mixture composition, temperature, and cross-linking time on film hardness and gel-fraction content has been established. The mixtures with optimal hardness and gel-fraction content were analyzed by IR spectroscopy regarding functional groups content during cross-linking at different temperatures. The reaction scheme of three-dimensional cross-linked structure formation has been suggested.

6.1 INTRODUCTION

Polyurethane materials, as well as materials based on epoxy resins are characterized by good operational properties. They are widely used for the production of high-quality coatings, polymer concrete compositions, gums, etc.[1–3] However, each of mentioned materials has own disadvantages. If we compare epoxy and polyurethane materials, the epoxy compounds have worse abrasion resistance, adhesive strength relative to nonferrous metals, and aromatic fuel tolerance. Polyurethane materials have low hardness and strength; they are not stable relative to acid and alkali influence. To avoid abovementioned disadvantages, the combination of both materials takes place.[4]

Modification of epoxy resins by isocyanate with epoxy–urethane compounds increases the epoxy resins functionality and significantly improves the adhesive and cohesive strength, as well as thermal resistance of the final materials.[5,6] This is a result of isocyanate and secondary hydroxy groups interaction and formation of urethane links with high cohesion energy.[7]

A series of scientific works[8–12] deals with the investigation of the materials based on epoxides and isocyanate mixtures formation. In work[8] it is shown that in epoxy–isocyanate systems without catalyst, the urethane is formed at 333K due to the reaction between isocyanate and secondary hydroxy groups of epoxy oligomers:

$$-O-CH_2-CH-CH_2- \ + \ O=C=N- \ \longrightarrow$$
$$\underset{OH}{|}$$

$$\longrightarrow \ -O-CH_2-CH-CH_2-$$
$$\underset{\underset{\underset{NH-}{|}}{\underset{C=O}{|}}}{\overset{|}{O}}$$

If the temperature is low, the system may be turned into glassy state that makes impossible formation of epoxy–urethane materials with necessary operational properties.

It is well known[9,13] that cross-linking of epoxy resins are mostly realized by using aliphatic amines. This fact complicates usage of compounds with isocyanate groups because under ambient conditions these groups immediately react with amino groups of hardener and form urea groups. Thus, it is impossible to obtain high-quality coatings. To prevent such disadvantages, the products of isocyanate interaction and compounds containing mobile hydrogen atom (blocked isocyanates) are used.[8,12,14]

On the other hand, it is known[15] that coatings based on epoxy resins and oligomers with unsaturated fragments are characterized by better performance to compare with coatings without them.

So, in the present work, we investigate the possibility of polymer films formation using Epidian-5 epoxy resin, diisocyanate, polyamine, and hydroxy-acrylic oligomer (HAO) of the formula:

(HAO)

Such oligomer with acrylic, primary and secondary hydroxy groups should influence the temperature of film formation and the process as itself.

6.2 EXPERIMENTAL

6.2.1 MATERIALS

Epidian-5 (Sarzyna-Ciech) with molecular weight (Mn) of 390 g/mol and epoxy groups content of 20.0% was used as an initial epoxy resin. 4,4'-Diphenylmethanediisocyanate (MDI, Suprasec 1306, Hunstman) was used as isocyanate. Triethylenetetramine (Z-1, CEDAR) was used as a hardener.

The modifier HAO was synthesized via the reaction between hydroxy derivative of Epidian-6 epoxy resin and acrylic acid.[16] It was found for HAO Mn 495 g/mol and hydroxyl number 351 mg KOH/g.

6.2.2 CROSS-LINKING OF EPOXY–ISOCYANATE MIXTURES

Epoxy–isocyanate mixtures, the compositions of which are given in Table 6.1, were prepared by mixing HAO with MDI at room temperature for 10–15 min. The Epidian-5 was added under stirring until homogeneous mixture was obtained. Then Z-1 hardener was added and the mixture was poured over previously degreased glass plates. Films formation was studied under the following conditions: first the compositions were cured at room temperature for 24 h and then at 323 K, 343K, 363K, 383K, 403K, or 423K for 15, 30, 60, or 90 min. The structural changes were controlled by gel-fraction content (G) of the grinded samples in Soxhlet apparatus during their extraction by chloroform for 10 h and films hardness (H) determined by M-3 pendulum device at room temperature (ISO 1522).

TABLE 6.1 Composition of Epoxy–Isocyanate Mixtures.

Component	Component content (mass parts)		
	I	II	III
Epidian-5	92.5	85	70
MDI	2.5	5	10
HAO	5	10	20
Z-1	13	12	11

Note: Content of Z-1 was calculated for total amount of epoxy groups based on the ratio 20 g of epoxy groups per 14 mass parts of Z-1 hardener.

6.2.3 IR-SPECTROSCOPIC INVESTIGATIONS

IR-spectra of epoxy–isocyanate mixtures were recorded using Therma Electron Corporation-Nicolet 8700 instrument at the spectroscopic laboratory of Gdansk Technical University. The instrument is equipped with Specac Golden Gate adapter and diamond crystal ATR. The epoxy–isocyanate mixtures were applied by thin layer over KBr plates and IR-spectra were recorded. Then, the plates were placed in desiccator, held for 24 h at room temperature and again IR-spectra were recorded. Then, the mixtures were heated gradually at 383K (30 min), 403K (30 min), and 423K (30 min). IR-spectra were recorded after each heating step.

6.3 RESULTS AND DISCUSSION

The results represented in Table 6.2 (mixtures I, II, and III) reveal that gel-fraction content in the samples and film hardness depends on the mixture composition, temperature, and cross-linking time.

TABLE 6.2 Dependence of Gel-Fraction Content and Film Hardness upon the Temperature and Cross-Linking Time.

Mixture in accordance with Table 6.1	T (K)	Symbol	Symbol values at cross-linking time (min)					
			24 h under ambient conditions	15	30	45	60	75
I	363	H	0.58	0.86	0.91	0.92	0.92	0.99
		G	77.9	92.8	94.3	93.0	93.1	94.4
	383	H	0.58	0.93	0.97	0.99	1.00	1.00
		G	77.9	93.3	93.4	94.8	95.5	96.5
II	323	H	0.59	0.70	0.80	0.84	0.84	0.87
		G	75.5	75.8	77.0	83.7	84.5	90.4
	343	H	0.59	0.83	0.92	0.92	0.91	0.99
		G	75.5	84.3	89.9	92.2	92.8	92.9
	363	H	0.59	0.99	1.00	0.91	0.93	1.00
		G	75.5	90.6	94.5	94.7	95.7	95.2
	383	H	0.59	0.78	0.87	0.82	0.84	0.86
		G	75.5	93.4	95.6	97.1	97.1	100.0
	403	H	0.59	0.51	0.73	0.75	0.78	0.83
		G	75.5	78.4	80.7	86.9	90.6	97.2
	423	H	0.59	0.73	0.75	0.80	0.83	0.91
		G	75.5	76.9	80.2	89.3	93.8	97.8 M
III	363	H	–	–	–	–	–	–
		G	70.7	78.7	87.4	88.3	89.9	91.3

Note: G—gel-fraction content, %; H—film hardness, relative units.

As one can see from Table 6.2, the increase of HAO amount from 5 to 10 mass parts in the mixtures I–III practically does not change film hardness at room temperature. At the same time, gel-fraction content slightly decreases. To our mind, the reason is that at room temperature the functional groups of HAO oligomer do not participate in the reactions which result in cross-linked structures formation. Moreover, three-dimensional structures are formed at room temperature due to the reaction of Epidian-5 epoxy groups with amino groups of Z-1 hardener in accordance to the scheme:

$$\text{(Epidian-5)} + H_2N\text{ww} \longrightarrow \text{wwC}\underset{H}{\overset{OH}{-}}C\overset{H_2}{-}NH \quad (1)$$

Therefore, for the more amount of HAO and the less amount of epoxy resin in the mixture, the less number of cross-linked structures is formed at room temperature and gel-fraction content decreases (Table 6.2). The decrease of hardness is not observed. Such compensation effect may be provided by the reactions of isocyanate with primary hydroxy groups of HAO, the amount of which increases from 2.5 to 10 mass parts in the mixtures I–III and leads to formation of long-chain linear structures:

$$HO\text{ww} + O{=}C{=}N\text{ww} \longrightarrow \text{ww}O-\overset{O}{\overset{\|}{C}}-NH\text{ww} \quad (2)$$
$$\text{(HAO)} \qquad \text{(MDI)}$$

Due to the long carbon chain, the structures provide high hardness of the films, that is why it does not decrease at smaller epoxy resin content in the mixtures. However, such structures do not compensate the decrease of gel-fraction content, since they are linear ones and do not take part in cross-linked structures formation.

While adding Z-1 hardener to epoxy–isocyanate mixture, the residual unreacted isocyanate MDI may react with amino groups of hardener in accordance to the reaction:

$$O{=}C{=}N\text{ww}R + H_2N\text{ww}R' \longrightarrow R\text{ww}\underset{H}{N}-\overset{O}{\overset{\|}{C}}-NH\text{ww}R' \quad (3)$$

Where R= fragment of MDI, R'= fragment of Z-1

This reaction is undesirable because it leads to instant formation of foamed porous structures inside the epoxy matrix. The structures are brittle and cause surface heterogeneity. The reaction (3) is improbable for the mixtures I and II since hydroxyl groups of MDI are mostly blocked by HAO molecules according to the reaction (2).

Moreover, the reaction (1) is exothermal process and causes the mixture heating. The result is possible reaction of HAO oligomer homopolymerization by double bonds. The formation of linear structures with long carbon chain also provides high hardness of the films but does not increase their gel-fraction content. It is the additional explanation of the fact of film hardness increasing with the increase of HAO amount in the mixtures from 5 to 10 mass parts.

Further increase of HAO amount to 20 mass parts (mixture III, Table 6.2) leads to the negative result. It is impossible to determine films hardness, because they become heterogeneous with nonuniform and hump-shaped surface. At the same time, the gel-fraction content of the films decreases. To our mind, all abovementioned may be explained by the inadmissible increase of MDI isocyanate amount to 10 mass parts caused by the increase of HAO amount. In other words such amount of MDI isocyanate groups could not be completely blocked [see the reaction (2)], therefore while adding hardener the unreacted isocyanate groups instantly react with Z-1 amino groups in accordance with the reaction (3). Moreover, the content of epoxy resin in the mixture (III) is much less to compare with the mixtures I and II that causes the decrease in gel-fraction content of obtained films. Thus, it is inexpedient to increase HAO and MDI amount in epoxy–isocyanate mixtures.

The heating of mixtures I–III to 363K intensifies the reaction (1). But at this temperature, the additional reactions involving Epidian-5 epoxy groups and secondary hydroxyl groups obtained via the reaction (1) as well as secondary hydroxyl groups of HAO and Epidian-5 become possible:

(4)

(Epidian-5)

(5)

(Epidian-5)   (HAO or Epidian-5)

The reactions (4) and (5) lead to the increase of cross-linked structures density in epoxy matrix. The result is higher film hardness and gel-fraction content of the films based on mixtures I–III (Table 6.2).

The hardness of the mixture II (5 mass parts of HAO oligomer) is equal to 0.99 relative units after only 15 min of heating at 363K and is practically unchanged with the temperature increase. At the same time, film hardness of the mixture I (2.5 mass parts of HAO) gradually increases from 0.86 relative units at heating for 15 min to 0.99 relative units after 75 min of heating. The reason is that under mentioned conditions, the intensive homopolymerization of HAO molecules by double bonds takes place in the mixture II. As a result, the formed long linear structures increase film hardness. At this stage, the content of functional unsaturated bonds in HAO oligomer is almost zero and further heating of mixture does not cause increasing of film hardness. With the increase in heating time, gel-fraction content increases for both mixtures (Table 6.2) confirming the partial proceeding of the reactions (4) and (5) in the mixture I and II.

The mixture III heating at 363K for 15–75 min also leads to the gradual increasing of gel-fraction content, but it is lower to compare with gel-fraction content of films based on mixtures II and I. The reason is the low content of epoxy resin (70 mass parts, Table 6.1) in this mixture to compare with that in mixtures I and II (92.5 and 85 mass parts, respectively).

Taking into account the obtained experimental data (H and G) for the mixtures I–III, the films based on mixtures I and II are characterized by best properties. Therefore, we used these mixtures to study cross-linking process at higher temperatures, at 383K in particular.

For the mixture I, the increase of heating temperature to 383K slightly increases gel-fraction content. To our mind, the mentioned effect is achieved due to the reactions (1), (4), and (5). Secondary hydroxyl groups of HAO oligomer are less active than those of epoxy resin. Under the same conditions for the mixture II, containing more HAO oligomer, the values of gel-fraction content are considerably higher. This fact confirms our earlier assumptions and indicates the greater role of HAO secondary hydroxyl groups (reaction 5). Higher HAO content in this mixture causes the increase of cross-linked structures density and, as a result, higher gel-fraction content of obtained films.

For the mixture I, the film hardness gradually increases at 383K. Moreover, after heating films for 45 min, we obtained almost maximal film hardness equal to 0.99 relative units. Further increase in heating time does not affect the film hardness.

Since the films based on mixture II are characterized by high gel-fraction content and sufficient hardness, we studied the temperature effect on cross-linking process using this mixture.

The decrease in temperature to 343K and 323K decreases both gel-fraction content and hardness (Table 6.2). Moreover, the decrease in hardness is slight, because such temperatures are high enough for the proceeding of the reactions (1), (2) and further HAO molecules homopolymerization, due to which the structures providing coatings hardness are formed. At the same time gel-fraction content sharply decreases, that indicates the intensity decrease of the reactions (4) and (5). It means that the temperature of 323K is not high enough to form cross-linked structures via the reactions (4) and (5).

The increase of cross-linking temperatures to 403K and 423K abnormally decreases both gel-fraction content and final coatings hardness. It may be caused by the decomposition of some structures in cross-linked matrix. It is known that under low temperature, MDI isocyanate molecules may react with Z-1 hardener molecules in accordance with the reaction (3). As we mentioned before, this reaction is undesirable since it leads to the formation of foamlike structures. But if the amount of such molecules in the reaction system is small, they form linear structures with reactive NH– groups. Such groups may react at 343–363K with the residual epoxy groups of three-dimensional cross-linked structures and graft in such a way to the general cross-linked matrix:

$$(6)$$

Cross-linked matrix with   Cross-linked matrix
residual epoxy groups

Where R= fragment of MDI, R'= fragment of Z-1

If such reactions proceed till 383K, they slightly provide the increase of gel-fraction content. However, at higher temperatures (400K and more) the decomposition of urea bonds is possible. As a result there is a partial detachment of short linear structures from general three-dimensional matrix. The structures may be washed out by a solvent and thus cause the decrease of gel-fraction content of obtained films.

To confirm the abovementioned assumptions and processing of the mentioned reactions, IR investigations have been conducted in accordance to the methodic described in Section 6.2.3. Mixture II (Table 6.1) was chosen as the most optimal one.

Structural changes in epoxy–isocyanate mixtures were controlled by the absorbance bands at 918 cm^{-1}, corresponding to asymmetric stretching vibrations of epoxy ring, at 3400 cm^{-1}, corresponding to the stretching vibrations of hydroxyl groups, and at 1625 cm^{-1}, corresponding to unsaturated double bonds.

IR spectra were recorded after mixtures preparing, after keeping mixture at room temperature for 24 h, and after heating for 0.5 h at 383K, 403K, or 424K (Figs. 6.1–6.3).

As we can see from Figure 6.1 right after mixture preparing, there is an intensive absorbance band at 918 cm^{-1} that indicates high concentration of epoxy groups in the mixture. Keeping mixture at room temperature for 24 h and its heating at 383K for 0.5 h sharply reduce the intensity of absorbance band that confirms occurrence of the reaction (1) with participation of epoxy groups. Further heating (Fig. 6.1) does not influence the intensity of absorbance band at 918 cm^{-1} that indicates almost total exhaustion of epoxy groups in the system.

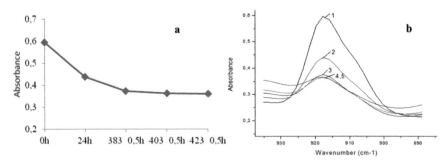

FIGURE 6.1 Intensity of absorption band at 918 cm^{-1} (a) and its change (b) after mixing the components of mixture II (1), after curing it at room temperature for 24 h (2), and after heating at 383K (0.5 h) (3), 403K (0.5 h) (4), and 423K (0.5 h) (5).

The possibility of the reaction (1) proceeding is confirmed by the results presented in Figure 6.2. While keeping the mixture at room temperature for 24 h, we observe the sharp increase of absorbance band intensity at 3400 cm^{-1} characterizing the changes of hydroxyl groups concentration in the mixture. This effect is caused by the reaction (1) at room temperature when opening of the epoxy rings and formation of secondary hydroxyl groups take place. Primary hydroxyl groups of HAO oligomer react with isocyanate according to the reaction (2) and form the product with NH– groups. Characteristic absorbance bands of such groups appear at 3310 cm^{-1} which are very close to absorbance bands of OH– groups.

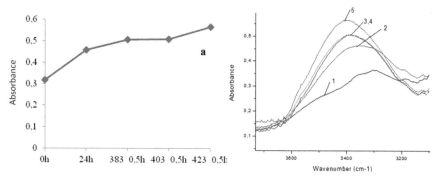

FIGURE 6.2 Intensity of absorption band at 3400 cm^{-1} (a) and its change (b) after mixing the components of mixture II (1), after curing it at room temperature for 24 h (2), and after heating at 383K (0.5 h) (3), 403K (0.5 h) (4), and 423K (0.5 h) (5).

The mixture heating to 383K leads to the increase of absorbance band intensity at 3400 cm^{-1} that confirms proceeding of the reactions (1), (4), and (5). Moreover, the intensity of absorbance bands at 918 cm^{-1}, characterizing epoxy groups content and absorbance bands at 3400 cm^{-1}, characterizing hydroxyl groups content is the same within the interval corresponding to the mixture heating from 383K to 403K for 0.5 h (Fig. 6.1a). But the increase in temperature to 423K slightly increases the intensity of absorbance bands at 3400 cm^{-1} indicating the occurrence of chemical reactions with formation of OH– groups.

Figure 6.3 represents the change of absorbance bands intensity characterizing the variation of unsaturated double bonds concentration. While keeping the mixture at room temperature for 24 h, the decrease of absorbance bands intensity at 1625 cm^{-1} is observed. It is the confirmation of our assumption that at room temperature, self-heating of the mixture takes place due to the reaction (1), followed by HAO homopolymerization relative to the double bonds. The mixture heating to 383K intensifies this process (Fig. 6.3). Further heating to 403K and 423K does not essentially change the absorbance band intensity at 1625 cm^{-1}. This fact indicates that double bonds of HAO oligomer are almost totally exhausted at 383K and at 403K, HAO homopolymerization is almost absent or is very slight.

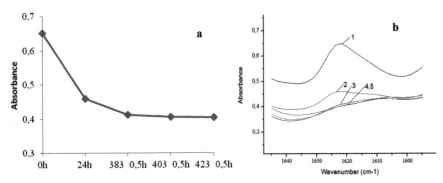

FIGURE 6.3   Intensity of absorption band at 1625 cm^{-1} (a) and its change (b) after mixing the components of mixture II (1), after curing it at room temperature for 24 h (2), and after heating at 383K (0.5 h) (3), 403K (0.5 h) (4), and 423K (0.5 h) (5).

6.4   CONCLUSIONS

We synthesized new epoxy–isocyanate mixtures, consisting of Epidian-5 epoxy resin, Z-1 hardener, MDI diisocyanate, and active additive on the basis of hydroxyl derivative of epoxy resin and acrylic acid. The Epidian-5:MDI:HAO:Z-1 ratio of 85:5:10:12 mass parts was found to be the optimum one. On the basis of investigated mixtures, we obtained epoxy–urethane coatings with gel-fraction content of 93–95% and hardness of 0.99–1.00 relative units (if the temperature of coating formation was 343–363K), and gel-fraction content of 100% and hardness of 0.86 relative units (formation temperature was 383K). It was confirmed that cross-linked structures were formed as a result of chemical interaction of functional groups of the mixture components.

KEYWORDS

- epoxy–isocyanate mixtures
- hydroxyl-acrylic oligomer
- triethylenetetramine
- cross-linking
- IR-spectroscopy

REFERENCES

1. Saunders, D.; Frish, K. *Khimiya poliuretanov*; Khimia: Moskva, 1968; p 470.
2. Kompozitsionnye materialy na osnove poliuretanov: Per. s angl./Pod red. J. M. Buista. M.: Khimia, 1982; p 240.
3. Lipatov, Y. S.; Kercha, Y. Y.; Sergeeva, L. M. *Structura i svoistva poliuretanov;* Naukova Dumka: Kuiv, 1970; p 288.
4. Kudriavtseva, B. B. Lakorasochnye marerialy i ih primenenie. **2003,** *7–8*, 24–28.
5. Lapitskii, V. A.; Krytsuk, A. A. *Fiziko-mekhanicheskie svoistva epoksidnyh polimerov I stekloplastikov*; Naukova Dumka: Kiev, 1986; p 96.
6. Szycher, M. *Szycher's Handbook of Polyurethanes*, 2nd ed.; CRC Press: Boca Raton, FL, USA, 2013; p 1112.
7. Smirnov, Y. N.; Valueva, L. F.; Lapitskii, V. A. Otverzshdenie epoksiuretanovyh oligo-merov. *Plasticheskie masy.* **1985,** *12*, 41–42.
8. Morev, F.; Prokopchuk, N.; Krutko, E. *Trudy BGTU* **2012,** *4*, 88.
9. Bratychak, M. M.; Ivashkiv, O. P.; Astakhova, O. T. *Dopovidi NAN Ukrainy* **2014,** *8*, 97.
10. Takeshita, Y.; Becker, E.; Sakata, S.; Miwa, T.; Sawada, T. *Polymer* **2014,** *55*, 2505.
11. Dhevi, D. M.; Jaisankar, S. N.; Pathak, M. *Eur. Polym. J.* **2013,** *49*, 3561.
12. Kirillov, A. N. *Polym. Sci. Series D* **2014,** *7*(1), 14.
13. Moshinsky, L. *Epoxy Resins and Hardeners. Structure, Properties, Chemistry and Topology of Curing*; Arcadia Press Ltd.: Tel Aviv, 1995; 370 p.
14. Bratychak, M. M.; Ivashkiv, O. P.; Astakhova, O. T.; Haponuk, Y. *Ukr. Khim. Zh.* **2015,** *81*(3), 59.
15. Jatsyshyn, O.; Astakhova, O.; Lazorko, O., et al. *Chem. Chem. Technol.* **2013,** *1*, 73.
16. Ivashkiv, O.; Bruzdziak, P.; Shyshchak, O., et al. *Chem. Chem. Technol.* **2015,** *4*, 411.

PART II
Computational Approaches

CHAPTER 7

THE SUPERPOSING SIGNIFICANT INTERACTION RULES (SSIR) METHOD

EMILI BESALÚ[1,*], LIONELLO POGLIANI[2], and
J. VICENTE JULIAN-ORTIZ[2]

[1]*Institut de Química Computacional i Catàlisi (IQCC) and
Departament de Química, Universitat de Girona, 17003 Girona,
Catalonia, Spain*

[2]*Unidad de Investigación de Diseño de Fàrmacos y Conectividad
Molecular, Departamento de Química Física, Facultad de Farmacia,
Universitat de València, Burjassot (València), Spain, and MOLware
SL, Valencia, Spain*

Corresponding author. E-mail: emili.besalu@udg.edu

CONTENTS

ABSTRACT

The Superposing Significant Interaction Rules method is a simple rule generator that detects molecular aggregations especially rich in molecules tagged as being of interest under certain criterion. A series of rules are used to rank molecular sets of congeneric families, and the procedure allows for proposing new structures in a kind of inverse Structure-Activity Relationship (SAR). An example of application is shown demonstrating how, due to the simple arbitrary and symbolic molecular codification needed, the method can serve to generate confidential models. As the method deals with dichotomic properties, a special balanced leave-two-out procedure is also considered.

7.1 INTRODUCTION

Nowadays, requirements present in both the industry and the scientific community have being growing as long as the chemical data are expanding, and more efficient tools are needed to mine and extract relevant information. (Quantitative) Structure-Activity Relationship, (Q)SAR, fields are in need to look for new, and, especially, for simple methods that give rise to reliable models. The Superposing Significant Interaction Rules (SSIR) methodology, first described in the work of Besalú,[1] constitutes a simple procedure that, dealing with a dichotomic molecular property, is able to rank a molecular series of congeneric compounds. As it will be shown, the procedure can act as an inverse (Q)SAR tool[2,3] proposing new structures of interest to the researcher. Additionally, the process allows to deal with confidential data because the symbolic codification permits a potential (Q)SAR customer to provide data in a masked way. The whole SAR analysis can be done without revealing the actual nature of the compounds being investigated. Privacy is an important issue when dealing with libraries provided by third parties or when implementing a Software as a Service (SaaS) host.[4] The secretiveness can be implemented across SSIR because the algorithm only needs minimum symbolic information to be run.

7.2 GENERAL FRAMEWORK

The basic details of SSIR procedure have been described elsewhere[1] and will not be repeated here. Instead, the main leading ideal will be sketched out, but to a detail enough to reproduce the example shown below. It must be said that the primary idea to design SSIR was based on a well-known technique in engineering: the design of experiments (DoE).[5] In brief, one of the authors had the necessity to optimize a chemical design process (peptides design) but without the restriction to have "a priori" knowledge of responses for a predefined set of levels combination along factors, that is, molecular residues combination along substitution sites.

7.2.1 MOLECULAR DESCRIPTION

The original field of SSIR application is the world of congeneric molecular series. A congeneric series is described by a common molecular structure presenting n sites of substitution that are known as factors in the language of DoE. Each one of these sites is able to accommodate two or more substituents or residues (levels per factor). If the distinct substituents per site are symbolically (and arbitrarily) denoted by A, B, C, ..., then each compound can be simply represented by a series of n-ordered letters or symbols as, for example, $CABD$. Using this notation, the mere position of each letter identifies the substitution site. It is worth noting that, without loss of generality, the series of codifying letters (the alphabet) can be repeated along the distinct sites. If each site presents $m_i > 1$ substituents (sites with a single substituent do not enter into the combinatorial reckoning), the number of possible residue combinations is

$$M = \prod_{i=1}^{n} m_i, \tag{7.1}$$

where M being the theoretical number of structures in the whole virtual database. Note that, normally, the researcher only holds a, usually small, fraction of these, say $N < M$.

7.2.2 RULES

The way SSIR works is built on rules. A rule (or a variable) is a simple, and a priori, random specification of molecules defining a subgroup. For instance, the rule $AXBX$ stands for all the structures presenting at the same time the residue labeled A of the set of ones attached to the site number 1 and *simultaneously* the residue labeled B in the anchorage point number 3. It is said that the rule condenses these structures. The X symbol stands here for a wild card or for just "any residue" symbol. A rule can also involve negative terms, as in the case $AX\bar{B}X$, that is, the set of molecules presenting residue A at site 1 and, at the same time, not presenting the residue B at site 3. The former rules are of order 2 because they inform about the allowed or not allowed residues in only two substitution sites. The rule of order 0 XXXX virtually stands for the whole database. The maximal rule of order is n, the number of sites. Each rule of order n specifies a single analogue.

The generation of the rules is computationally systematic and relies on combinatorics.[6–8] The pseudocode for the generation of all the described rules follows:

- An external single loop identifies the rules of order k with $k = 1$, 2, 3, ... n. The theoretical final value of k is n, but usually it is not compulsive nor necessary to explore all the possible orders.
- Next loop generates all the combinations of n sites taken within the groups of k. A total of $C(n, k)$ combinations are generated identifying the sites involved in the next loop.
- The third-level loop generates all the variations of residues attached to the previously selected combinations of sites.
- Finally, the innermost loop eventually generates the negation terms (the above described bar notation) for each selected residue in the previous step. As it will be seen, this loop may generate up to 2^k terms.

7.2.3   QUANTIFYING THE RULE SIGNIFICANCE

In practice, despite a rule points to many virtual compounds, SSIR focuses the attention on $a < M$ compounds being studied by the researcher and for those where a dichotomic property has been defined. A simple probabilistic formula provides a factor that is treated as a p-value[9,10] attached to each rule. Let us suppose that, along known compounds, b of them are labeled as being of interest (i.e., be active, be a drug, present a small IC_{50} value, and so on). Then, if a rule condenses c analogues and d of these are of interest, the hypergeometric probability is attached to this event (under equal or uniform circumstances).[9,10]

$$P(d,c;b,a) = \frac{\binom{b}{d}\binom{a-b}{c-d}}{\binom{a}{c}} \text{ with } d \leq c \leq a \leq M \text{ and } d \leq b \leq a \leq M. \quad (7.2)$$

In this context, one can define the statistical significance of the rule in terms of a p-value as the probability to condense d or more (denoted $d+$) structures of interest:

$$p\,(d+,c;a) = \sum_{i=d}^{min(b,c)} P(i,c;b,a) = 1 - \sum_{i=max(0,c+b-a)}^{d-1} P(i,c;b,a). \qquad (7.3)$$

The basic version of SSIR procedure aims to generate all the possible rules of orders of 1, 2, 3, ... and evaluates for each one the attached p-value. Only the rules bearing a p-value lesser than or equal to a predefined threshold or cut value, say p_c, are accounted for. Once the rules are filtered by the respective p-values, the selected ones can bear a positive or a negative vote. A selected rule is provided with a positive vote if it condenses compounds of interest in a higher ratio than the, $p \leq p_c < 50\%$, neutral mode. If the rule condenses more analogues of noninterest than the expected neutral mode (i.e., $p > 50\%$), there is a possibility to attach it to a negative vote if the complementary probability is significant ($1-p \leq p_c < 50\%$). This means that the rule performs well-condensing structures of noninterest (this also constitutes a relevant information), and that the probability factor p-value becomes

$$p(d-,c;b,a) = 1 - p(d+,c;b,a) + P(d,c;b,a), \qquad (7.4)$$

where the d symbol stands for "d or less actives."

7.2.4 THE (Q)SAR SSIR MODEL

The full series of rules that are being assigned to a positive vote constitute a sort of *bonus* list. The *malus* list is formed by the set of rules having a negative vote. Despite there are several ways to combine the rules and votes, these possibilities will not be explored here. The basic SSIR program version[11] simply considers the direct sum of integer votes.

Once the selected rules (and their votes) are known, each molecular structure will cumulate the votes for all the significant variables condensing it. The ultimate goal is to apply this voting scheme also over new compounds not present in the sublibrary in order to rank them properly. For ranked or continuous molecular property values, the resulting series of votes per molecule is expected to be correlated with them. For categorical or pure dichotomic properties, it is assumed that the higher the number of votes a structure is collecting, the higher probability of being of a particular group.

7.2.5 CROSS-VALIDATION

If integer votes are being considered in SSIR, some cross-validation routines are easy to be implemented. This is so because the rankings are obtained by a mere addition of integer values. In particular, the leave-one-out (L1O) procedure can be simulated in a fast way[1]: first, it is only necessary to perform the full training and keep in memory the number of relevant items attached to each rule, that is, the number of condensed structures and the number of these that are of interest. Then, the L1O footage has not to be explicitly reproduced. Instead, for each left-out structure, it is only necessary to loop over the rules and, per each one, redo a simple count of condensed and active structures that would be obtained if the model was rebuild from scratch leaving the cross-validated structure apart. This process allows recalculating the rule significance probability, and decide for its vote assignation.

To the best of our knowledge, balanced leave-two-out (BL2O) cross-validation algorithm was presented for the first time.[1] The adjective balanced comes from the fact that, at every simulated cross-validation loop, two structures are left out simultaneously, one active and another being nonactive. This prevents to generate all the $n(n-1)/2$ combinations of molecular pairs but only $n_a \times n_i$, that is, the product of the number of active (n_a) ones by the nonactives (n_i). For every generated molecular pair, a couple of prediction votes is given that is added up. At the end of the procedure, the sum of votes for the active and the nonactive compounds has to be divided by n_i and n_a, respectively (or multiplied by n_a and n_i, respectively), setting an homogeneous scaling in both cases. This gives a series of comparable (and also balanced) votes that conforms a ranking. Due to the above comments, the BL2O procedure can be also implemented in a fast way without explicitly generating training/testing calculations for each left-out pair, but getting the equivalent result.

Along the BL2O cycles, it is counted for how many times an active molecule had more votes than the nonactive companion (correct internal classification), the times both structures received the same number of votes (ties) and the times the active item received less votes than the nonactive one (incorrect internal classification). As it will be seen below, this information constitutes a classification performance parameter.

All SSIR procedures, including cross-validation protocols, are implemented in our computer program SSIR.[11]

7.3 APPLICATION EXAMPLE: TIBO DERIVATIVES

The main goal of presenting this classical application example is to show the basic SSIR footage and reveal how it is simple to obtain molecular rankings in a systematic and reproducible way. Even more, SSIR requires a minimum and very condensed amount of information: Apart from a dichotomic molecular partition, only an arbitrary and symbolic codification of the residues is needed. These niggling requirements encompass the ability to share the data in a confidential manner. The reduced information needed by SSIR makes very difficult the revelation of some of the database compounds or similar analogues. This constitutes a valuable feature, as it is evident that there exists an awareness for sharing chemical information in a secure way, at least at the level of sharing databases in a protected mode from third parties.[12,13] On the contrary, if some, even partial, chemical information is given (even codified in the form of calculated indices) or some residues or additional features are revealed, it is possible to give enough clues to disclose some of the compounds.[14]

7.3.1 THE MOLECULAR SET

The molecular family chosen is a well-known set of tetrahydroimidazo[4,5,1-jk][1,4]benzodiazepinone (TIBO) analogues consisting of 90 molecules, the merge of training compounds listed in the work of Huuskonen[15] (which is the same as the ones in Hannongbua et al.[16]), and Zhou and Madura.[17] In these lists, six compounds were repeated, that is, compounds 1, 2, 36, 37, 38, and 45 in Huuskonen's list that are same as compounds 44, 45, 3, 47, 2, and 50 of Zhou and Madura's article, respectively. Note that in Zhou and Madura's article, compounds 38 and 50 appear to be the same. A personal communication of the authors revealed the real structure of compound 38 (X = H, Z = O, R = 2-Methylallyl, Y = 4-CHMe$_2$, pIC$_{50}$ = 4.90). Table 7.1 describes the codification dictionary of all the nonredundant training compounds. Note that this dictionary is only known by the database owner. The information that has to be transferred (eventually to a third party) to apply SSIR is only the set of symbols (not the residues) and the dichotomized property.

Table 7.2 lists the compounds codification according to the original numbering of Huuskonen, Zhou, and Madura's articles. It is worth noting that the only information needed by the method proposed here is the

TABLE 7.1 Owner's Dictionary Used to Codify the Compounds of References Huuskonen, Hannongbua,[15-17] Molecules Present Four Substitution Sites (Labeled R_1, X, R_2, and R_3 in Huuskonen's and Hannongbua et al., Articles, Which Correspond to Sites Y, Z, X, and R in Zhou and Madura's Article, Respectively). Note that the Symbol Codification is Arbitrary, and that the Same Set of Symbols can be Repeated Along Substitution Sites.

R_1		X		R_2		R_3	
Symbol	Residue	Symbol	Residue	Symbol	Residue	Symbol	Residue
A	H	O	O	A	8-Cl	A	DMA^a
B	5-Et	S	S	B	9-Cl	B	2-MA^b
C	5-i-Pr			C	H	C	$CH_2CH(CH_3)_2$
D	5,5-di-Me			D	8-F	D	n-Pr
E	4-Me			E	8-SMe	E	H
F	4-i-Pr			F	8-OMe	F	CPM^c
G	4-n-Pr			G	$8-OC_2H_5$	G	CH_2CHdCH_2
H	7-Me			H	8-CN	H	CH_2CO_2Me
I	4,5-di-Me (cis)			I	8-CHO	I	CH_2-2-furanyl
J	4,5-di-Me (trans)			J	$8-CONH_2$	J	$CH_2CH_2CH=CH_2$
K	4-keto-5-Me			K	8-Br	K	$CH_2CH_2CH_3$
L	4,5-di-benzo			L	8-I	L	$CH_2CH=CHMe$ (E)
M	5,7-di-Me (trans)			M	8-C≡CH	M	$CH_2CH=CHMe$ (Z)
N	5,7-di-Me (cis)			N	8-Me	N	$CH_2CH_2CH_2Me$
O	5,7-di-Me (R,R;trans)			O	$8-NMe_2$	O	$CH_2C(Br)=CH_2$
P	5,7-di-Me (S,S;trans)			P	$9-NH_2$	P	$CH_2C(Me)=CHMe$ (E)

TABLE 7.1 *(Continued)*

R$_1$		X		R$_2$		R$_3$	
Symbol	Residue	Symbol	Residue	Symbol	Residue	Symbol	Residue
Q	4,7-di-Me (trans)			Q	9-NMe$_2$	Q	CH$_2$C(C$_2$H$_5$)=CH$_2$
R	5,6-CH$_2$C(=CHCH$_3$)CH$_2$(S)			R	9-NHCOMe	R	CH$_2$CH=CHC$_6$H$_5$ (Z)
S	6,7-(CH$_2$)$_4$			S	9-NO$_2$	S	CH$_2$C(CH=CH$_2$)=CH$_2$
T	5-Me (S)			T	9-F		
U	5-Me			U	9-CF$_3$		
V	4-CHMe$_2$			V	10-OMe		
W	4-Me (R)			W	10-Br		

[a] 3,3-dimethylallyl.
[b] 2-methylallyl.
[c] cyclopropylmethyl.

TABLE 7.2 SSIR Codification of the Mixture List of Compounds of References Huuskonen, Hannongbua et al., and Zhou and Madura.[15-17] Molecules Show Four Substitution Sites (Sites Labeled R_1, X, R_2, and R_3 in Huuskonen's and Hannongbua et al. Articles, Which Correspond to Sites Y, Z, X, and R in Zhou and Madura's Article, Respectively) that Here Have Been Codified from Left to Right.

Compound number	Original numbering[a]	SSIR codification	pIC_{50}	Compound number	Original numbering[b]	SSIR codification	$logIC_{50}$
1	1	ASAA	7.34*	47	1	TSCA	7.36*
2	2	ASBA	6.80*	48	4	TSDA	8.24*
3	3	BOCB	4.30	49	5	TSEA	8.30*
4	4	COCB	5.00	50	6	TSFA	7.47*
5	5	COCA	5.00	51	7	TSGA	7.02*
6	6	DOCB	4.64	52	8	TSHA	7.25*
7	7	EOCB	4.49	53	9	TSIA	6.73
8	8	ESBB	6.17	54	10	TOJA	5.20
9	9	ESBC	5.66	55	11	TOKA	7.33*
10	10	FOCD	4.13	56	12	TSKA	8.52*
11	11	FOCB	4.90	57	13	TOLA	7.06*
12	12	GOCD	3.74	58	14	TSMA	7.53*
13	13	GOCB	4.32	59	15	TONA	6.00
14	14	HOCD	4.08	60	16	TSNA	7.87*
15	15	HOCA	4.92	61	17	TOOF	5.18
16	16	HOAA	6.84*	62	18	TOPF	4.22
17	17	HOBA	6.79	63	19	TOQF	5.18

TABLE 7.2 (Continued)

Compound number	Original numbering[a]	SSIR codification	pIC_{50}	Compound number	Original numbering[b]	SSIR codification	$logIC_{50}$
18	18	HSCD	5.61	64	20	TORF	3.80
19	19	HSCA	7.11*	65	21	TSSF	5.61
20	20	HSAA	7.92*	66	22	TSTA	7.60*
21	21	HSBA	7.64*	67	23	TOUA	5.23
22	22	IOCA	4.25	68	24	TSUA	6.31
23	23	ISCA	5.65	69	25	TOVA	5.18
24	24	JSCC	4.87	70	26	TSVA	5.33
25	25	JSCA	4.84	71	27	TSWA	5.97
26	26	KSBD	4.30	72	28	TOCG	4.15
27	27	LSCC	5.00	73	29	TOCB	4.33
28	28	MSCA	7.38*	74	30	TOCH	3.04
29	29	NSCA	5.94	75	31	TOCI	3.97
30	30	OOBA	6.64	76	32	TOCJ	4.30
31	31	OSBA	6.32	77	33	TOCK	4.05
32	32	POBA	5.30	78	34	TOCF	4.36
33	33	QSCA	4.59	79	35	TOCL	4.24
34	34	RSBE	5.42	80	36	TOCM	4.46
35	35	SSBE	5.70	81	37	TOCN	4.00

TABLE 7.2 (Continued)

Compound number	Original numbering[a]	SSIR codification	pIC$_{50}$	Compound number	Original numbering[b]	SSIR codification	logIC$_{50}$
36	36	TSAA	8.34*	82	38	VOCB	4.90
37	37	TOBA	6.74	83	39	TOCO	4.21
38	38	TSBA	7.42*	84	40	TOCP	4.54
39	39	TSBC	7.47*	85	41	TOCQ	4.43
40	40	TSCC	7.22*	86	42	TOCR	3.91
41	41	UOCD	4.22	87	43	TOCS	4.15
42	42	USCD	5.78	88	46	WSBF	5.66
43	43	UOCB	4.46	89	48	TSBF	7.47*
44	44	USCA	7.01*	90	49	TSCF	7.22*
45	45	TOCA	5.48				
46	46	TSCB	7.58*				

[a]References: Huuskonen[15] and Hannongbua et al.[16]
[b]Reference: Zhou and Madura.[17]
*Activities marked with an asterisk (pIC$_{50}$ > 6.79) mean molecule of interest (i.e., active).

symbolic codification and the characteristic of a compound of being or not of interest. The pIC_{50} values are not needed and this also means that the particular property or activity being studied is not revealed. The database owner keeps the information presented in Table 7.1 (the real molecules) and the property values, whereas the (Q)SAR modeler only needs a reduced part of the information presented in Table 7.2 (the molecular codification and the dichotomic labeling). SSIR procedure allows getting a model from this minimal information, even if the modeler does not know which kind of compounds has been dealing with. Even more, this also applies at the final stage when new structures are being proposed: the modeler only knows about generic "new compounds" codified in the same manner as it was done in Table 7.2. Only the database owner is able to decode the new proposed analogues in terms of real chemical compounds.

The four molecular sites allow for 23, 2, 23, and 19 residues (levels) each. SSIR program can operate in two modes, either allowing or not allowing the generation of negation terms in rules, and this is equivalent to either consider or not consider the pseudocode innermost loop described in Section 2. If negation terms are not allowed, there are 67, 1533, 12,857, and 20,102 definable rules of orders 1–4, respectively. The number of rules of maximal order is the same as the number of analogues definable in the full database, M. If negation terms are allowed, then the number of rules increases. For instance, there are 132 rules of order 1, as 23 + 23 + 19 rules come from three residues, each one multiplied by 2 if individual negations are also considered. This gives 130 rules. The two remaining rules are the two residues of the second site. Note that for binary sites, there is no need to generate negations because one level is the natural negation of the other one. Even more, for rules of order 1 involving a binary site, only one of the levels has to be considered. This is so because this level, if significant, will bear a positive or a negative vote, retaining all the relevant information. This information is enough because the other level will automatically bear a negative or a positive vote, respectively, being the information wholly redundant if both rules are to be kept.

The number of complete rules of orders 2, 3, and 4 is 5872, 91,632, and 160,816, respectively. Figure 7.1 shows the distribution of p-values (eq 7.3) attached to the rules of order 4 (negation terms allowed).

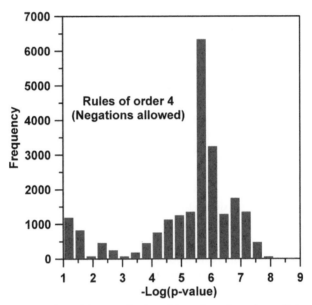

FIGURE 7.1   Distribution of the negative logarithm of p-values along all the rules of order 4 for the TIBO family (representation starts at $p = 0.1$).

7.3.2 TRAINING AND CROSS-VALIDATION

Several calculations of training, L1O, BL2O, and predictions over the external set have been conducted changing the cutoff p-value (p_c) from 10^{-2} to 10^{-7}. It is interesting to look at rules of low order, as some of them can reveal general trends and features found in the database. For instance, regarding the rules of order 1, only five bear p-values lesser than 0.01, if negations are not allowed. All these rules focus on the presence of a particular residue in specific sites. Ranging from most to less significant, these rules are: *XOXX, XSXX, XXXA, XXCX,* and *XXAX* with significances equal to $4\ 10^{-7}$, $4 \cdot 10^{-7}$, $2.6 \cdot 10^{-5}$, $1.7 \cdot 10^{-3}$, and $6.9 \cdot 10^{-3}$, respectively. For the first and third cases, the vote is negative, that is, the rule points to avoid the specified residues at the corresponding sites.

As said above, note that the rules *XOXX* and *XSXX*, as being of order 1 for a binary (O, S) site, have the same probabilistic significance and the only difference among them is the sign vote. Strictly, only one of the rules is to be kept in SSIR model. Let us see it in detail. In training, rule *XOXX* condenses 46 of the 90 molecules, and only 3 of these 46 are actives or of interest. According to eq 7.3, this gives a probability $p(3+, 46; 27, 90) = 0.99999998$

to condense three or more active analogues. This is not significant. On the contrary, the complementary success corresponds to point the attention to the 63 (i.e. 90 − 27) nonactive analogues. The *XOXX* rule condenses 43 (i.e. 46 − 3) of these nonactive compounds. For this case, eq 7.3 gives a significant probability of p (43+,46;63,90) = 4 · 10^{-7}, as stated above. The conclusion is that rule *XOXX* is efficient to point to nonactive analogues and, hence, must bear a negative vote. But as the rule is of order 1 and it involves a binary site (*O*, *S*), the complementary rule is *XSXX* for which the same significant and redundant probability values are reproduced, p ((27−3)+,90−46;27,90) = p (24+,44;27,90) = 4 · 10^{-7}, and points to active compounds, that is, its vote assignation must be positive.

The variables of order 2 without negative terms generate a model involving only 6 rules (for $p_c \leq 0.01$): *TSXX, XSXA, XOCX, TXXA, TOXX,* and *XXAA*. The respective p-values are 1.2 · 10^{-7}, 3.6 · 10^{-7}, 7 · 10^{-7}, 6.9 · 10^{-4}, 1.4 · 10^{-3}, and 7.6 · 10^{-3}; all positive votes but the third and fifth. Consistently, many of these rules involve combinations of terms appearing in previous rules of order 1. Here, the moiety *T* at the first site appears to be relevant. By the way, concerning the confidentiality issue notes that, in a real example, only the database owner will know that this residue is a 5-Me(S) group. This is the essence of SSIR method: the automation of search rules. For some almost trivial cases, the rules could be deduced by inspection but this can require endeavor. In general, the algorithm performs the inspection systematically for more complicated cases involving larger databases, extra molecular substitution points, and available residues.

Table 7.3 summarizes the representative results obtained for $p_c = 10^{-5}$. The reported parameters are the Area Under ROC (AU-ROC)[18,19] curves and Spearman's rank correlation coefficient,[20] ρ, that is shown in brackets. The procedure does not take into account any continuous experimental activity but only the labels of being active or nonactive attached to every molecule. Despite of that, it is worthy of mention that Spearman's ρ presents a quite high correlation between the molecule category and the cumulated votes.

The SAR models were obtained in batches of prefixed rule orders. The table also shows how many variables are generated in each case and, between brackets, how many of them are below the cutoff p-value. There are no significant rules of order 3 or 4 if negations are not allowed. For the case of order 4, as expected, none of such rules is significant because each rule only embraces a single compound. Figure 7.2 shows the ROC for the training model involving rules of order 3, when negation terms are allowed.

TABLE 7.3 Exploration Results Obtained for the TIBO Family of Analogues. Rank Correlation Coefficients (in Brackets) and AU-ROC Curves are Given. For the BL2O the Number of Good, Tie, and Bad Classifications are Given. For all Calculations, the p-Value Cutoff was $p_c = 0.00001$.

Rules order	Negations allowed	Rules (significant)[a]	Cutoff p-value = 0.00001		AU-ROC (Spearman ρ)[b]	External set prediction[e]	
			TRN	L1O	BL2O[c]	TRN	BL2O
1	No	67 (2)	0.786 (0.682)	0.786 (0.682)	0.712 (0.657) 1032/564/105	0.794	0.829 (0.710) (0.801)
	Yes	132 (2)	0.786 (0.682)	0.756 (0.641)	0.695 (0.636) 972/624/105	0.794	0.829 (0.710) (0.801)
2	No	1533 (3)	0.899 (0.827)	0.899 (0.827)	0.899 (0.827) 1437/186/78	0.836	0.836 (0.809) (0.809)
	Yes	5872 (126)	0.919 (0.832)	0.767 (0.698)	0.780 (0.709) 1459/95/147	0.888	0.937 (0.852) (0.892)
3	No	12,857 (0)	—	—	—	—	
	Yes	91,632 (2501)	0.928 (0.825)	0.764 (0.678)	0.796 (0.708) 1526/22/153	0.958	0.958 (0.909) (0.909)
4	No	20,102 (0)	—	—	—	—	
	Yes	1,60,816 (16,112)	0.927 (0.819)	0.755 (0.642)	0.767 (0.657) (0.757) 1495/47/159	0.958	0.958 (0.905) (0.905)

[a]Number of rules giving p-values equal or under the cutoff along the single fitting of the training set. This number of significant ones can vary along L1O or BL2O iterations.

[b]Spearman's rank correlation coefficient between molecule class and number of collected votes.

[c]BL2O procedure needs $27 \times 63 = 17,071$ cycles.

[d]For BL2O calculations, the number of good, tie, and bad classifications along the cycles is given in this order.

[e]Predictions obtained by application of the training model (TRN) and all the collected rules along iterations in BL2O processes.

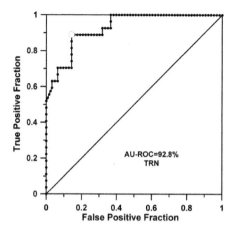

FIGURE 7.2 ROC curve for the training of TIBO molecules using rules of order 3 (negation terms allowed, $p_c = 10^{-5}$). The small empty circle signals the best point found from the probabilistic point of view, that is, having the lesser p-value computed according to formula (3). The AU-ROC value is also indicated.

From the ranking imposed by previous training model, Figure 7.3 shows the successive p-values obtained when tracing back the sorted list of molecules and counting the number of active ones that are being collected in groups of molecules sharing the same number of votes. The value inside each circle gives the number of cumulated collected active structures.

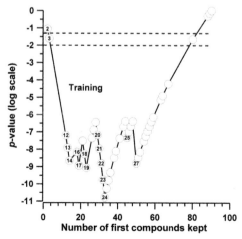

FIGURE 7.3 Positions of 27 active TIBOs appearing in the sorted database by the model of order 3 (negations allowed, $p_c = 10^{-5}$). Note the logarithmic scale in the vertical axis. The two horizontal discontinuous lines stand for 1% and 5% significance levels.

For this case, 24 active structures are found in the first bloc of 33 ranked compounds and, according to formula (3), the attached probability is $p(24+,33;27,90) = 1.7 \cdot 10^{-11}$, which is the depicted value for the point labeled "24." This "optimal" point of minimal p-value corresponds to the one marked with an empty circle in Figure 7.2. Similarly, all the 27 active structures are found in the first bloc of 50 ranked compounds with $p(27+,50;27,90) = 1.6 \cdot 10^{-9}$. Note that this efficiency has been obtained from no knowledge of the chemical residues' nature present in each compound.

As said above, a fast BL2O procedure has been designed and implemented throughout our SSIR program. For this case, it gives 1701 left-out pairs or cycles (27 actives × 63 nonactive). After cumulating and scaling the votes of rules of order 3 (negations allowed), the model is resulted to have $\rho = 0.708$ and AU-ROC = 0.796. Better results are obtained for other p_c values or order rules, but this was not accompanied with better predictions over the external set (see Section 3.3). Along the BL2O cycles, for 1526 active/ nonactive pairs, the internal classification was correct; for 22 cases, a tie in votes was reproduced and for the remaining 153 couples, the active item received fewer votes than the nonactive one. Note that the ratio of correct explicit pair classifications is 0.897, an interesting value that can be treated as an AU-ROC or efficiency parameter. This is so because it is well known that, for a single fitting calculation, given a couple of molecules (one of interest and the other of noninterest), the AU-ROC corresponds to the *a posteriori* probability that the classifier correctly sorts the pair.[21] Figure 7.4 shows the ROC for the model involving significant rules or order 3 (negation terms allowed) collected during all the BL2O iterations. The first ranked molecules were erroneous, but the next series resulted to be very rich in active analogues.

L1O and BL2O procedures seem to reveal that the training model is quite unstable because the rank parameters decrease. A deeper investigation showed that this was due to spurious borderline rules which are not significant in training but become significant in some cross-validation cycles. These rules became significant because the momentary extraction of some compounds during cross-validation changed the probability assignations. During cross-validation cycles, if only original significant training rules are allowed to enter into local models (and, eventually, several of them will become nonsignificant), the correlation parameters increase substantially and artificially.

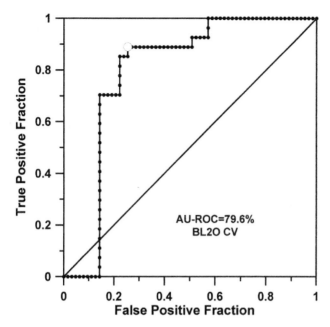

FIGURE 7.4 ROC curve for the BL2O cross-validation procedure of TIBO molecules using rules of order 3 (negation terms allowed, $p_c = 10^{-5}$). The small empty circle signals the best point found from the probabilistic point of view, that is, having the lesser p-value computed according to formula (3). The AU-ROC value is also indicated.

7.3.3 TEST FOR VALIDATION

Once a ranking model is defined, prediction for new structures can be formulated by accumulation of votes coming from the selected significant rules. Each tested structure collects the votes of the corresponding matching rules. This demands the training set to be provided with all the residues that will appear in future predictions. If this condition is not met, compounds presenting new substitutions are not properly evaluated by SSIR method. The absence of residues will lead to the application of less rules, leaving the counters of votes near the conservative zone of zero votes. For the particular case of Huuskonen's or Hannongbua et al. test set (Table 7.4), structures T6 and T7 have been treated here as equivalent because the fourth residue is not present in training and, hence, is transparent to the training rules.

In general, the inclusion of negation terms in rules gives the best results in both training and test/validation batches. From the collected significant rules of orders 3 and 4 in training (Table 7.3), the predictions over the external test set gave in both cases AU-ROC = 0.958. This last value was also obtained with other tests involving $p_c \leq 10^{-4}$ values. In all these cases, the test compounds of interest are ranked first, but systematically three compounds of noninterest (*TSIA*, *TSWA*, and *TSVA*) are placed in between at positions 10–12. Figure 7.5 visualizes the result obtained when the training model (2501 significant rules of order 3, including negations) is applied over the external test set.

TABLE 7.4 SSIR Codification for the 24 Test Set Molecules of Reference Huuskonen[15]. Lowercase Letters (a, b, c, and d) Stand for Residues Not Present in Training Set.

Compound number	SSIR codification	pIC_{50}
T1	*AOCA*	4.9
T2	*AOCB*	4.33
T3	*AOCD*	4.05
T4	*AOCa*	4.43
T5	*TSCA*	7.355*
T6	*TOCb*	4.154
T7	*TOCc*	3.999
T8	*TSDA*	8.235*
T9	*TOKA*	7.324*
T10	*TSKA*	8.521*
T11	*TSNA*	7.865*
T12	*TSFA*	7.468*
T13	*TSdA*	7.592*
T14	*TOHA*	5.94
T15	*TSHA*	7.25*
T16	*TONA*	6
T17	*TSVA*	5.33
T18	*TOVA*	5.18
T19	*TSWA*	5.97
T20	*TSIA*	6.73
T21	*TOLA*	7.06*
T22	*TSLA*	7.32*
T23	*TOMA*	6.36
T24	*TSMA*	7.53*

*The 11 activities marked with an asterisk ($pIC_{50} > 6.79$) mean molecule of interest (i.e., active).

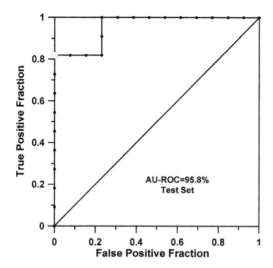

FIGURE 7.5 ROC curve for the TIBO external test set. The model included rules of order 3 (negation terms allowed, $p_c = 10^{-5}$). The small circle signals the best point found from the probabilistic point of view, according to formula (3). The AU-ROC value is also indicated.

It is expected that SSIR rules generated during training have some information content in order to (probabilistically) point to active compounds when external sets are being considered. As an illustrative example, for this TIBO family, the calculation with rules of order 3 without negations ($p_c = 0.001$) only generates a single rule during training: give a positive vote to the substructure *TSXA*. Along the training set, the visual detection of such a rule is not immediate. Noticeably, this single rule suffices to establish two categories in the external test set: the analogues fulfilling or not fulfilling this pattern. Hence, from this rule, 12 of the 24 test molecules receive a positive vote and, among them, 9 are analogues of interest. According to eq 7.3, this results to show a predictive significance of $p(9+,12;11,24) = 0.006$ for the external set.

7.3.4 INVERSE SAR: NEW ANALOGUES PROPOSALS

For the case of combinatorial analogues treatment, SSIR procedure also acts as an inverse SAR tool[22] because of its ability to suggest for new compounds. Despite it is envisaged a direct algorithm able to generate good candidate analogues directly from the knowledge of the selected significant rules, we here are simply referring to a brute force procedure consisting to generate new substituent combinations, apply the rules and select the analogues bearing a maximum number of positive votes. Within this context, in general, and in order to prevent the combinatorial explosion of possible derivatives, not all the M analogues of eq 7.1 are to be generated in silico. In order to preserve affinity to the multidimensional space spanned by the training compounds, many times it only suffices to generate analogues that differ from several training items only by a few substitutions, as it will be explained now.

The whole training database expands 20,102 TIBO analogues. For the model of rules of order 3 ($p_c = 10^{-5}$, negations allowed), SSIR program was asked for predictions of new combinatorial analogues having a high number of votes. For exploratory purposes, only 68 compounds have been initially considered: those having at most only one single substitution difference with (in any site) respect to at least six training compounds. As said, these kinds of criteria are implemented in SSIR program in order to prevent extrapolations out of the molecular training domain. The SSIR answer was to synthesize as good candidates 24 analogues which, listed in decreasing preference are (the brackets collect analogues with the same number of votes) *TSLA*, (*TSQA, TSRA, TSPA, TSOA*), *TSSA, TSJA, ASCA*, (*SSCA, VSCA, WSCA, RSCA, KSCA, DSCA, ESCA, GSCA, BSCA, FSCA*), *LSCA*, (*PSCA, CSCA*), *ESBA, OSCA*, and *TSCD*. Note that the first ranked compound *TSLA* is an active analogue present in Huuskonen's external test set. The remaining compounds presumably will show low activity, as all of them have negative votes. This *malus* list included three compounds of noninterest (*TOHA, TOMA*, and *AOCB*) which are already present in Huuskonen's test set. The other low-activity test analogues *AOCA* and *AOCD* also presented negative votes, but these compounds are generated beyond the above restriction considered to prevent extrapolations.

The generation of the training model involving rules of order 3 or 4 and the generation of new analogues proposals only takes a few seconds on an Intel-i7 CPU (4 GB memory) running under Windows 7. Running times can increase substantially if the number of rules reaches the combinatorial explosion.

7.3.5   SOME ADVANTAGES AND DRAWBACKS OF SSIR METHOD FUTURE PROSPECTS

One of the main advantages of SSIR method relies on the fact that the data can be processed in a systematic way and that preprocessing tasks are minimal. The setup from scratch is fast because no conformational analysis or no elaborated indices calculations are needed to start a computational run. The symbolic treatment allows for several fields of application beyond the molecular one. Once the symbolic descriptors and the running parameters (property-cut parameter if dichotomization is needed, threshold p-value, rule orders being considered, the option to include negation terms, ...) are fixed, the method can be easily reproduced. As it has been shown, simple leave-out and replace cross-validation procedures (L1O, BL2O) can be rapidly executed.

The main drawbacks of the method open new doors for exploration and future analysis of the SSIR characteristics and performance. As it occurs in other methodologies, the results are depending on the balancing of the database. Despite the method is feasible in a variety of circumstances, the substituents' population should be well represented along the library. Another factor to be taken into account is the degree of library dilution: as more sites or substitution options are present, more training molecules are presumably needed in order to reach meaningful results. Regarding the problem of the combinatorial explosion for the number of available rules, work is done now in our laboratory in order to implement nonexhaustive searches of significant rules. As it has been seen above, the test or external sets presenting new substituents never appearing in training will presumably force conservative/useless predictions. Furthermore, the method cannot deal with libraries of analogues that only have one single substitution site.

There are several aspects of the method which demand more attention as there is room for additional refinements not explored in this work. For instance, the set of collected votes per rule can be analyzed numerically using linear or nonlinear procedures. The collected votes per molecule and rule (and also per range of p-values) can be interpreted as new molecular descriptors. This idea is in turn connected with the option to merge rules of distinct order, belonging to distinct ranges of significances or having distinct votes sign. For instance, the series of votes can be combined by means of the construction of linear combinations or multilinear, discriminant or partial least squares derived weights. This panorama opens the door to transform the SAR model into a QSAR one and related work is now in progress in our laboratory. Additionally, votes of distinct sign can be merged but with distinct weights.

Another important aspect to be investigated is the possible implementation of the false discovery rate (FDR)[23] test. Standard scientific protocols recommend filtering the selected "significant" rules according to Bonferroni's test or similar. FDR filters will reduce the number of selected rules (this is equivalent to force the reduction of the cutoff p_c value) and it is expected to increase the quality of the model. Notwithstanding this logical proposal, it has been stated that the implementation of FDR filters in many cases dramatically reduces the number of rules, leading to nonpredictive models. In all the calculations presented here no FDR sieve has been used. Despite of that, in all the tests and calculations done, the performance over the test sets was more than satisfactory. It seems that the inclusion of many rules tends to stabilize the models, perhaps including compensatory effects. More investigations are to be performed regarding the possibility to filter individual rules according to its performance in cross-validation.

7.4   CONCLUSIONS

It has been described SSIR, a systematic procedure useful to rank series of combinatorial analogues. The mathematical foundations and the algorithm have been revisited and a dataset has been explored in order to reveal the main characteristics and performance of SSIR. The method is fast for rules of low order and can also be implemented for database ranking or filtering purposes, and even as an inverse SAR tool. The symbolic nature of the needed information allows manipulating the database in a confidential manner, if needed. The BL2O procedure for cross-validation has been also considered to check for model robustness at a cross-validation level.

ACKNOWLEDGMENT

The author acknowledges the Generalitat de Catalunya (Departament d'Innovació, Universitats i Empresa) for the financial support given to the QTMEM (Química teòrica i Modelatge i Enginyeria Molecular) research group of the University of Girona (code 2014-SGR-1202). A personal communication of J. Madura and Z. Zhou regarding a correction of a typographical error in a reference is also acknowledged.

KEYWORDS

- **Superposing Significant Interaction Rules**
- **Structure-Activity Relationships**
- **congeneric families**
- **molecular codification**
- **dichotomic properties**

REFERENCES

1. Besalú, E. Fast Modeling of Binding Affinities by Means of Superposing Significant Interaction Rules (SSIR) Method. *Int. J. Mol. Sci.* **2016**, *17*, 827. DOI: 10.3390/ijms 17060827.
2. Miyao, T.; Arakawa, M.; Funatsu, K. Exhaustive Structure Generation for Inverse-QSPR/QSAR. *Mol. Inform.* **2010**, *29*(1–2), 111–125. DOI: 10.1002/minf.200900038.
3. Wong, W. W. L.; Burkowski, F. J. A Constructive Approach for Discovering New Drug Leads: Using a Kernel Methodology for the Inverse-QSAR Problem. *J. Cheminform.* **2009**, *1*, 4. DOI: 10.1186/1758-2946-1-4.
4. http://en.wikipedia.org/wiki/Software_as_a_service (accessed June 3, 2016).
5. Eriksson, L.; Johansson, E.; Kettaneh-Wold, N.; Wikström, C.; Wold, S. *Design of Experiments. Principles and Applications;* Umetrics Academy: Umea, Sweden, 2000.
6. Carbó, R.; Besalú, E. Nested Summation Symbols and Perturbation Theory. *J. Math. Chem.* **1993**, *13*, 331–342.
7. Besalú, E.; Carbó, R. Generalized Rayleigh-Schrödinger Perturbation Theory in Matrix Form. *J. Math. Chem.* **1994**, *15*, 397–406.
8. Carbó, R.; Besalú, E. Definition, Mathematical Examples and Quantum Chemical Applications of Nested Summation Symbols and Logical Kronecker Deltas. *Comput. Chem.* **1994**, *18*(2), 117–126.
9. Besalú, E.; Ponec, R.; de Julián-Ortiz, J. V. Virtual Generation of Agents Against *Mycobacterium tuberculosis*. A QSAR Study. *Mol. Divers.* **2003**, *6*(2), 107–120.
10. Barroso, J. M.; Besalú, E. Design of Experiments Applied to QSAR: Ranking a Set of Compounds and Establishing a Statistical Significance Test. *Theochemistry* **2005**, *727*(1–3), 89–96.
11. Besalú, E. SSIR program v1.0, Girona, Spain, 2015.
12. Bologa, C.; Allu, T. K.; Olah, M.; Kappler, M. A.; Oprea, T. I. Descriptor Collision and Confusion: Toward the Design of Descriptors to Mask Chemical Structures. *J. Comput. Aided Mol. Des.* **2005**, *19*, 625–635. DOI: 10.1007/s10822-005-9020-4.
13. Tetko, I. V.; Abagyan, R.; Oprea, T. I. Surrogate Data—A Secure Way to Share Corporate Data. *J. Comput. Aided Mol. Des.* **2005**, *19*, 749–764. DOI 10.1007/s10822-005-9013-3.
14. Filimonov, D.; Poroikov, V. Why Relevant Chemical Information Cannot be Exchanged Without Disclosing Structures. *J. Comput. Aided Molec. Des.* **2005**, *19*, 705–713. DOI 10.1007/s10822-005-9014-2.
15. Huuskonen, J. QSAR Modeling with the Electrotopological State: TIBO Derivatives. *J. Chem. Inf. Comput. Sci.* **2001**, *41*, 425–429.
16. Hannongbua, S.; Pungpo, P.; Limtrakul, J.; Wolschann, P. Quantitative Structure-Activity Relationships and Comparative Molecular Field Analysis of TIBO Derivatised HIV-1 Reverse Transcriptase Inhibitors. *J. Comput. Aided Mol. Des.* **1999**, *13*, 563–577.
17. Zhou, Z.; Madura, J. D. CoMFA 3D-QSAR Analysis of HIV-1 RT Nonnucleoside Inhibitors, TIBO Derivatives Based on Docking Conformation and Alignment. *J. Chem. Inf. Comput. Sci. 44*, 2167–2178.
18. Egan, J. P. *Signal Detection Theory and ROC Analysis*; Academic Press: New York, 1975.
19. Besalú, E.; De Julián Ortiz, J. V.; Pogliani, L. *Quantum Frontiers of Atoms and Molecules*; Putz, M. V., Ed.; NOVA Publishing Inc.: New York, 2010.

20. Forlay-Frick, P.; Van Gyseghem, E.; Héberger, K.; Vander Heyden, Y. Selection of Orthogonal Chromatographic Systems Based on Parametric and Non-parametric Statistical Tests. *Anal. Chim. Acta* **2005,** *539*, 1–10.

21. Mason, S. J.; Graham, N. E. Areas Beneath the Relative Operating Characteristics (ROC) and Relative Operating Levels (ROL) Curves: Statistical Significance and Interpretation. *Q. J. R. Meteorol. Soc.* **2002,** *128*, 2145–2166.

22. Jin Cho, S.; Zheng, W.; Tropsha, A. Rational Combinatorial Library Design. 2. Rational Design of Targeted Combinatorial Peptide Libraries Using Chemical Similarity Probe and the Inverse QSAR Approaches. *J. Chem. Inf. Comput. Sci.* **1998,** *38*, 259–268.

23. Holm, S. A Simple Sequentially Rejective Multiple Test Procedure. *Scand. J. Statist.* **1979,** *6*, 65–70.

CHAPTER 8

THE VISION OF APPLICATION OF MULTIOBJECTIVE OPTIMIZATION AND GENETIC ALGORITHM IN MODELING AND SIMULATION OF THE RISER REACTOR OF A FLUIDIZED CATALYTIC CRACKING UNIT: A CRITICAL REVIEW

SUKANCHAN PALIT[*]

Department of Chemical Engineering, University of Petroleum and Energy Studies, Energy Acres, Bidholi via Premnagar, Dehradun 248007, Uttarakhand, India

[*]*Corresponding author. E-mail: sukanchan68@gmail.com; sukanchan92@gmail.com*

CONTENTS

ABSTRACT

The world of science and technology is moving toward a newer visionary frontier. Man's vision, mankind's prowess, and civilization's progress all will lead a long way in the true emancipation of human scientific endeavor. In today's scientific world, petroleum engineering science is facing an inimitable global crisis with the depletion of fossil fuels. Technological initiative is the only answer to this immense crisis. Depletion of fossil-fuel resources is a bane to human civilization. Today, human research pursuit is at its helm at every step of human life and human civilization. Chemical process engineering, chemical process modeling, and petroleum engineering science are ushering in a new era of true realization of science. In such a crucial juncture, mathematical tools such as multiobjective optimization and genetic algorithm (GA) are opening new windows of innovation and scientific instinct in decades to come. Fluidized catalytic cracking unit (FCCU) is a major component in the future of petroleum refining science. Difficulties, barriers, and challenges in petroleum refining and design of an FCCU are overwhelming. The author with deep comprehension and unending insight brings to the scientific forefront the barriers and difficulties in designing a fluid catalytic cracking (FCC) unit. Scientific regeneration and scientific fortitude are needed in the research pursuit today. The author delves deep into the nitty-gritty's of the design of the riser section of the FCC unit. The riser section deserves immense importance in the pursuit of science and engineering of petroleum refining. History of petroleum refining is ushering in a new eon of science in today's scientific world. The author of this treatise deeply comprehends and focuses on the FCC process and instinctively reviews recent developments in its modeling, monitoring, control, and optimization. The riser unit and its design evolve immense challenges and difficulties. The author pointedly focuses on modeling, simulation, control, and optimization of this key downstream unit.

8.1 INTRODUCTION

Human civilization and scientific rigor are moving toward a newer realm and visionary frontier. Petroleum engineering science and chemical process engineering are witnessing immense changes and drastic challenges. Science, engineering, and technology are in the path of renewed glory and immense greatness. Depletion of fossil fuels has renewed scientific concern and is devastating the petroleum engineering scenario. Scientific research pursuit, the scientific cognizance, and deep scientific introspection will go a long and visionary way in the true realization and emancipation of petroleum engineering science and chemical process modeling. FCCU is a part of the petroleum refinery. Energy sustainability and the global energy scenario are at a deep distress. The need for technology initiative and application of mathematical tools are reshaping the chemical process modeling scenario. The author with deep and cogent insight brings to the scientific forefront the application of multiobjective optimization (MOO) science and also the emerging area of GA. Scientific rigor, immense scientific vision and insight are the torchbearers of this detailed treatise. GA and its applications are the pallbearers toward a greater emancipation of applied mathematics, chemical process engineering, and chemical process modeling. Scientific vision, scientific rigor, and the world of deep scientific understanding in the field of chemical process modeling and simulation are opening up new avenues in the wide path of engineering science today. In this treatise, the author repeatedly points out the difficulties and barriers in the design of riser unit of fluidized catalytic cracking (FCC) in a petroleum refinery with deep scientific vision and immense technological introspection.[24,25]

8.2 VISION AND AIM OF THE PRESENT TREATISE

The author with precision deals deeply with the MOO of the riser reactor of the fluidized catalytic cracking unit (FCCU). The scientific truth, the scientific vision, and the scientific forbearance need to be reenvisioned at every step of its immense application. Human scientific endeavor and scientific rigor are witnessing devastating and drastic challenges in the present decade. The vision of petroleum engineering science is undergoing drastic changes with the grave concern of depletion of fossil-fuel resources. Energy sustainability, global energy scenario, and the progress of science and engineering will all lead a long way in the true realization and effective emancipation of petroleum engineering paradigm. In this present treatise, the world of challenges and difficulties are brought to the forefront of scientific horizon with visionary appeal and unending effectivity. The author details the basic fundamentals of MOO, its varied applications in chemical engineering and petroleum engineering systems, the future recommendations of the study, and the holistic vision of the application of GA. The vision of this treatise is wide and bright. The difficulties, the barriers, and the challenges of design of an FCCU are delineated in minute details. The science of modeling and simulation of an FCC unit, through this treatise, surpasses visionary boundaries. Fluidized catalytic cracking and catalytic reactions are the heart of a petroleum refinery. Scientific vision, the challenge to move forward, and the instinct of scientific innovation are the pallbearers of greater advancements in research in petroleum refining.[24,25]

Scientific vision and deep scientific understanding in the field of modeling, simulation, control, and optimization of a riser unit of FCCU is totally unsolved. In this treatise, the author deeply comprehends the success of mainly the control and simulation of the riser reactor with an aim toward global energy sustainability.[24,25]

8.3   SCOPE OF THE STUDY

The vision of science in today's world and in today's human civilization is inspiring and in the similar vein groundbreaking. Science and engineering is surpassing barriers of truth and understanding in today's world. Energy sustainability is ushering in a new eon of scientific truth and scientific forbearance. The scope of the vision of MOO is vast and versatile. The author deeply comprehends the success and difficulties in the application of MOO in designing riser reactor of an FCC unit. Technological vision in today's world is in the midst of scientific challenges, scientific understanding, and deep scientific questions. Scientific candor today stands in the midst of deep distress. Modeling, simulation, and optimization of the riser reactor stand as an important content in the success of chemical process modeling and petroleum engineering science today. The scope of the study in the field of application of MOO and GA is surpassing visionary boundaries. The challenge, the vision, and the application stand in the midst of deep introspection. Human science is in a state of distress. Energy sustainability and the future of science are connected by an umbilical cord. The application areas of GA will surely open new windows of innovation and scientific instinct in years to come. Science is slowly and steadily opening up new frontiers of scientific cognizance. In a similar vein, GA and bioinspired optimization are gearing up for newer challenges and newer visionary realm.[24,25]

8.4   THE NEED AND THE RATIONALE OF THIS STUDY

Petroleum engineering science is moving from one drastic change toward another. The paradigm of global energy sustainability needs to be envisaged at every step of human life. Depletion of fossil fuel resources has changed the scientific astuteness of scientific research pursuit in petroleum engineering science. Vision of science, the progress of technology, and the visionary era of technological validation all will lead a long way in the true emancipation and great realization of global energy sustainability. The need and the rationale of this study are immense and far-reaching. The author gleans and delineates the world of scientific validation in the design of the riser reactor in an FCCU in a petroleum refinery. The challenge, the motive, and the paradigm of modeling, simulation, and optimization of a petroleum refining unit are ushering in a newer visionary scientific domain in years to come. Depletion of fossil-fuel resources is a bane to human civilization and a disaster to scientific research pursuit. The author skillfully delineates with deep introspection the immense difficulties and unimaginable barriers in the modeling, simulation, control, and optimization of a petroleum refining unit.

8.5   FUTURE RECOMMENDATIONS OF THE STUDY

Designing chemical engineering systems and petroleum refining units is difficult and arduous task. The author with deep insight and wide vision informs in a narrative the challenges in designing FCCU and futuristic vision behind it. The scientific vision, the scientific fortitude, and the immense scientific challenges will go a long way in the true emancipation of application of MOO. Scientific validation and scale-up of the FCC model stands as a backbone to the futuristic vision of petroleum refining paradigm today. Scientific research pursuit and immense scientific determination should be at that angle. The world of challenges in the field of chemical process modeling and simulation are revolutionizing and restructuring the avenues of science today. Future research endeavor, the progress of science, and the forays in the domain of chemical process engineering are opening up new eons of science and engineering.

Future of chemical process engineering and petroleum engineering science is extremely bright. Scientific endeavor needs to be reenvisioned with deep foresight. Future recommendations of the study should be targeted toward greater scientific understanding in the field of applications of MOO, bioinspired optimization, and GA. Chemical process modeling, the world of chemical process design, and the future progress of engineering science need to be reshaped at every step of scientific research pursuit. Future of chemical process engineering and petroleum engineering science is bright and far-reaching. Human scientific endeavor and scientific cognizance are in the path of immense regeneration. Energy sustainability, ecological imbalance, and global concern for fossil-fuel resources have urged the global scientific community to devise new green technologies and newer innovations. Vision of science in today's world is unimaginable. Future areas of research and future endeavor should be targeted toward the direction of holistic sustainable development.[24,25]

8.6   SCIENTIFIC DOCTRINE AND SCIENTIFIC COGNIZANCE BEHIND MOO

Scientific doctrine and the wide world of scientific cognizance are in the path of a newer realm and a new visionary future. MOO is opening up newer future dimensions in the field of chemical process modeling and petroleum refining unit designing. The vision, the challenges, and the scientific forbearance are inspiring and far-reaching. MOO is revolutionizing the scientific panorama and the ardent scientific horizon of scientific research pursuit. Human scientific progress, immense academic rigor, and the unimaginable challenges in the design of a riser unit in FCC unit in a petroleum refinery are changing the scientific scenario of chemical process modeling. Scientific doctrine, scientific cognizance, and immense scientific sagacity are in a state of distress and deep turmoil. The challenge of science and engineering needs to be reenvisioned. Scientific doctrine in today's world of engineering science is in a state of distress. Environmental engineering and petroleum engineering science are ushering in a new era of scientific cognizance and scientific introspection. The challenge, the scientific inspiration, and the scientific doctrine need to be reshaped at each step of human life.

Scientific doctrine and scientific cognizance in today's world of scientific research pursuit stand in the confluence of deep scientific comprehension and immense scientific sagacity. The challenge of human scientific research pursuit is awesome and visionary. MOO and its application are the heart of this treatise. The author pointedly focuses on the difficulties, barriers, and hurdles in the successful application of MOO and GA in modeling and optimization of an FCCU.

8.7   DEFINITION OF MOO, THE UNDERLYING PRINCIPLES, AND FUTURISTIC VISION

MOO is a wide branch of applied mathematics which is surpassing vast boundaries of science and engineering. Man's progress, mankind's prowess, and the achievements of science are leading a long way in the true realization of applied mathematics, chemical process modeling, and the wide world of optimization. The visionary realms of science need to be realized at every step of human life and wide scientific forays. The underlying principles of MOO need to be reenvisioned at every stride of scientific research pursuit. Science and engineering of MOO is extremely complicated yet very effective. The futuristic vision of GA and bioinspired optimization is wide and far-reaching. Science and technology are moving very fast. Chemical process modeling and optimization is entering a new era of scientific vision and scientific challenge. Application areas need to be restructured and reenvisioned to the fullest scientific details. GA and bioinspired optimization are ushering in a new era and gearing toward a newer realm in decades to come.

8.8 WHAT IS FCC AND WHAT ARE ITS FUNDAMENTALS?

FCCU stands as an important and decisive component in the design and operation of a petroleum refinery. Scientific vision, the unmitigated difficulties, and the deep scientific introspection are the forerunners toward a greater understanding in the design, modeling, simulation, and control of an FCC unit.[24,25]

Growing demand for refinery products combined with the decreasing quality of crude oil and tighter product specifications due to drastic environmental constraints is forcing refiners to make definite investments and devise innovative technologies. Fluid catalytic cracking (FCC) remains a key unit in many refineries; it consists of a three step process: reaction, product separation, and regeneration.[24,25]

Risers are considered vital parts in FCC conversion units. It is inside the riser reactor that the heavy hydrocarbon molecules are cracked into lighter petroleum fractions such as liquefied petroleum gas (LPG) and gasoline. The FCC process is considered a key process in the world of petroleum industry, since it is the main unit responsible for the profitable conversion of heavy gas oil into commercial valuable products. There has been immense research work done in the field of riser design and riser fundamentals. In this treatise, the author also discusses the recent work done in the control and optimization of an FCCU primarily the riser reactor.[24,25]

8.9 WHAT IS MOO?

MOO and its scientific principles in today's scientific world stand between deep scientific comprehension and scientific vision. The challenge of the human scientific endeavor is slowly unfolding with each step of scientific rigor. The vision of science, the progress of technology, and the ardent scientific and academic rigor will lead a long and visionary way in the true realization of technology and engineering.

Multivariable optimization algorithms are used in algorithms for optimizing functions having multiple design and decision variable. A deep and proper understanding of the single-variable function optimization algorithms would be immensely helpful in appreciating multivariable function optimization algorithms.[24,25]

8.9.1 WHAT IS GA, BIOINSPIRED OPTIMIZATION, SIMULATED ANNEALING, AND OTHER NONTRADITIONAL OPTIMIZATION ALGORITHMS?

GAs mimic the principles of natural genetics and natural selection to constitute search and optimization procedures. Simulated annealing mimics the cooling phenomenon of molten metals to constitute a search procedure. Professor John Holland of University of Michigan, Ann Arbor, USA envisaged the concept of these GAs in the mid-1960s and published his phenomenal work. Bioinspired optimization mimics traditional biological sciences. Bioinspired, short form for biologically inspired, computing is a field of study that integrates subfields related to the topics of connectionism, social behavior, and emergence. It is often related to the field of artificial intelligence. Biologically inspired computing is a branch of natural computation. Nontraditional optimization involves GA, simulated annealing, and bioinspired optimization.[24-26]

8.10   THE SCIENTIFIC VISION AND DEEP SCIENTIFIC UNDERSTANDING BEHIND APPLICATION OF GA

GA and its applications are visionary yet immature. Scientific rigor and academic prowess are at its helm at each step of human progress. GA and bioinspired optimization in today's scientific world are witnessing tremendous changes and deep introspection. The challenge of human scientific endeavor needs to be readdressed today. GA and application of MOO are widely restructuring the face of chemical process engineering and chemical process modeling.[24-26]

8.11 THE SCIENTIFIC DOCTRINE AND THE SCIENTIFIC VISION BEHIND DESIGN OF RISER REACTOR OF AN FCC UNIT

The scientific doctrine and the deep scientific vision behind the design of a riser reactor are awesome and groundbreaking. Scientific endeavor along with scientific fortitude need to be readdressed at each step of human civilization. Petroleum engineering science in today's scientific landscape is witnessing drastic changes with the evergrowing concern of depletion of fossil-fuel resources. Environmental engineering science is the other face of the civilization's coin today. Scientific sagacity today is at the helm of human scientific research pursuit. Scientific vision and scientific doctrine are unfailingly leading a long and visionary way in the true realization of energy sustainability today. Petroleum refining needs to be reenvisioned with each step of human scientific endeavor. Depletion of fossil-fuel resources is changing the paradigm of research pursuit.

8.12 THE VISION, THE CHALLENGE, AND THE DEEP SCIENTIFIC INTROSPECTION IN THE DESIGN OF AN FCC UNIT

The challenge and the vision of science in today's scientific research pursuit in the design of an FCC unit are widening and surpassing vast boundaries. Today's technology thrusts and technology introspection should be targeted toward scientific validation. The challenge and vision of science are veritably inspiring. This treatise targets the success of application of MOO in the visionary and emerging world of optimization of the riser reactor of FCCU. Scientific vision, scientific candor, and the world of scientific sagacity are unfolding a newer scientific generation and scientific horizon in decades to come.

8.13 PETROLEUM REFINING FUTURE, ENERGY SUSTAINABILITY, AND THE VISION FOR THE FUTURE

The future of petroleum refining is moving from one paradigmatic shift over another. Depletion of fossil-fuel resources are the vexing and burning issues of our times. Mankind's prowess is witnessing the test of our times as civilization moves toward a newer world of energy sustainability. Both, energy and environmental sustainability stand in the midst of unmitigated crisis and immense introspection. The success of human scientific endeavor is at a state of deep scientific distress. Future of petroleum refining needs to be readdressed at each step of scientific endeavor and each step of scientific sagacity. The vision for the future in the field of petroleum refining and design of petroleum refining unit should be targeted toward efficiency and effectivity of the process. Catalytic cracking is an energy-intensive process and the vision of modeling and simulation should be toward immense efficiency of the process.

8.14   THE SCIENCE OF APPLICATION OF GA IN DESIGNING RISER REACTOR OF A FCC UNIT

The science of application of GA in designing riser reactor is the crux of this well-informed and well-observed treatise. Application of MOO and GA needs to be readdressed and reenvisioned at each step of scientific advancement. Riser design is the heart of petroleum refining unit. The innovative vision and the deep challenge behind application of GA need to be deeply understood in the path toward design, modeling, simulation, and optimization of an FCCU.

8.15   APPLICATION AREAS IN THE FIELD OF MOO

Application areas in the field of optimization are vast, versatile, and visionary. Scientific research pursuit is at its helm at every step of human endeavor. Scientific validation and science of scale-up need to be envisioned at each stride of human scientific rigor. MOO has innumerable applications in today's world of science and engineering. Science, technology, and engineering are reshaping and restructuring themselves with immediate and urgent needs. At this critical juxtaposition of scientific reshaping, MOO and other tools of science such as evolutionary computation are changing the face of human scientific endeavor and civilization's forays. Application areas in the field of MOO and GA are immense, wide, and versatile. Chemical process modeling paradigm is ushering in a new era of immense scientific cognizance. Scientific history and scientific candor are witnessing revolutionary changes with the passage of human history and time.

8.16   GA AND THE WORLD OF IMMENSE CHALLENGES

GA is regenerating itself with each step of application of optimization. The immense challenges and barriers are in the path of reshaping. Advancements of science and engineering are changing the scenario of applications of MOO. Human endeavor, the progress of technology, and the visionary world of applied mathematics and optimization will go a long way in the true emancipation of the world of today's scientific fortitude. GA and bioinspired optimization are ushering in a new world of scientific achievement. The success, the vision, and the immense scientific challenges will lead a long way in the true realization of MOO. The hidden questions of computational time and the efficiency stand as strong backbone to the futuristic vision of MOO. GA, evolutionary MOO, and bioinspired optimization are the cruxes of the science of optimization. The challenge needs to be reenvisioned. GA and the world of bioinspired optimization are ushering in a new eon of scientific challenge, scientific vision, and deep scientific understanding. The candor of science, the vision to move forward, and the immense urge to excel are the effective pallbearers of a new paradigm in the domain of optimization.

8.17 APPLICATION AREAS OF GA

Application areas of GA are wide and far-reaching. Vision of science and engineering in today's human civilization are crossing wide boundaries. The world of MOO is ushering in a new era of scientific profundity. The difficulties and the barriers of application of bioinspired optimization are challenged and reenvisioned at each step of research pursuit. The challenge, the vision, and the scientific urge to excel in the area of GA are opening up new vistas of scientific perspicuity in years to come. Bioinspired optimization and evolutionary MOO are the hallmarks of a new scientific generation and a newer realm.

8.18 THE TRUE VISION OF FCCU AND ITS MODELING

FCC and its modeling are moving toward a newer phase of scientific vision and scientific understanding. The definitions of applied mathematics are changing with each step of scientific forays. Chemical process modeling, simulation, and optimization are being reenvisioned with each defining moments of the advancements of human civilization. Scientific challenges in today's world are at its helm and barriers need to be redefined. The research work in the field of design of FCCU is not new yet in a latent stage. Scientific sagacity and the world of visionary challenges are scientifically inspiring and are surpassing wide boundaries. The true vision of

FCC is witnessing a newer realm and a purposeful vision. Modeling, simulation, optimization, and control of an FCC unit stand at the heart of design of a petroleum refining unit. Process optimization of a petroleum refining unit in today's world of science is surpassing visionary frontiers and is at the helm of scientific perspicuity. The challenge and vision of optimization and control need to be reenvisioned at each step of design of an FCC unit in petroleum refinery. Riser modeling optimization and control stands as a major component of the efficiency and success of catalytic cracking process.

8.19   RISER DESIGN IN AN FCC AND THE PROGRESS OF SCIENCE

Riser design in an FCC is witnessing immense challenges. Human scientific research pursuit today stands in the midst of multitudes of crossroads. Science and engineering of optimization are passing through an eon of widespread changes. The vision of energy sustainability, the grave concerns of future of fossil-fuel resources, and the futuristic progress of science will all go a long way in unraveling the scientific truth and scientific vision. Scientific and academic rigor in the field of riser design is opening up new areas of innovation and definitive vision. Science is a massive colossus without a will of its own. In such a crucial juncture of scientific history and scientific cognizance, man's vision as well as a scientist's prowess are reshaping and reenvisioning the domain of petroleum engineering science.

Riser design is of vital importance in the design, modeling, and simulation of an FCCU. Depletion of fossil-fuel resources and the grave concerns for environmental and energy sustainability have urged human scientific endeavor to devise newer innovations and newer technologies. The success of human sustainable development will never be a mirage with the passage of human history and time. Energy sustainability in today's human civilization is in the path of immense reform. Riser design is the heart of petroleum refining process. Scientific vision and deep scientific understanding needs to be rebuilt at each step of designing, modeling, simulation, and optimization of a riser reactor.[24–26]

8.20   MOO: ITS VISION AND DIFFICULTIES

MOO and application of GA are moving toward a newer phase in this decade of human scientific endeavor. Scientific adjudication and scientific judgment are entering a newer visionary era. The ultimate vision and aim of application of MOO is to reduce the computational time in optimization. Bioinspired optimization and GA are changing the scientific horizon in decades to come. MOO and applications of GA are surpassing vast and versatile visionary frontiers. The world of challenges rests upon scientific validation, scientific candor, and immense scientific understanding. The true defining moment of MOO and bioinspired optimization has led to a new world of scientific challenges and vision.

8.21 EVOLUTIONARY MOO AND THE SCIENTIFIC DOCTRINE

Evolutionary MOO and its scientific doctrine need to be reenvisioned at each step of civilization's destiny. The grave concern for energy and environmental sustainability and the world of immense challenges are changing the face of human civilization. Scientific sagacity is witnessing severe challenge. Evolutionary MOO and bioinspired optimization are moving in the newer futuristic direction of scientific optimism and immense scientific hope. Human scientific rigor in today's world is in the state of definite scientific vision and deep scientific understanding.

8.22 FLUIDIZED CATALYTIC CRACKING UNIT IN A PETROLEUM REFINERY AND APPLICATION OF EVOLUTIONARY MOO

FCC and the world of energy sustainability are surpassing wide and visionary horizons. In today's human civilization, energy and environmental sustainability are connected by an unsevered umbilical cord. The present challenge of the human civilization lies on the successful energy and environmental sustainability. Scientific research rigor and academic prowess in the field of design of the riser in an FCC unit need to be reenvisioned. Petroleum refining is witnessing dramatic and drastic changes. The fascinating world of science and technology is witnessing new rigor at each turn of the present century. Petroleum refining is moving toward a newer and visionary scientific paradigm and a newer scientific cognizance. The challenge of design, modeling, simulation, and optimization of a petroleum refining unit is the vision of today and the paradigm of tomorrow. Scientific forbearance and the difficulties and barriers are at its helm with each step of human endeavor.

8.23 VISIONARY SCIENTIFIC ENDEAVOUR IN THE FIELD OF DESIGN OF A RISER IN A FLUIDIZED CATALYTIC CRACKING UNIT

Scientific endeavor in today's human civilization is arduous as well path-breaking. Petroleum refining is the backbone of many developed and developing economies. The grave concern of depletion of fossil-fuel resources and the uncertainties of petroleum engineering science have changed the world of scientific research pursuit. The challenges, the vision, and the progress of science are ushering a new era in chemical process modeling and the vast domain of optimization.

Souza et al. (2007)[1] discussed with lucid details a modeling and simulation procedure of industrial FCC risers.[1] The importance of the FCC utilities for the petroleum industry is growing steadily and drastically as the demand for the utilization of heavy oil increases. Therefore, there is a greater interest of the petroleum refining industry in the development of new technologies, which may increase the conversion, and also the quality of the gas oil into noble products such as gasoline and LPG. For these reasons, several works have been found in technical literature related to the catalytic cracking of petroleum fractions. Science and engineering of the design of a riser unit of FCC are crossing wide and visionary frontiers.[1] Such studies can be divided in two major research avenues: the fluid flow and the catalytic cracking reaction schemes, both of great importance for the modeling and simulation of an FCC unit. It can also be observed from scientific advances that there exist countless types of models for both fluid flow and cracking kinetics, varying from simple models to three-dimensional and three-phase models. This work deeply focuses its attention on the riser reactor.[1] It is inside the reactor that all the cracking reactions responsible for the heavy gas oil conversion into lighter petroleum fractions take place.[1] The hot catalyst (around 700°C) coming from the regenerator enters at the bottom of riser and is brought into contact with a liquid stream of gas oil (around 240°C) flowing from a number of nozzles and, almost instantaneously, is vaporized.[1] This inlet zone is characterized by the presence of turbulence and high temperature and concentration gradients. This turbulent zone is important, but it happens only in the few first meters of the riser, and in some sense the presence of turbulence is not considered in its simulation.[1]

Another visionary area not attempted in the area of riser models is the hypothesis that the gas oil injected in the bottom of riser is well mixed with the catalysts.[1]

In general, the fluid flow models can be classified in three categories. First, the one-dimensional models that are normally simple to formulate and to solve.[1] They are more suitable when the interest is to explore the influence of operating conditions, test a kinetic model, or when the simulation includes not only the riser but also other equipment such as the regenerator and the stripper.[1]

The second type comprises the semiempirical models, which are usually described as core-annulus models.[1] Normally the particle fall velocity and particle concentration are determined empirically. These models cannot predict results for different operational conditions from those of the model parameter estimation.[1]

Fernandes et al. (2005)[2] discussed in details mechanistic dynamic modeling of an industrial FCC unit. The aim of this study is to obtain a model that can simulate the performance of an industrial FCCU in steady and dynamic state, and will invariably be used in studies of control and real-time optimization.[2] The challenge of science and engineering in FCC design is unfolding at each step of this scientific pursuit. The model includes the riser, the stripper/disengage, the regeneration system, and the catalyst transport lines.[2]

Llanes et al. (2008)[3] drastically in a visionary review validated the use of modeling to fine-tune cracking operations.[3] Rigorous nonlinear reactor kinetic models have traditionally been used to support refinery planning, especially in defining linear programming (LP) submodels and operating constraints for reactors.[3] According to the authors, application of kinetic models in engineering, operations support, and optimization are less common due to intense efforts required to produce accurate results for each operating scenario.[3] This is an example of a refinery which uses next-generation simulation tools to cost-effectively select new catalysts. This is a case history which is written with deep comprehension and instinctive scientific imagination. CEPSA's La Rabida refinery in Spain has used nonlinear kinetic FCC model in combination with other advanced simulation tools for several operational studies.[3] The visionary business objective of these applications is to improve FCC unit performance in various areas and to include[3]:

- New catalyst selection
- Unit monitoring, troubleshooting, and optimization
- LP planning model updates
- Examining new feed scenarios

The modeling of the FCC unit is the backbone of the entire petroleum refinery. The challenge and vision of today's science are unimaginable. The

FCC model comprises a reactor–regenerator section block where all the data related to the reaction section is introduced.[3] The FCC reactor–regenerator model seamlessly integrates with other segments of the simulation model where downstream separation units are constructed with standard simulator unit operations such as distillation columns, heat exchangers, drums, pumps, compressors, and more.[3]

Dagde et al. (2012)[4] discussed lucidly modeling and simulation of industrial FCC unit and did analysis based on five-lump kinetic scheme for gas oil cracking. The challenge and the deep scientific understanding of chemical kinetics are slowly unfolding with each step of instinctive and innovative scientific pursuit. Models which describe the performance of riser and regenerator reactors of FCC unit are presented in this treatise. The riser–reactor is modeled as a plug-flow reactor operating adiabatically, using five-lump kinetics for the cracking reactions. Chemical kinetics and chemical reaction engineering are the forerunners toward a visionary domain of chemical process design and chemical process modeling.[4] Vision of science, the scientific and academic rigor, and the march of technology will all lead a long way in the true emancipation of holistic energy sustainability. The efficacy of scientific modeling and simulation of an FCC unit will widen the frontiers of science and engineering.[4]

Bhende et al. (2014)[5] deeply discuss with cogent insight modeling and simulation for FCC unit for the estimation of gasoline production.[5] Fluidized Catalytic Cracking

(FCC) is one of the most important processes in the petroleum refining industry for the conversion of heavy gas oil to gasoline and diesel. Further, valuable gases such as ethylene, propylene, and isobutylene are produced.[5] This work targets the development of a mathematical model that can simulate the behavior of the FCC unit, which consists of feed and preheat system, riser, stripper, reactor, regenerator, and the main fractionators.[5]

Ahari et al. (2008)[6] discuss a mathematical model of the riser reactor in industrial FCC unit. In this article, the author discusses and develops a one-dimensional adiabatic model for riser reactor of an FCC unit.[6] The hydro-dynamic model was described by one general model. Simulation studies are performed to investigate the effect of changing process variables such as input catalyst temperature and catalyst-to-oil ratio.[6]

Bispo et al. (2014)[7] discussed with lucid depths modeling, optimization, and control of an FCC unit using neural networks and evolutionary methods. Artificial neural network (ANN) stands as a major component in control and optimization of a fluidized bed catalytic cracking reactor–regenerator system. This innovative case study, whose phenomenological model was validated with industrial data, is a multivariable and nonlinear process with strong interactions with operating variables.[7]

Rao et al. (2004)[8] discussed industrial experience with object-oriented modeling with an FCC case study. Process modeling and simulation has emerged as an important tool in the advancement of chemical process engineering science. There is an immediate need for refinement of advanced modeling and simulation studies. This work researches on the framework of one such multipurpose simulator, MPROSIM, an object-oriented process modeling and simulation environment.[8]

Dasila et al. (2012)[9] deeply comprehend parametric sensitivity studies in a commercial FCC unit. A steady-state model was developed for simulating the performance of an industrial FCC unit which was next applied to parametric sensitivity studies. The simulator includes kinetic models for the riser reactor and regenerator systems.[9]

Souza et al. (2006)[10] discussed and developed a two-dimensional (2D) model for simulation, control, and optimization of FCC risers. A simple 2D formulation for FCC risers has been developed. The model delineates and approximates the mixture of gas oil, steam, and solid catalyst that flows inside the riser reactor by an equivalent fluid with average properties. The cracking equations are modeled by a six-lump kinetic model combined with two energy equations.[10]

Pahwa et al. (2016)[11] described lucidly computational fluid dynamics (CFD) modeling of FCC riser reactor. The authors of this work present a two-phase flow FCC riser model incorporating a four-lump kinetic scheme. The two-phase flow (gas–solid) in the riser is modeled using the Eulerian–Eulerian multiphase flow model.[11]

Varshney et al. (2014)[12] effectively modeled a riser reactor in a Resid FCCU using a multigrain model for an active matrix-zeolite catalyst. The riser reactor is simulated using a multigrain model of the catalyst in which the amorphous matrix as well as the embedded Y-zeolite crystals are porous and reactive. The technology of FCC design is today entering a visionary era. The scientific vision and the scientific cognizance are moving from one paradigm over another.

8.24 SCIENTIFIC RESEARCH PURSUIT IN THE FIELD OF MOO

Research vision is the utmost need of the hour. Sustainability and sustainable development are the unanswered branches of human scientific endeavor. MOO and its research domain are ushering in new future thoughts and newer pragmatic solutions. Science of optimization and bioinspired optimization are gaining new grounds. The purpose, the vision, and the immediate mission of optimization need to be reshaped and reenvisioned at each step of futuristic scientific endeavor.

MOO and evolutionary techniques are changing the face of chemical process modeling and simulation. The challenge and vision of chemical process modeling need to be reenvisioned.

Deb et al. (2002)[13] discussed a fast and elitist multiobjective GA (NSGA-II) in a phenomenal paper. Multiobjective evolutionary algorithms (MOEAs) that use nondominated sorting and sharing have been criticized mainly for (1) computational complexity, (2) nonelitism approach, and (3) the need for specifying a sorting sharing parameter. The paper deals with a nondominated sorting-based MOEA, called nondominated sorting GA II (NSGA-II), which alleviates all the above three difficulties.[13]

Yapo et al. (1998)[14] discussed with deep insight multiobjective global optimization for hydrologic models. The development of automated (computer-based) calibration methods has focused mainly on the selection of a single-objective measure of the distance between the model-simulated output and the data and the selection of an automatic optimization algorithm to search for the parameter values which minimize the distance.[14] To calibrate a hydrologic model, the hydrologist must specify values of its parameters in such a way that the model's behavior closely matches that of the real system it represents. In some cases, the appropriate values for a model parameter can be determined through direct measurements conducted on the real system.[14] Because of the time-consuming nature of manual trial-and-error calibration, there has been a great deal of research into the development of automated calibration methods.[14] The author rigorously brings into the scientific forefront the multiobjective global optimization for hydrological models.[14]

Marler et al. (2004)[15] surveyed MOO methods for engineering. Current continuous nonlinear MOO concepts and methods are presented. It consolidates and relates seemingly different terminology and methods. In the paper, GAs are also presented in lucid details. Discussion is provided on three fronts, concerning the advantages and loopholes of individual methods, the different classes of methods, and the field of MOO as a whole.[15]

Wang et al. (2006)[16] reviewed metamodeling techniques in support of engineering design optimization. Computation-intensive design problems are becoming increasingly relevant in manufacturing industries. Expensive analysis and simulation processes are the hallmarks of these processes, thus the need of metamodeling techniques.[16]

Gupta (2012)[17] discussed in a phenomenal paper MOO and biomimetic adaptations of GA. Scientific vision and scientific imagination of GA are veritably emboldened with this endeavor. Optimization using more than one objective function (MOO) has become quite relevant in engineering science today. The challenge and vision of biomimetic optimization are reenvisioned and re-enshrined in this major work. MOO provides a decision maker with an idea of the trade-offs between noncommensurate objectives. An example is the maximization of the profit and minimization of pollutants. Two ideas of biological sciences, namely jumping genes (transposons) and the altruism of honeybees, have been biomimicked to give faster adaptations of multiobjective GAs as well as simulated annealing.[17]

Kasat et al. (2002)[18] deeply comprehends MOO of industrial FCC units using elitist nondominated sorting GA. This treatise deeply discusses optimal operation of fluidized bed catalytic cracking unit. A five-lump model is used to characterize the feed and the products. The challenge and deep vision of science are widening. The model is tuned using industrial data. The elitist nondominated sorting GA (NSGA-II) is used to solve a three-objective function optimization problem.[18]

Kasat et al. (2003)[19] dealt lucidly with MOO of an industrial fluidized bed catalytic cracking unit (FCCU) using GA with the jumping genes operator. The MOO of industrial operations using GAs and its variants, often requires inordinately large amounts of computational (LPU) time. The vision, the challenge, and the rigor of science and engineering need to be rebuilt in such a critical situation. An adaptation to speed up the solution procedure was the need of the hour. In this treatise, an adaptation from natural genetics is presented in lucid details.[19]

Gupta et al. (2013)[20] in a well-informed and well-observed treatise discussed MOO using GA. GA is among the more popular evolutionary optimization techniques. Its multiobjective domains and relevant versions are useful for solving industrial problems that are more meaningful and groundbreaking. The scientific rigor of this treatise unfolds greater visionary areas of engineering of GA.[20]

8.25   THE WORLD OF CHALLENGES, THE FUTURISTIC VISION, AND THE AVENUES OF RESEARCH PURSUIT IN THE FIELD OF GA

GA and bioinspired optimization are witnessing new challenges and surpassing visionary scientific frontiers. The avenues of scientific research pursuit need to have a clear-cut vision and a rigorous and purposeful mission. Scientific and academic rigor is in the path of immense restructuring. Research endeavor and research forays in the field of GA and bioinspired optimization are seeing the newer vision of the day. Sustainability research, sustainable development, and the wide world of chemical process modeling are veritably ushering in a new genre of science and engineering in years to come. Challenges are immense but options are many in the field of chemical process engineering. GA and evolutionary MOO are facing tremendous challenges with the passage of human history and time.[21–23]

8.26   IMMENSE DIFFICULTIES AND SCIENCE OF OPTIMIZATION

Science and engineering of optimization is moving toward a visionary realm and a world of greater scientific understanding. The barriers and difficulties in its application are immense and thought-provoking. The author reenvisions the loopholes in the difficult terrains of optimization. These terrains need to be overcome in the attainment and realization of global energy sustainability. Bioinspired optimization in today's scientific world is witnessing new and unimaginable challenges. GA and bioinspired optimization are the new-generation innovative technologies which needs to be reenvisioned and re-enshrined at each step of scientific endeavor. Design of an FCCU is an extremely difficult scientific pursuit. The challenge, the vision, and the efficacy needs to be scientifically understood at each stride of research endeavour.[21–23]

8.26.1   DIFFICULTIES OF COMPUTATIONAL TIME

Computational time stands as a major component in the success of bioinspired optimization and application of GA. Evolutionary MOO also stands in the midst of immense barriers with respect to computational time. Traditional optimization techniques are beset with immense difficulties such as computational time. Thus, there is the need of improved version and non-traditional bioinspired techniques such as GA or evolutionary MOO. The challenge and wide vision of technology and engineering need to be re-enshrined and rebuilt with each step of future science.[21–23]

8.27   VISION OF SCIENCE, ENERGY SUSTAINABILITY, AND THE FUTURE PROGRESS OF HUMAN CIVILIZATION

The progress of human civilization today stands in the midst of immense distress and deep scientific innovation. The challenge of sustainable development is revealing at each turn of decade in the present century. Social sustainability, economic sustainability, and environmental sustainability stand in the crucial juncture of human scientific rigor. Scientific rigor and true scientific vision of bioinspired optimization and GA are the pallbearers of a greater emancipation of energy and environmental sustainability. Human scientific cognizance is witnessing the inevitable test of our times. The author repeatedly delineates the effect and success of bioinspired optimization and application of GA in designing FCCU.

The future progress of human civilization depends on the progress of energy as well as environmental sustainability. Petroleum engineering science today stands in the midst of immense disaster and intense introspection. Depletion of fossil-fuel resources and the grave concerns of ecological balance have plunged human civilization to a deep crisis. In such a situation, human civilization and progress of human mankind are the ultimate parameters toward a greater vision of tomorrow. Sustainable development of human civilization and human progress are in the path of immense reshaping. The progress of science and technology is in a state of great distress with the passage of history and time. The challenge, the vision, and the purpose of research pursuit need to be reshaped at every stroke of a decade in the present century in the field of energy and environmental sustainability.[21–23]

8.28 VALIDATION OF SCIENCE AND DESIGN OF A RISER REACTOR IN AN FCC UNIT

Validation of science and engineering has moved to a new phase of scientific regeneration. GA and its application in the design of the riser reactor in an FCC unit have revolutionized and reframed the visible world of chemical process modeling and applied mathematics. Science has moved toward a newer path of scientific regeneration and scientific forbearance. The scientific horizon of chemical process engineering and bioinspired optimization is reframing and restructuring the world of science, engineering, and technology. The true vision of GA still needs to be explored. Human scientific endurance is in the avenue of greater emancipation. Validation of science, the futuristic vision of engineering, and immense scientific fortitude are the need of the hour in the futuristic path of scientific research endeavor. Deep scientific understanding and immense foresight are the true parameters of human civilization today. Validation of science, the rigor of technology, and the immense scientific foresight are the foremost parameters of human research pursuit today.[21–23]

8.29   FUTURISTIC VISION OF CHEMICAL PROCESS MODELING AND PROGRESS TOWARD FUTURE

Chemical process modeling is witnessing a newer visionary future. Human scientific endeavor, the scientific urge to excel, and human mankind's prowess will all lead a long way in true realization of energy and environmental sustainability. Vision of science and progress of chemical process engineering in today's world are linked by an unsevered umbilical cord. Engineering science in today's scientific horizon is connected to energy sustainability in a similar umbilical cord. The world of scientific progress, the concern for ecological balance, and the eon of scientific research pursuit are the pallbearers of a newer visionary future.[21–23]

8.30   GA APPLICATIONS, THE SCIENCE OF OPTIMIZATION, AND THE VISION FOR THE FUTURE

GA applications and its innovative scientific research pursuit are witnessing a new restructuring and newer reenvisioning. The challenge of science, the vast and versatile world of chemical process modeling, and the success of realization of energy sustainability are the torchbearers of tomorrow's scientific research pursuit. Thus, the science of optimization and its immense innovation are opening up new vistas of scientific learning and true scientific understanding in years to come. A scientist's visionary prowess, the innovation and greatness of engineering science, and the futuristic world of optimization in today's world are leading a long way in successful designing of a petroleum refining unit. MOO and application of GA are reframing the scientific horizon of petroleum engineering science.[21–23]

8.31   SCIENTIFIC COGNIZANCE, SCIENTIFIC VISION, AND TRUE SCIENTIFIC UNDERSTANDING

Scientific cognizance and scientific understanding are in the phase of new restructuring and reenvisioning. Human scientific vision, the march of science, and the scientific challenges are opening up wide vistas and vast boundaries of scientific endeavor. Scientific cognizance is witnessing a real test of times in its path toward scientific validation. Scientific validation and scale-up are reorganizing the chemical process design scenario. The challenge of chemical process design and modeling lies in the success of scientific validation and scale-up. True scientific vision and deep scientific understanding are the bedrock of today's pursuit of science and engineering.

8.32 SCIENCE OF SCALE-UP AND THE PATH TOWARD SCIENTIFIC VALIDATION

Science of scale-up is entering a new phase of chemical process modeling and simulation. Scientific validation today stands in the bedrock of immense vision and introspection. Scale-up of a chemical process is the backbone of chemical process engineering. Scientific validation is the crux of the entire domain of chemical process engineering and petroleum engineering science. Scientific truth, the immense vision, and the multitude of scientific barriers will all lead a long way in true emancipation of petroleum engineering science. Scientific validation and scale-up are the today's visionary avenues of science and engineering. The challenges of scale-up of chemical processes are immense and the vision is vast and versatile. The path toward scientific validation is opening up a new world of scientific imagination and scientific candor in years to come.

8.33 DEPLETION OF FOSSIL-FUEL RESOURCES AND A NEW VISIONARY FUTURE

Depletion of fossil-fuel resources is changing the face of global energy sustainability. The definitive challenge, the scientific astuteness, and the deep scientific introspection are widening the futuristic vision of petroleum engineering science and chemical process modeling. The future is arduous and catastrophic. Petroleum refining is witnessing deep crisis with respect to growing concern of depletion of fossil-fuel resources. Upgradation of petroleum refining technology is the immediate need of an hour. Scientific vision and technological motivation need to be reenvisioned at each step of scientific endeavor. In this treatise, the author deeply comprehends the success of petroleum refining techniques, the difficulties and the road toward energy sustainability.

8.34 THE CHALLENGE, THE VISION, AND THE SCIENTIFIC WISDOM IN MODELING AND SIMULATION OF AN FCC UNIT

Modeling, simulation, optimization, and control of an FCC unit are the upshot of this treatise. The author skillfully delineates the success and the definitive vision in mathematical modeling and application of applied mathematics in chemical process engineering. Growing concern of depletion of fossil-fuel resources, the future of environmental engineering science, and the futuristic status of global energy sustainability are the pallbearers of a greater visionary tomorrow in the field of chemical process modeling and simulation.

8.35 SCIENTIFIC ACUITY, THE DEEP SCIENTIFIC UNDERSTANDING, AND SCIENTIFIC INTROSPECTION

Scientific acuity and deep scientific understanding are in today's world in a state of immense distress and deep insight. The scientific introspection, the grave concerns of energy sustainability, and the future of fossil-fuel resources are dramatically changing the global scientific scenario. Human challenges today are in the grip of immense introspection. The author repeatedly stresses on the future of modeling, simulation, and optimization of an FCC unit in a petroleum refinery with a vision toward energy sustainability. Science is a huge colossus with immense visionary implications. Future of modeling, simulation, and optimization is showing a new path in the research pursuit in FCC unit. Modeling, simulation, and control of riser reactor are at the heart of refinery optimization. The author brings positively to the scientific forefront the challenge and vision of design of FCC unit.

8.36 FUTURE TRENDS AND FUTURE FLOW OF THOUGHTS IN THE FIELD EVOLUTIONARY MOO

Science and engineering of optimization is gearing for newer challenges and newer vision. Bioinspired optimization is ushering in a new era of scientific vision and scientific candor. Scientific research pursuit in today's world stands in the midst of deep foresight and immense introspection. The scientific reality and the deep scientific understanding in the field of application of MOO need to have a futuristic overhauling. Science, technology and engineering science needs to gear toward a newer beginning.

8.37 CONCLUSION

Human civilization and forays into applied science are ushering in a new era of scientific vision and deep scientific understanding. Applied mathematics, chemical process modeling, and science of optimization are entering a new phase of scientific research pursuit. Global petroleum engineering concerns and the grave future of depletion of fossil-fuel resources have urged scientists and technologists to gear toward innovation and vision. The author deeply delineates the mathematical tools in designing chemical engineering systems and the future trends in research endeavor. Technological vision is at the forefront of human scientific research pursuit. FCC and application of GA in its design are the pallbearers of a new visionary era of the science of optimization. MOO with the help of GA has a definite and purposeful vision and that vision needs to be realized and harnessed. The difficulties, the barriers, and the vision of the science of optimization will all lead a long way in the true emancipation of MOO and GA. The vision of science is an immense colossus without a definite will of its own. Petroleum engineering science today stands in the midst of immense distress and an unending catastrophe. The author skillfully delineates the success, the vision, and the future trends in the research pursuit in the domain of design of FCCU. Modeling, simulation, and optimization of petroleum refining processes stand as a major barrier and an unimaginable difficulty in the futuristic vision of science and engineering. This treatise delineates the major hurdles in the design of a FCCU and opens up new windows of innovation and scientific vision in decades to come.

KEYWORDS

- **fluidized catalytic cracking**
- **petroleum**
- **vision**
- **modeling**
- **science**
- **engineering**

REFERENCES

1. Souza, J. A.; Vargas, J. V. C.; Von Melen, O. F.; Martignoni, W. P. Modeling and Simulation of Industrial FCC Risers. *Eng. Termica* **2007,** *6*(1), 19–25.

2. Fernandes, J.; Verstraete, J. J.; Pinheiro, C. C.; Oliveira, N.; Ramoa Ribeiro, F. In *Mechanistic Dynamic Modeling of an Industrial FCC Unit*, European Symposium on Computer Aided Process Engineering, Barcelona, 2005.

3. Llanes, J. M.; Miranda, M.; Mullick, S. Use Modeling to Fine Tune Cracking Operations. *Hydrocarb. Process.* **2008**, *87*, 123–132.

4. Dagde, K. K.; Puyate, Y. T. Modeling and Simulation of Industrial FCC Unit: Analysis Based on Five-Lump Kinetic Scheme for Gas-oil Cracking. *Int. J. Eng. Res. Appl.* **2012**, *2*(5), 698–714.

5. Bhende, S. G.; Patil, K. D. Modeling and Simulation for FCC Unit for the Estimation of Gasoline Production. *Int. J. Chem. Sci. Appl.* **2014**, *5*(2), 38–45.

6. Ahari, J. S.; Farshi, F.; Forsat, K. A Mathematical Modeling of the Riser Reactor in Industrial FCC Unit. *Pet. Coal* **2008**, *50*(2), 15–24.

7. Bispo, V. D. S.; Silva, E. S. R. L.; Meleiro, L. A. C. Modeling, Optimization and Control of a FCC Unit Using Neural Networks and Evolutionary Methods. *Engevista* **2014**, *16*(1), 70–90.

8. Rao, R. M.; Rengaswamy, R.; Suresh, A. K.; Balaraman, K. S. Industrial Experience with Object Oriented Modeling-FCC Case Study. *Chem. Eng. Res. Des.* **2004**, *84*(A4), 527–552.

9. Dasila, P. K.; Choudhury, I.; Saraf, D.; Chopra, S.; Dalai, A. Parametric Sensitivity Studies in a Commercial FCC Unit. *Adv. Chem. Eng. Sci.* **2012**, *2*, 136–149.

10. Souza, J. A.; Vargas, J. V. C.; Von Meien, O. F.; Martignoni, W.; Amico, S. C. A Two Dimensional Model for Simulation, Control and Optimization of FCC Risers. *Am. Inst. Chem. Eng. J.* **2006**, *52*(5), 1895–1905.

11. Pahwa, R.; Gupta, R. K. CFD Modeling of FCC Riser Reactor. *Int. Res. J. Eng. Technol.* **2016**, *3*(2), 206–209.

12. Varshney, P.; Kunzru, D.; Gupta, S. K. Modelling of the Riser Reactor in a Resid Fluidised Bed Catalytic Cracking Unit Using a Multi-grain Model for an Active Matrix-zeolite Catalyst. *Indian Chem. Eng.* **2014**, *57*(2), 115–135.

13. Deb, K.; Pratap, A.; Agarwal, S.; Meyarivan, T. A Fast and Elitist Multiobjective Genetic Algorithm: NSGA-II. *IEEE Trans. Evol. Comput.* **2002**, *6*(2), 182–197.

14. Yapo, P. O.; Gupta, H. V.; Sorooshian, S. Multi-objective Global Optimization for Hydrologic Models. *J. Hydrol.* **1998**, *204*, 83–97.

15. Marler, R. T.; Arora, J. S. Survey of Multi-objective Optimization Methods for Engineering. *Struct. Multidiscip. Optim.* **2004**, *26*, 369–395.

16. Wang, G. G.; Shan, S. Review of Metamodeling Techniques in Support of Engineering Design Optimization. *J. Mech. Des. ASME Trans.* **2006**, *129*(4), 370–380.

17. Gupta, S. K. Multi-objective Optimization: Biomimetic Adaptations of Genetic Algorithm. *Indian Chem. Eng.* **2012**, *54*(1), 1–11.

18. Kasat, R. B.; Kunzru, D.; Saraf, D. N.; Gupta, S. K. Multi-objective Optimization of Industrial FCC Units Using Elitist Nondominated Sorting Genetic Algorithm. *Ind. Eng. Chem. Res.* **2002**, *41*, 4765–4776.

19. Kasat, R. B.; Gupta, S. K. Multiobjective Optimization of an Industrial Fluidized Bed Catalytic Cracking Unit (FCCU) Using Genetic Algorithm (GA) with the Jumping Genes Operator. *Comput. Chem. Eng.* **2003**, *27*, 1785–1800.

20. Gupta, S. K.; Garg, S. Multi-objective Optimization Using Genetic Algorithm. In *Control and Optimization of Process Systems, Advances in Chemical Engineering*; Elsevier: New York, 2013; Vol. 43, pp 205–245.

21. Palit, S. Application of Evolutionary Multi-objective Optimization in Designing Fluidised Catalytic Cracking Unit and Chemical Engineering Systems—A Scientific Perspective and a Critical Overview. *Int. J. Comput. Intell. Res.* **2016,** *12*(1), 17–34.

22. Palit, S. The Future Vision of the Application of Genetic Algorithm in Designing a Fluidized Catalytic Cracking Unit and Chemical Engineering Systems. *Int. J. Chem. Technol. Res.* **2014–2015,** *7*(4), 1665–1674.

23. Palit, S. Modelling, Simulation and Optimization of a Riser Reactor of Fluidised Catalytic Cracking Unit with the Help of Genetic Algorithm and Multi-objective Optimization—A Scientific Perspective and a Far-reaching Review. *Int. J. Chem. Eng.* **2016,** *9*(1), 89–102.

24. Deb, K. *Multi-objective Optimization Using Evolutionary Algorithm. Wiley Interscience Series in Systems and Optimization*; John Wiley and Sons, 2010.

25. Elnashaie, S. S. E. H.; Elshishini, S. S. *Modelling, Simulation and Optimization of Industrial Fixed Bed Catalytic Reactors*; Gordon and Breach Science Publishers: UK, 1993.

26. www.wikipedia.com (accessed Jan, 2017).

CHAPTER 9

A THEORETICAL STUDY OF BIMETALLIC CuAu$_n$ (n = 1–7) NANOALLOY CLUSTERS INVOKING CONCEPTUAL DFT-BASED DESCRIPTORS

PRABHAT RANJAN[1], TANMOY CHAKRABORTY[2,*], and AJAY KUMAR[1]

[1]*Department of Mechatronics Engineering, Manipal University Jaipur, Dehmi-Kalan, Jaipur 303007, India*

[2]*Department of Chemistry, Manipal University Jaipur, Dehmi-Kalan, Jaipur 303007, India*

Corresponding author. E-mail: tanmoy.chakraborty@jaipur.manipal.edu; tanmoychem@gmail.com

CONTENTS

ABSTRACT

Due to diverse nature of applications in the field of science and technology, nowadays study of bimetallic nanoalloy clusters is of immense importance. A deep insight is required to explore the structure and physicochemical properties of such compounds. Among such nanoalloy clusters, the compound formed between Cu and Au has received a lot of attention because of their unique electronic, optical, mechanical, and magnetic properties, which have large application in the area of nanotechnology, material science, and medicine. Density functional theory (DFT) is one of the successful approaches of quantum mechanics to study the electronic properties of materials. Conceptual DFT-based descriptors have been used to reveal experimental properties qualitatively. In this report, we have investigated the geometrical structures and physicochemical properties of $CuAu_n$ ($n = 1–7$) invoking DFT methodology. The result predicts that low-lying ground state structure of copper–gold clusters have planar structure. The computed highest occupied-lowest unoccupied molecular orbital (HOMO-LUMO) energy gap of Cu–Au shows interesting odd–even oscillation behavior. A close agreement between experimental and our computed bond length is also observed.

9.1   INTRODUCTION

In the recent years, nanomaterials and nanotechnology have emerged as a new research domain in science and technology. Nanomaterials with at least one of its dimensions in the nanometer range of 1–100 nm have various applications in the area of health care, medicine, energy, electronics, and space industries.[1] The materials with this particular size range have considerably different property from the atomic/molecular and bulk material. However, there are still some cases of nonlinear transition of certain physicochemical properties, which may vary depending on their size, shape, and composition.[2,3] A large number of scientific reports are available which describes change in the optical, electronic, magnetic, mechanical, and other physicochemical properties of nanomaterials due to change in size and structure.[4-6] Advancement of methodologies and characterization techniques has a dependence on utilization of different compositions of nanoalloys.[7-9] A deep insight into the research of nanoalloys with proper control of their nanoscale structures may lead to some other alternatives for high-performance devices and technologies.[1]

The noble metal clusters can be extensively applied in various strategic and technological areas due to its unique catalytic, magnetic, mechanical, electronic, and optical properties.[7,10–15] There are a number of examples, where it has been seen that the combination of two or more metals enhances the abovementioned properties.[16,17] Bimetallic nanoclusters have been identified as important catalyst in fuel cell electrode reactions.[18] The study based on core–shell structure of nanocompounds is very much popular because its properties can be tuned through suitable control of other structural and chemical parameters. During the last few years, bimetallic nanoalloy clusters such as copper, silver, and gold, have been real field of research from experimental and theoretical point of view.[19–25] Group 11 metal (Cu, Ag, and Au) clusters exhibit the filled inner d orbitals with one unpaired electron in the valence s shell. This electronic arrangement is responsible for reproduction of exactly similar shell effects[26–30] which are experimentally observed in the alkali metal clusters.[31–33] Gold nanoclusters are very much popular due to their potential applications in the area of fabrication technology, material science, catalysis, biology, and medicine.[34–45] In the periodic table, among the nanoalloy clusters of group 11 elements, the compound formed between copper and gold is very much popular due to its potential technological applications. It has been already reported that s–d hybridization is strong in gold as compared to copper and silver due to relativistic contraction of the d orbitals.[46–49] Studies based on first principle calculation have shown that gold clusters with more than 10 atoms have planar structure.[50–51] However, no direct experiment has been performed on neutral gold clusters.[19] It has been already established that impurity atoms can enhance the physical and chemical properties of doped gold clusters that are sensitive toward the nature of dopant atoms. In recent years, many studies have been done on doped or mixed gold clusters to enhance the stability and physicochemical properties of gold clusters.[52–57] The nanoalloy clusters of copper–gold as a catalyst can enhance the reaction efficiency and selectivity.[58–62] Though some studies have been performed on CuAu$_n$ nanoalloy clusters, a theoretical analysis invoking Conceptual Density Functional Theory (CDFT)-based descriptors is still unexplored in this domain.

DFT is one of the most popular approaches for the quantum mechanical computation of periodic systems. Due to its computational friendly behavior, DFT is widely accepted method to study the many-body systems. DFT, which includes electron correlation properties, has been commonly described as a practical and effective computational tool, especially for metallic nanoclusters. In the domain of material science research, particularly in superconductivity of metal-based alloys, magnetic properties of nanoalloy clusters, quantum fluid dynamics, molecular dynamics, nuclear physics, DFT has gained a huge importance.[63] The study of DFT covers three major domains such as theoretical, conceptual, and computational.[64–66] Conceptual DFT is established as an important approach to study the chemical reactivity of materials.[67–69] The conceptual DFT is highlighted following Parr's dictum "Accurate calculation is not synonymous with useful interpretation. To calculate a molecule is not to understand it."[70] We have rigorously applied conceptual DFT-based global and local descriptors for studying physicochemical properties of nanoengineering materials and drug-designing process.[71–77]

In this report, we have studied a number of bimetallic nanoalloy clusters containing copper and gold, invoking DFT-based global descriptors such as hardness, HOMO (highest occupied molecular orbital)-LUMO (lowest unoccupied molecular orbital) gap, softness, electrophilicity index, and electronegativity. An attempt has been made to correlate the properties of the compounds with their computational counterparts.

9.2   COMPUTATIONAL DETAILS

In this study, we have made a theoretical analysis on the bimetallic nanoalloy clusters of $CuAu_n$, where ($n = 1$–7). The 3d modeling and structural optimization of all the compounds have been performed using Gaussian 03[78] within DFT framework. For optimization purpose, the local spin density approximation (LSDA) exchange correlation with basis set LanL2dz has been adopted. Although LDA functionals are not so much complex but their effectiveness is already established in manifold applications, particularly for solid-state physics,[79] where accurate phase transitions in solids[80] and liquid metals[81,82] are predicted, and for lattice crystals, in which 1% precision are successfully achieved.[83] Energy minimization has been done without imposing any restriction on molecular spin. We have adopted Z-axis as spin polarization axis. Symmetrized fragment orbitals (SFOs) are combined with auxiliary core functions (CFs) to ensure orthogonalization on the (frozen) core orbitals (COs).

Invoking Koopmans' approximation,[67] ionization energy (I) and electron affinity (A) of all the nanoalloys have been computed using the following ansatz:

$$I = -\varepsilon_{HOMO} \tag{9.1}$$

$$A = -\varepsilon_{LUMO} \tag{9.2}$$

Thereafter, using I and A, the conceptual DFT-based descriptors, such as electronegativity (χ), global hardness (η), molecular softness (S), and electrophilicity index (ω), have been computed. The equations used for such calculations are as follows:

$$\chi = -\mu = \frac{I + A}{2},$$ (9.3)

where μ represents the chemical potential of the system.

$$\eta = \frac{I - A}{2}$$ (9.4)

$$S = \frac{1}{2\eta}$$ (9.5)

$$\omega = \frac{\mu^2}{2\eta}.$$ (9.6)

9.3 RESULTS AND DISCUSSION

9.3.1 STRUCTURE OF BIMETALLIC COPPER–GOLD NANOALLOY CLUSTERS

In this section, low-lying geometrical structure of small copper–gold CuAu$_n$ (n = 1–7) nanoalloy clusters has been studied. A large number of isomeric structures of copper–gold clusters have been optimized and the most stable structures with minimum energy are shown in Figure 9.1. The most stable structure is placed in the left and total energy of the cluster increases from left to right. The low-lying triangular structure of CuAu$_2$ with an apex angle of 69.58° has C_s symmetry. A linear structure has C_{2v} symmetry which is having high energy than the triangular structure. CuAu$_3$ clusters have three isomers, showing rhombus structure is energetically more favorable than the Y-shaped structure. We have optimized four low-lying isomers for CuAu$_4$, CuAu$_5$, and CuAu$_6$. The structure of CuAu$_4$ indicates an early presence of 3d geometries with C_s symmetry but it is less stable than the planar structure.

The structure of CuAu$_5$ and CuAu$_6$ also predicts that planar geometry is more stable as compared to the 3d structure. For octamer of CuAu$_7$, we have optimized five low-lying isomers; a starlike planar structure with C_{2v} symmetry has been identified as the lowest energy.

For *n* = 1

For *n* = 2

For *n* = 3

For *n* = 4

For *n* = 5

For *n* = 6

For *n* = 7

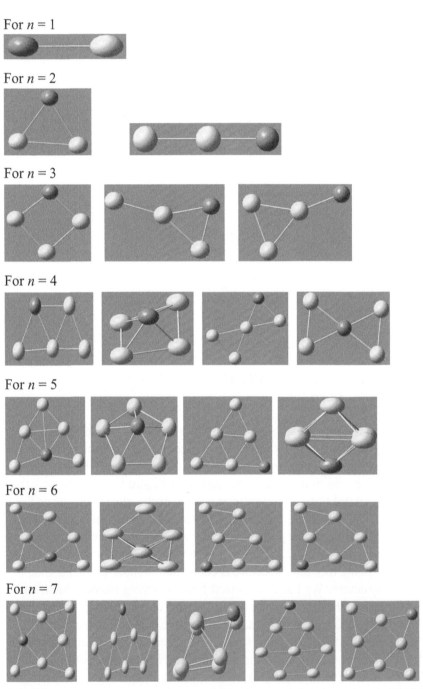

FIGURE 9.1   The lowest energy and low-lying structures of CuAu$_n$ (n = 1–7) nanoalloy clusters.

9.3.2 ELECTRONIC PROPERTIES AND DFT-BASED DESCRIPTORS

Theoretical analysis of copper–gold bimetallic nanoalloy clusters has been done invoking DFT methodology. The orbital energies in form of HOMO-LUMO gap along with computed DFT-based descriptors, such as molecular electronegativity, global hardness, global softness, and global electrophilicity index, have been reported in the Table 9.1. It is well known that there is a direct relationship between HOMO-LUMO energy gap and stability of the nanoalloy clusters. Table 9.1 reveals that HOMO-LUMO gap of the nanoclusters runs hand in hand along with their evaluated global hardness values. As the frontier orbital energy gap increases, the hardness value increases. The molecule having highest HOMO-LUMO gap will be least prone to response against any external perturbation. Result from Table 9.1 reveals that CuAu$_5$ with C_{2v} symmetry has maximum HOMO-LUMO gap whereas CuAu$_2$ with C_s symmetry possesses the lowest energy gap. Though there is no such available quantitative data for optical properties of aforesaid clusters, we can tacitly assume that there must be a direct qualitative relationship between optical properties of Cu–Au nanoalloy clusters with their computed HOMO-LUMO gap. The assumption is based on the fact that optical properties of materials are interrelated with flow of electrons within the systems, which in turn depends on the difference between the distance of valence and conduction band. A linear relationship between HOMO-LUMO gap with the difference in the energy of valence and conduction band is already reported.[81] Keeping it in view, we may conclude that optical properties of the bimetallic nanoalloy clusters increase with increase in their hardness values. Similarly, the softness data exhibit an inverse relationship with the experimental optical properties. Similar relationships are also observed between other descriptors and HOMO-LUMO gap. The linear correlation between HOMO-LUMO gap along with their computed electrophilicity index is lucidly plotted in Figure 9.2. The value of regression coefficient (R^2 = 0.955) supports our predicted model.

As per the theory of cluster physics, dissociation energy and second difference of total energy have a strong influence on the relative stability of nanoalloy clusters. These two energies are highly sensitive quantities and they exhibit pronounced even–odd alternation behavior for neutral and charged clusters, as a function of cluster size.[19] The similar type of oscillation behavior is also exhibited by the HOMO-LUMO gap of any particular compound. It has already been reported that cluster with an even number of total atoms exhibits high value of HOMO-LUMO gap as compared to their neighbor cluster with odd number of atoms. The stability of the even number electronic cluster is actually an outcome of their closed electronic configuration which always produces extra stability. We have reported HOMO-LUMO gap as a function of cluster size in Figure 9.3. It is distinct from Figure 9.3 that the Cu–Au clusters containing even number of total atoms possess higher HOMO-LUMO gap as compared to the clusters having odd number of total atoms. So, our computed data successfully reveal the experimental odd–even alternation behavior of bimetallic nano-alloy clusters.

TABLE 9.1   Computed DFT-based Descriptors of $CuAu_n$ ($n = 1$–7) Nanoalloy Clusters.

Species	HOMO-LUMO gap	Electronega-tivity	Hardness	Softness	Electrophilicity index	Symmetry
CuAu	1.959	5.224	0.979	0.510	13.931	C_{2v}
$CuAu_2$	0.952	4.938	0.476	1.051	25.610	C_s
$CuAu_3$	1.768	5.591	0.884	0.565	17.678	C_1
$CuAu_4$	1.033	5.331	0.516	0.967	27.507	C_s
$CuAu_5$	2.040	5.564	1.02	0.490	15.172	C_{2v}
$CuAu_6$	1.306	5.496	0.653	0.765	23.130	C_s
$CuAu_7$	1.958	5.686	0.959	0.510	16.507	C_1

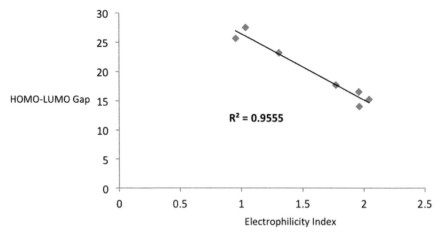

FIGURE 9.2   A linear correlation plot of electrophilicity index versus HOMO-LUMO gap.

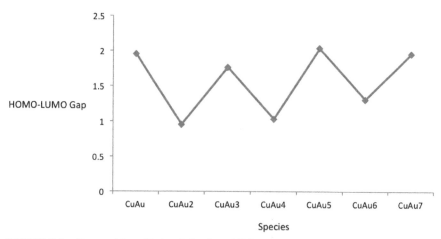

FIGURE 9.3   Even–odd oscillation behavior of HOMO-LUMO gap and CuAu$_n$ (n = 1–7) nanoalloy clusters.

A comparative analysis has been made between experimental bond length[84–86] and our computed data of the species, namely Au$_2$, Cu$_2$, and CuAu. The result is reported in Table 9.2. A close agreement between experimental data and our computed bond length is reflected in Table 9.2.

TABLE 9.2   The Calculated Bond Lengths (Å) for the Au$_2$, Cu$_2$, and CuAu Species.

Species	Theoretical bond length	Experimental bond length
Au$_2$	2.48	2.47[84]
Cu$_2$	2.16	2.22[85]
CuAu	2.40	2.42[86]

9.4   CONCLUSION

Bimetallic nanoalloy clusters have received a lot of attention due to their wide range of technological applications. In this report, geometrical structure and physicochemical properties of copper–gold nanoalloy clusters have been systematically investigated by using DFT methodology. The study predicts that lowest energy structure of most stable isomers has planar structure. The result reveals that HOMO-LUMO gap of this nanoclusters runs hand in hand along with the computed hardness. Due to nonavailability of any quantitative benchmark, the optical property of copper–gold nanoalloy clusters has been assumed to be exactly equivalence of its HOMO-LUMO gap. Here, our evaluated data reveal that optical property of these nanocompounds maintains a direct relationship with computed global hardness and inverse relationship with computed softness. This trend is consistent with other experimental facts also. The value of regression coefficients of model, obtained between electrophilicity index and HOMO-LUMO gap, successfully supports our approach. The nanoalloy clusters also exhibit interesting odd–even oscillation behavior, showing that even number of atoms in clusters is more stable as compared to their neighbor odd number of clusters. From our study, we can claim that $CuAu_5$ is the most stable cluster among the reported systems. The calculated bond lengths for the species Au_2, Cu_2, and $CuAu$ are numerically very close to the experimental value, which supports efficacy of our analysis.

KEYWORDS

- **electronic properties**
- **nanoalloy clusters**
- **copper–gold**
- **odd–even oscillation**

REFERENCES

1. Rao, C. N. R.; Cheetham, A. K. *J. Mater. Chem.* **2001,** *11,* 2887–2894.
2. Kastner, M. A. *Phys. Today* 1993, *46,* 24–31.
3. Haruta, M. *Cattech* **2002,** *6,* 102–115.
4. Khosousi, A. Z.; Dhirani, A. A. *Chem. Rev.* **2008,** *108,* 4072–4124.
5. Ghosh, S. K.; Pal, T. *Chem. Rev.* **2007,** *107,* 4797–4862.
6. Chaudhuri, R. G.; Paria, S. *Chem. Rev.* **2012,** *112,* 2373–2433.
7. Ferrando, R.; Jellinek, J.; Johnston, R. L. *Chem. Rev.* **2008,** *108,* 845–910.
8. Oderji, H. Y.; Ding, H. *Chem. Phys.* **2011,** *388,* 23–30.
9. Baletto, F.; Ferrando, R. *Rev. Mod. Phys.* **2005,** *77,* 371–423.
10. Teng, X.; Wang, Q.; Liu, P.; Han, W.; Frenkel, A. I.; Wen, W.; Marinkovic, N.; Hanson, J. C.; Rodriguez, J. A. *J. Am. Chem. Soc.* **2008,** *130,* 1093–1101.
11. Henglein, A. *J. Phys. Chem.* **1993,** *97,* 5457–5471.
12. Davis, S. C.; Klabunde, K. J. *Chem. Rev.* **1982,** *82,* 153–208.
13. Lewis, L. N. *Chem. Rev.* **1993,** *93,* 2693–2730.
14. Schmid, G. *Chem. Rev.* 1992, *92,* 1709–1727.

15. Schon, G.; Simon, U. *Colloid Polym. Sci.* 1995, *273*, 101–117.
16. Oderji, H. Y.; Ding, H. *Chem. Phys.* 2011, *388*, 23–30.
17. Liu, H. B.; Pal, U.; Medina, A.; Maldonado, C.; Ascencio, J. A. *Phys. Rev. B* 2005, *71*, 075403.
18. Hsu, P. J.; Lai, S. K. *J. Chem. Phys.* 2006, *124*, 044711.
19. Wang, H. Q.; Kuang, X. Y.; Li, H. F. *Phys. Chem. Chem. Phys.* 2010, *12*, 5156–5165.
20. Taylor, K. J.; Hall, C. L. P.; Cheshnovsky, O.; Smalley, R. E. *J. Chem. Phys.* 1992, *96*, 3319–3329.
21. Cheeseman, M. A.; Eyler, J. R. *J. Phys. Chem.* 1992, *96*, 1082–1087.
22. Fernandez, E. M.; Soler, J. M.; Garzon, I. L.; Balbas, L. C. *Phys. Rev. B: Condens. Matter Mater. Phys.* 2004, *70*, 165403.
23. Quinn, B. M.; Dekker, C.; Lemay, S. G. *J. Am. Chem. Soc.* 2005, *127*, 6146–6147.
24. Eustis, S.; El-Sayed, M. A. *Chem. Soc. Rev.* 2006, *35*, 209–217.
25. Cottancin, E.; Celep, G.; Lerme, J.; Pellarin, M.; Huntzinger, J. R.; Vialle, J. L.; Broyer, M. *Theor. Chem. Acc.* 2006, *116*, 514–523.
26. Katakuse, I.; Ichihara, T.; Fujita, Y.; Matsuo, T.; Sakurai, T.; Matsuda, H. *Int. J. Mass Spectrom. Ion Process.* 1985, *67*, 229–236.
27. Katakuse, I.; Ichihara, T.; Fujita, Y.; Matsuo, T.; Sakurai, T.; Matsuda, H. *Int. J. Mass Spectrom. Ion Process.* 1986, *74*, 33–41.
28. Heer, W. A. D. *Rev. Mod. Phys.* 1993, *65*, 611–676.
29. Gantefor, G.; Gausa, M.; Meiwes-Broer, K. H.; Lutz, H. O. *J. Chem. Soc. Faraday Trans.* 1990, *86*, 2483–2488.
30. Leopold, D. G.; Ho, J.; Lineberger, W. C. *J. Chem. Phys.* 1987, *86*, 1715–1726.
31. Lattes, A.; Rico, I.; Savignac, A. D.; Samii, A. A. Z. *Tetrahedron* 1987, *43*, 1725–1735.
32. Chen, F.; Xu, G. Q.; Hor, T. S. A. *Mater. Lett.* 2003, *57*, 3282–3286.
33. Taleb, A.; Petit, C.; Pileni, M. P. *J. Phys. Chem. B* 1998, *102*, 2214–2220.
34. Liu, H. Q.; Tian, Y.; Xia, P. P. *Langmuir* 2008, *24*, 6359–6366.
35. Noonan, K. J. T.; Gillon, B. H.; Cappello, V.; Gates, D. P. *J. Am. Chem. Soc.* 2008, *130*, 12876–12877.
36. Wang, T.; Hu, X. G.; Dong, S. J. *J. Phys. Chem. B* 2006, *110*, 16930–16936.
37. Majumder, C.; Kandalam, A. K.; Jena, P. *Phys. Rev. B Condens. Matter Mater. Phys.* 2006, *74*, 205437.
38. Li, X.; Kiran, B.; Cui, L. F.; Wan, L. S. *Phys. Rev. Lett.* 2005, *95*, 253401–253404.
39. Torres, M. B.; Fernandez, E. M.; Balbas, L. C. *J. Phys. Chem. A* 2008, *112*, 6678–6689.
40. Teles, J. H.; Brode, S.; Chabanas, M. *Angew. Chem.* 1998, *110*, 1475–1478.
41. Valden, M.; Lai, X.; Goodman, D. W. *Science* 1998, *281*, 1647–1650.
42. Yoon, B.; Hakkinen, H.; Landman, U.; Wörz, A. S.; Antonietti, J. M.; Abbet, S.; Judai, K.; Heiz, U. *Science* 2005, *307*, 403–407.
43. McRae, R.; Lai, B.; Vogt, S.; Fahrni, C. J. *J. Struct. Biol.* 2006, *155*, 22–29.
44. Ackerson, C. J.; Jadzinsky,P. D.; Jensen, G. J.; Kornberg, R. D. *J. Am. Chem. Soc.* 2006, *128*, 2635–2640.
45. Shaw III, C. F. Chem. Rev. 1999, *99*, 2589–2600.
46. Kabir, M.; Mookerjee, A.; Bhattacharya, A. K. *Eur. Phys. J. D* 2004, *31*, 477–485.
47. Heard, C. J.; Johnston, R. L. *Eur. Phys. J. D* 2013, *67*, 37–53.
48. Hakkinen, H.; Moseler, M.; Landman, U. *Phys. Rev. Lett.* 2002, *89*, 176103.
49. Massobrio, C.; Pasquarello, A.; Car, R. *Chem. Phys. Lett.* 1995, *238*, 215–221.
50. Valden, M.; Lai, X.; Goodman, D. W. *Science* 1968, *281*, 1647–1650.

51. Alonso, J. A. *Chem. Rev.* **2000,** *100,* 637–678.
52. Autschbach, J.; Hess, B. A.; Johansson, M. P.; Neugebauer, J.; Patzschke, M.; Pyykko, P.; Reiher, M.; Sundholm, D. *Phys. Chem. Chem. Phys.* **2004,** *6,* 11–22.
53. Walter, M.; Hakkinen, H. *Phys. Chem. Chem. Phys.* **2006,** *8,* 5407–5411.
54. Zhao, Y.; Li, Z.; Yang, J. *Phys. Chem. Chem. Phys.* **2009,** *11,* 2329–2334.
55. Tanaka, H.; Neukermans, S.; Janssens, E.; Silverans, R. E.; Lievens, P. *J. Chem. Phys.* **2003,** *119,* 7115–7123.
56. Tanaka, H.; Neukermans, S.; Janssens, E.; Silverans, R. E.; Lievens, P. *J. Am. Chem. Soc.* **2003,** *125,* 2862–2863.
57. Zhang, M.; He, L. M.; Zhao, L. X.; Feng, X. J.; Luo, Y. H. *J. Phys. Chem. C* **2009,** *113,* 6491–6496.
58. Yin, F.; Wang, Z. W.; Palmer, R. E. *J. Nanopart. Res.* **2012,** *14,* 1124.
59. Bauer, J. C.; Mullins, D.; Li, M.; Wu, Z.; Payzant, E. A.; Overbury, S. H.; Dai, S. *Phys. Chem. Chem. Phys.* **2011,** *13,* 2571–2581.
60. Bracey, C. L.; Ellis, P. R.; Hutchings, G. J. *Chem. Soc. Rev.* **2009,** *38,* 2231–2243.
61. Rapallo, A.; Rossi, G.; Ferrando, R.; Fortunelli, A.; Curley, B. C.; Lloyd, L. D.; Tarbuck, G. M.; Johnston, R. L. *J. Chem. Phys.* **2005,** *122,* 194308-1-13.
62. Molenbroek, A. M.; Norskov, J. K.; Clausen, B. S. *J. Phys. Chem. B* **2001,** *105,* 5450–5458.
63. Hafner, J.; Wolverton, C.; Ceder, G. *MRS Bull.* **2006,** *31,* 659–668.
64. Parr, R. G.; Yang, W. *Annu. Rev. Phys. Chem.* **1995,** *46,* 701–728.
65. Liu, S.; Parr, R. G. *J. Chem. Phys.* **1997,** *106,* 5578.
66. Ziegler, T. *Chem. Rev.***1991,** *91,* 651–667.
67. Parr, R. G.; Yang, W. *Density Functional Theory of Atoms and Molecules*; Oxford University Press: Oxford, 1989.
68. Chermette, H. *J. Comput. Chem.* **1999,** *20,* 129–154.
69. Geerlings, P.; Proft, F. D.; Langenaeker, W. *Chem. Rev.* **2003,** *103,* 1793 (Washington, D.C.).
70. Geerlings, P.; Proft, F. D. *Int. J. Mol. Sci.* **2002,** *3,* 276–309.
71. Ranjan, P.; Dhail, S.; Venigalla, S.; Kumar, A.; Ledwani, L.; Chakraborty, T. *Mater. Sci. Pol.* **2015,** *33,* 719–724.
72. Ranjan, P.; Venigalla, S.; Kumar, A, Chakraborty, T. *New Front. Chem.* **2014,** *23,* 111–122.
73. Venigalla, S.; Dhail, S.; Ranjan, P.; Jain, S.; Chakraborty, T. *New Front. Chem.* **2014,** *23,* 123–130.
74. Ranjan, P.; Kumar, A.; Chakraborty, T. AIP Conference Proceedings, 1724, 020072, 2016.
75. Ranjan, P.; Kumar, A.; Chakraborty, T. *Mater. Today Proc.* **2016,** *3,* 1563–1568.
76. Ranjan, P.; Kumar, A.; Chakraborty, T. *Environmental Sustainability: Concepts, Principles, Evidences and Innovations;* Mishra, G. C., Ed.; Excellent Publishing House: New Delhi, 2014; pp 239–242.
77. Ranjan, P.; Venigalla, S.; Kumar, A.; Chakraborty, T. *Recent Methodology in Chemical Sciences: Experimental and Theoretical Approaches*; Chakraborty, T., Ledwani, L., Eds.; Apple Academic Press and CRC Press: USA, 2015; pp 337–346.
78. Gaussian 03, Revision C.02, Frisch, M. J.; Trucks, G. W.; Schlegel, H. B.; Scuseria, G. E.; Robb, M. A.; Cheeseman, J. R.; Montgomery, J. A.; Vreven, J. T.; Kudin, K. N.; Burant, J. C.; Millam, J. M.; Iyengar, S. S.; Tomasi, J.; Barone, V.; Mennucci, B.;

Cossi, M.; Scalmani, G.; Rega, N.; Petersson, G. A.; Nakatsuji, H.; Hada, M.; Ehara, M.; Toyota, K.; Fukuda, R.; Hasegawa, J.; Ishida, M.; Nakajima, T.; Honda, Y.; Kitao, O.; Nakai, H.; Klene, M.; Li, X.; Knox, J. E.; Hratchian, H. P.; Cross, J. B.; Bakken, V.; Adamo, C.; Jaramillo, J.; Gomperts, R.; Stratmann, R. E.; Yazyev, O.; Austin, A. J.; Cammi, R.; Pomelli, C.; Ochterski, J. W.; Ayala, P. Y.; Morokuma, K.; Voth, G. A.; Salvador, P.; Dannenberg, J. J.; Zakrzewski, V. G.; Dapprich, S.; Daniels, A. D.; Strain, M. C.; Farkas, O.; Malick, D. K.; Rabuck, A. D.; Raghavachari, K.; Foresman, J. B.; Ortiz, J. V.; Cui, Q.; Baboul, A. G.; Clifford, S.; Cioslowski, J.; Stefanov, B. B.; Liu, G.; Liashenko, A.; Piskorz, P.; Komaromi, I.; Martin, R. L.; Fox, D. J.; Keith, T.; Al-Laham, M. A.; Peng, C. Y.; Nanayakkara, A.; Challacombe, M.; Gill, P. M. W.; Johnson, B.; Chen, W.; Wong, M. W.; Gonzalez, C.; Pople, J. A. Gaussian, Inc., Wallingford CT, 2004.

79. Jones, R. O.; Gunnarsson, O. *Rev. Mod. Phys.* **1989,** *61,* 689–746.
80. Zupan, A.; Blaha, P.; Schwarz, K.; Perdew, J. P. *Phys. Rev. B.* **1998,** *58,* 11266–11272.
81. Theilhaber, J. *Phys. Fluids B,* **1992,** *4,* 2044–2051.
82. Stadler, R.; Gillan, M. J. *J. Phys.: Condens. Matter.* **2000,** *12,* 6053–6061.
83. Argaman, N.; Makov, G. *Am. J. Phys.: Condens. Matter.* **2000,** *85,* 69–79.
84. Huber, K. P.; Herzberg, G. *Constraints of Diatomic Molecules*; Van Nostrand Reinhold Company: New York, 1979.
85. Beutel, V.; Kramer, H. G.; Bhale, G. L.; Kuhn, M.; Weyers, K.; Demtroder, W. *J. Chem. Phys.* **1993,** *98,* 2699–2708.
86. Bishea, G. A.; Pinegar, J. C.; Morse, M. D. *J. Chem. Phys.* **1991,** *95,* 5630–5645.

PART III
Clinical Chemistry and Bioinformatics

CHAPTER 10

BIOCHEMISTRY APPS AS ENABLER OF COMPOUND AND DNA COMPUTATIONAL: NEXT-GENERATION COMPUTING TECHNOLOGY

HERU SUSANTO[1,2,*]

[1]Department of Information Management, College of Management, Tunghai University, Taichung, Taiwan

[2]Computational Science, Research Center for Informatics, The Indonesian Institute of Sciences, Jakarta, Indonesia

[*]Corresponding author. E-mail: heru.susanto@lipi.go.id; susanto.net@ gmail.com

CONTENTS

ABSTRACT

In today's society, bioinformatics research is putting a great emphasis on answering "when," "what if," and "why" questions with the help of information technology. Researchers had argued that it could be the key factor in facilitating and attaining an efficient decision-making in medical research. With the current deluge of data, computational methods have become indispensable to biological investigations. Originally developed for the analysis of biological sequences, bioinformatics now encompasses a wide range of subject areas, including structural biology, genomics, and gene expression studies. Additionally, nowadays biological data are proliferating rapidly. With the advent of the World Wide Web and fast Internet connections, the data contained in these databases and a considerable amount of special-purpose programs can be accessed quickly and efficiently from any location in the world. As a consequence, computer-based tools now play an increasingly significant role in the advancement and development of biological research. This study investigates the role of information technology in enabling science as well as the consequences and implication of technology in supporting medical research and bioinformatics.

10.1   INFORMATION SYSTEM

Information system is a part of information technology. It has been well defined in terms of two perspectives: one relating to its purpose, the other relating to its structure. From a functional perspective, an information system (IS) is a technologically implemented medium for the purpose of recording, storing, and disseminating linguistic expressions, as well as for supporting inference-making. While from a structural perspective, an IS consists of a collection of people, processes, data, models, technology, and partly formalized language, forming a cohesive structure which serves some organizational purpose or function. In an organization, a convenient and fastest way to deal with large amount of data to organize and analyze the enormous information is by using IS. Many big companies, such as eBay, Amazon, and Alibaba, also use information system. Information is widely used in any organization nowadays. IS is a software that is used to collect, store, and process data. It is also used to spread information and feedback will be provided for it to meet an objective. Data are raw facts that are used for reference. Information, however, is a collection of data that can be used to answer question as well as for problem-solving. The primary purpose of

IS is to collect raw data and transform it to information that will be beneficial for decision-making.

The three main roles of an IS are to support competitive advantage decision-making, business decision-making, and business processes operations. For each role, it has its own level where it operates. For supporting competitive advantage and decision-making, it works on a strategic level, while to support business decision-making, it works on tactical level and to support business process and operations, it works on operational level. IS does not have only one type. In fact, it varies depending on the usage of the information in an organization. Decision support system is one type of IS. Often, it is used for decision-making where the system analyzes and works on existing data which will project statistical prediction and data model. Transaction process systems allow multiple transactions such as collecting, processing, storing, displaying, modifying, or canceling transaction at one time. Management information systems mainly focus on processing the data from transaction process, in order to make it useful for business decisions according to specific problems. Office automation systems are concerned with office tasks as it will control the organization's flow of information. Executive information systems create conceptual information, such as strategic and tactical decisions, to satisfy the senior management. Expert system is said to be a set of computer programs that can imitate human expert. In science, it is difficult to find the results of relationships between genetic variability, diseases, and treatment responses, where it usually comes out as uncertain and inconsistent.

However, there are systems such as Fuzzy Arden Syntax or the probabilistic OWL reasoner Pronto that can be used to increase the size of the classical rule engines. Such systems are examples of decision support systems. It must be associated with specialists and international bodies so it would affect the clinical practices efficiently. Based on these findings, the system we visualized must be directly interconnected with the IS of the hospital so there is existing workflow when handling information gathered from electronic patient records and clinical laboratories.

10.2 INFORMATION TECHNOLOGY

Information technology (IT) is a computer technology that consists of hardware, software, computer network, users, and the internet. IT enables organization to collect data, make their work organized, and analyze data that help them achieve their objectives. IT personnel mainly focus on certain area, such as business computer network, database management, information security, software development, and science fields. In bioscience, IT is generally used for automated data collection, statistical study of data, internet-accessible shared databases, modeling and simulation, imaging and visualization of data and investigation, internet-based communication among researchers, and electronic dissemination of research results. For instance, the use of IT for automated gene sequencers, which utilizes robotics to process models and computers to manage, store, and retrieve data, has made possible the rapid sequencing of the human genome, which in turn has resulted in first time expansion of genomic databases. Shared internet-accessible databases are important in paleontology; and models as well as databases are significantly used in population biology and ecology; and genomics is influencing many fields in biology. Furthermore, IT can be a distinct tool from the other scientific tools, for instance microscopes or physics accelerators, which are commonly used in the scientific process, such as data gathering. Additionally, IT supports hypothesis formation that is the first stage to gather observations about the problem examined in a biological study, and also research design, collection of data, data analysis, and communication of scientific result.[7]

In other science fields, IT also helps in analyzing subsurface creations, mapping, and modeling complex systems. For example, seismic data used to measure earthquakes were traditionally recorded on paper or film but today they are recorded digitally, making it possible for the researchers to analyze the data swiftly. Furthermore, internet-connected system allows many researchers to obtain and contribute data to address a large problem. In several areas of sciences, imaging and visualization has become important because it can give clear modeling that helps the researchers understand biological system such as tissues, organisms, and cells.

The role of IT toward software is that it helps the software to communicate with the hardware to input data, process data, and give the output. The communication and interaction between software and hardware can be defined when collecting and processing a signal detected by laboratory equipment, for example, charged-couple devices (CDD) in spectrophotometer that is used to measure the amount of light reflected or absorbed from a sample object and other devices that can be connected to a computer via an analog to digital connector. Computer-aided algorithms are being used to examine the behavior of thousands of genes at a time and are creating a foundation of data for building integrated models of cellular processes. Molecular biologists and computer scientists have experimented with various computational methods established in artificial intellect, including knowledge-based and expert systems, qualitative simulation, artificial neural networks, and other mechanical learning techniques. These methods have been applied to problems in data analysis, creation of databases with advanced retrieval capabilities, and modeling of biological systems. Practical outcomes have been obtained in finding active genes in genomic sequences, assembling physical and genetic maps, and predicting protein structure. IT has been proven to have conspicuously great impacts on different areas of bioscience.

10.3 BIOINFORMATICS

Bioinformatics is the application of computer technology in managing any biological information and their managements. This application is extensively being used to analyze specifically biological and genetic information based on living things, both in plants and animals. Essentially, it has been used in many applications such as studying human disease, managing biological information, information systems for molecular biology management, and other related fields. In other words, in biology, it conceptualizes the study of molecules and constitutes informatics techniques, a combination of computer science, applied math, and statistics in the sense of physical chemistry.

Previously, the use of traditional methods for studying diseases mostly observed single factors, but the present technologies help in studying multiple factors at the same time, together with each and every possible thousands of variables. Bioinformatics methods are also being used to study the fundamental molecular biology of the disease and move toward personalized medicine, which involves understanding these diseases and shifting to new and more fitting treatment regimes.

The purpose of studying bioinformatics is to organize any biological data in a specified database which will help researchers to access or even enter any new related information. Bioinformatics is also a useful tool in developing the resources with the aid of data analysis, and this tool is able to develop, implement computational algorithms, and other software tools that help in understanding, interpreting, and analyzing any biological data that serve the humankind in a meaningful manner. Moreover, the computational tools are efficient in interpreting the results of biological research or can be applied during protein, cell, and gene research, in discovery and development of any new drugs or even during developing herbicide-resistant crop combination.[1,6]

The essence of bioinformatics application simply involves melding biology with computer science, the use of genomics information in understanding human disease, and the identification of new molecular targets for drug discovery on a large scale. The uses of bioinformatics are mostly in analyzing biological and genetic information which is associated with biomolecules on a large scale, disciplines related to molecular biology, from structural biology, genomics to gene expression studies applied to gene-based drug discovery, and development of genomic information resulting from the Human Genome Project. Experts and researchers, furthermore, apply bioinformatics in studying the sequences of genomes which appeared both in plants and animals in the field of agricultural studies with the advancement in genomics technologies throughout the years from 1950s until now. Genome sequencing apparently involves genetic databases for patients; next-generation sequencing (NGS) technology allows researchers to study complex genomics research, analyze genome-wide methylation or DNA–protein interactions, and study microbial diversity in the environment and in humans.[8,11,15]

10.4   BIOINFORMATICS APPLICATION IN PROGRESS

Grid infrastructures and next-generation computing technology played an important role in the recent decade in supporting scientific, computer-based analysis. However, the increasing complexity of bioinformatics resulted in finding new solutions to speed up computational time. Grid infrastructure is not completely satisfactory in terms of providing services and managing data that are reliable for presenting bioinformatics.

Another key issue is represented by the fact that the grid is offering poor chances to customize the computational environment. In fact, it is quite common in computational biology to make use of relational databases and/ or web-oriented tools to perform analyses, store output files, and visualize results, which are difficult to exploit without having administration rights on the used resources. Another related problem derives from the huge amount of bioinformatics packages available in different programming environments that typically require many dependencies and fine-tuned customizations for the various users. Here, the cloud computing offers a new solution to this limitation.[18]

The reason why the cloud computing is the best solution is that unlike in-house computing infrastructure, cloud computing is delivered over the internet. Cloud computing provides cheap, reliable, and large-scale data where a small organization can get the same information as a well-funded organization. Bioinformatics grew with an increasing use of the internet, which allows creation and sharing of large biological data and offers rapid publication of research results. The internet also provides the researchers an access to a complex supercomputing system such as grid infrastructure.

10.4.1 GENOMICS, BIOMEDICINE, AND MICROBIOLOGY

Genomics is a large-scale data acquisition using technological advancements which involves genome structures, evolution, and variations.[12] Genomics origin can be traced back as far as to the 19th century to the work of Gregor Mendel. However, in the middle of the 19th century, the progress of IS and IT was not as advanced as it is today. It is important to remember that genomics is an essential area of bioinformatics, as well as understanding its roles in the milestones of biological and molecular discovery, for instance, the Human Genome Project, an international scientific research project with goals to determine what makes up human DNA and its physical and functional characteristics, understanding heredity, understanding diseases, its role in pushing the innovation in genomic technologies, and many more. Another view on genomics is that the main concept of genome informatics was to analyze, process, and interpret all aspects of DNA in order to come up with a more defined and accurate information on biological structure and components of DNA.[3] All things considered, genomics is evolving duly because IT and IS keep on improving throughout the years. Owing to this, the world is progressing at a much faster pace, namely, in biomedicine and microbiology, and the knowledge that it brought, had been or is still being used to broaden our views on molecular mechanisms in spreading, treating, and curing diseases, and preventing the development of diseases.

Almost every year, new drugs are being discovered or improved to better serve their purpose in curing, treating, and preventing health issues all around the world. Bioinformatics acts as the main agent for its progression due to its vast collection of advanced tools for managing a large volume of data, as well as to help interpret, predict, and analyze clinical and preclinical data. As biological technologies progress, so do the data they produce and this often leads to a massive boom in database collection. Useful data could be proven to be useless without a proper system or tools to access them accurately and thus it is imperative to have computational tools that can search and integrate significant information. Development of bioinformatics resources has shown that it is essential in screening valuable data for effective drug solution in a profitable and timely manner. As an example, array comparative genomic hybridization (ACGH) method has been used globally for DNA analysis on normal and pathological clinical samples to check for the DNA copy number gain and loss across the chromosomes.[10] In addition, the current proteomic analysis that uses mass spectrometry (MS)-based technology is progressively used for identifying molecular network targets and is responsible for many discoveries in profiling correlations in the pathogenesis of certain human illnesses. Through this analysis and method, integration of the connections between proteomics and metabolomics platforms can increase the dynamic and potency of the drug treatment solution.[17]

Adverse drug reactions (ADRs) are often caused by all-purpose drugs that exist in today's market. Most people would prefer this type of drugs due to their economic status and often due to their state of living. Personalized medicine could provide the needed solution to these circumstances. Yan argued that through the development of pharmacogenomics and systems biology, personalized medicine could aid in the advancement of reductionism-based and disease-centered curative methods to systems-based, correlative, and human-centered care.[19] Developing further understanding between genotypes and phenotypes from analyzed data of genomic studies through data integration methods helps connect an efficient clinical and laboratory data flow. By implementing a translational informatics support into data mining techniques, knowledge discovery, and electronic health records (EHRs), better diagnostic and treatment selection can lead to a more suitable medication for the right people. A good translational bioinformatics will aid in establishing a powerful platform to connect various knowledge scopes for translating numerous biomedical data into predictive and preventive medicine.[19] Altogether, this will bring about an ideal personalized medicine that is less costly, with reduced errors and risks, diminished ADRs, and that overcomes the therapeutic obstruction.

According to Wu, Rice, and Wang, cancer is one of the most prevalent and profound diseases that can occur at anytime and anywhere in the body.[19] Its development is explained as an uncontrollable genetic mutation of cells in the body of organisms whereby it drastically affects the metabolism, loss in genes, and promotion of invasive tumor growth, metastases, and angiogenesis.[14] Multiple factors such as the period, severities, drug resistance, cell origins, locations, and affectability can be the causes for poor diagnoses and therapy results for the patient with cancer. However, over the years, the results from advanced and accurate clinical bioinformatics and uses of new systems clinical medicine have helped improve the results of cancer treatment and diagnosis all around the world. Adoptive immunotherapy (gene therapy) is commonly used for cancer treatment, it uses the technology of genetic modification whereby T cells with antitumor antigen receptors (TCR) or chimeric antigen receptors (CAR) are used, duly because they can target antigens expressed on tumor cells.[13] In recent times, the use of semantic web technology has enabled better understanding of high throughput clinical data and established quantitative semantic models gathered from Corvus (data warehouse providing systematic interface to numerous forms of omics data) rooted from systematic biological knowledge and by application of SPARQL endpoint.[4] In addition, application of new biomarker strategies in cancer bioinformatics has become more popular in monitoring progress of the disease and its response to therapy. It is expected to coordinate with clinical informatics which includes patient inputs, such as complaints, history, therapies, symptoms and signs, medical examinations, biochemical analysis, imaging profiles, and other valid inputs. On the whole, the expected result would provide more accurate interpretable signatures and therefore helps in better diagnosis and cancer solutions for specific patients.

The world's population is continuously booming and is expected to reach 11 billion by the year 2100 and this brings about new challenges in managing disease outbreaks. A rise of infectious diseases by new viruses and drug-resistant bacteria are the tendencies of disease outbreaks. Over the last two decades alone, new virus strains have kept the world in constant fear of deadly outbreak threats: swine flu pandemic, SARS (severe acute respiratory syndrome), HIV (human immunodeficiency virus) and AIDS (acquired immune deficiency syndrome), malaria, tuberculosis, and, more recently, Zika virus. Steps in bottling more outbreaks have been initiated throughout the globe but more importantly, the sharing of knowledge on how it came to be and the proper exploitation methods to discover the viruses' weaknesses has proven to be a better front in battling outbreaks. Next-generation genome sequencing has aided the advancement of biotechnologies and tools by providing new insight into viral distinctiveness, allowing in-depth sampling and providing bigger capacity for automation, and thus enabling new data interpretation on what could be done or changed to the characterization of viral quasi-species.[9] Kijak et al. named their bioinformatics package Nautilus, which runs on several operating systems, and it represents new sets of tools to support better data analysis, facilitating the application of next-generation sequencing and allowing better insight on HIV genome characterizations throughout the population and its evolution.[9] Nevertheless, the rapid occurrences of antimicrobial resistance in microorganisms cannot be diminished simply by continuous biological studies but this requires a global understanding in public to always keep a hygienic environment, in practice, as well as awareness, wherever and whenever they are. This method is the main preventative method for, generally, all kinds of microbial infections, duly because producing a constantly evolving medicine and treatment to fight against rapidly evolving antimicrobial-resistant illnesses will take higher health care expenditures and is time-consuming for all sides. The risk of death from resistant microorganisms is much higher than that of the same nonresistant microorganisms.

10.4.2 BIOLOGICAL DATABASE

Database is seen as a major tool for storing biological data for public use. Relational database concept of computer science and information retrieval concept of digital libraries are implemented to fully interpret biological database. Gene sequence, attributes, textual descriptions, and ontology classification are stored in biological database. The data mentioned are categorized as semistructured data that later can be displayed in table form, key delimited record, and XML structure. The common method of cross-referencing is often used by database accession number.

A biological database can be defined as the collection of biological data obtained during live-experiment and computation operation and analysis. Biological data should be organized properly to enable easy data operations such as manipulation, deletion, and calculation. The aim of the database must follow two principles: data should be accessible and it can be used in both single- and multiuser system environment.

Databases, generally, can be grouped into primary, secondary, and composite databases. Primary database includes data that are gathered during experiment such as nucleotide sequences and three-dimensional structures. It is often called archival database. The data gathered are resulted from experiments by researchers all around the world. Primary database includes GenBank, DDBJ, and SWISS-PROT. Secondary database is derived from analysis of the first-hand (primary) data such as sequences and secondary structure. The results of secondary database are usually in the form of conserved sequences, signature sequences, and active site residues of protein. Curated database is another term for secondary database. Some of the databases were created and hosted by the researchers themselves at their own laboratory such as SCOP, CATH, and PROSITE, which were developed in Cambridge University, University College of London, and Stanford University, respectively.

The first database was created in 1956, after the insulin protein sequence became available; insulin was the first protein to be sequenced. The content of the insulin sequence includes just 51 residues which characterized the sequence. Four years after the discovery of insulin protein sequence, the first nucleic acid sequence of yeast tRNA with 77 bases was discovered in 1960. Three-dimensional structure was studied as well as the first protein structure database with only 10 entries was created in 1972. Currently, the Protein Data Bank (PDB) has grown to store more than 10,000 entries.

Since protein sequence databases were maintained by individual laboratories, SWISS-PROT protein sequence, which was categorized as consolidated format database, was started in 1986. Database functionality was expanded not only with the capabilities of handling data and sophisticated queries facilities but also by implementing bioinformatics analysis function in modern databases.

Similar to general databases, biological database can also be categorized into two groups: sequence structure databases and pathway databases. Sequence structure databases mainly focus on nucleic acid sequence and protein sequence, whereas pathway databases focus on protein.

10.4.3 SEQUENCE DATABASES

Sequence databases are categorized as the most frequently used databases and some of them are marked as the best biological databases. GenBank is one of the examples of widely used sequence databases. GenBank focuses on DNA and protein sequence.

GenBank is classified as one of the most widely used sequence biological databases. The name refers to the DNA sequence databases of National Center for Biotechnology Information (NCBI). GenBank data are mainly made up of sequences submitted by individual laboratories and data interchange from international nucleotide sequence databases, European Molecular Biology Laboratory (EMBL) and DNA database of Japan (DDBJ).[16]

10.4.4 STRUCTURE DATABASES

To be able to completely understand the protein function, knowledge of protein structure and molecular interaction and mechanism must be understood thoroughly. The PDB is the worldwide repository of experimentally classified protein structures, nucleic acids, and complex assemblies, including drug–target complexes.[5] The PDB was created in Brookhaven National Laboratories in 1971. It mainly contains information on molecular structure of macromolecules from x-ray crystallography and nuclear magnetic resonance (NMR) method. Currently, Research Collaboratory for Structural Bioinformatics plays a huge role in maintaining PDB. One of the features of PDB is that it mainly allows user to display and present data either in plain text or through a molecular viewer using JMOL.

10.4.5   PATHWAY DATABASES

The growth of metabolic databases through metabolic study pathway will fulfill the need and enhance the development of system biology. One of the popular pathway databases is KEGG. KEGG refers to the Kyoto Encyclopedia of Genes and Genomes. KEGG database is the center of information for system analysis of gene function and connecting genomic information with higher order functional information. KEGG consists of three databases, namely, PATHWAY, GENES, and LIGAND. PATHWAY database is responsible for storing the higher order functional information. This information includes the computerized knowledge on molecular interaction networks. These data are often encoded by coupled genes on the chromosome which are crucial for predicting gene functions. GENES database consists of the collection of gene catalogs and sequences of genes and proteins produced by the Genome Project. The third database, LIGAND, keeps the information regarding chemical compounds and chemical reactions which are important for cellular processes.

10.4.6   PREVENTION AND TREATMENT OF DISEASES

Bioinformatics is a scientific discipline that deals with obtaining, analyzing, distributing, processing, and storing biological information. It uses scientific knowledge, such as algorithm and computer science, in order to understand the biological significance of a wide variety of data. With this, it enables researchers to find new strategies to look for clues in the prevention and treatment of diseases. Bioinformatics has turned into a key ingredient with the alliance of genomics, proteomics, and drugs in today's world.

In fact, bioinformatics owes its creation to the need to handle large amounts of data produced by these "-omic" technologies (genomics, proteomics, and, more recently, metabolomics). The information is generated by high-performance methods such as gene sequencing, DNA microarrays, and mass spectroscopy. Bioinformatics can be called a transverse activity because it is applicable to all the subsectors of biotechnology and life sciences. However, its main application is biomedicine. Bioinformatics manages and decodes "-omic" data and it facilitates the translational medicine concept by helping to distribute information throughout the entire health care value chain. This covers the discovery and analyzing of genes, the protein structures coded by these genes, and the design of molecules and drugs to counter these proteins, up to their clinical application, which is where bioinformatics plays a leading role in the development of specific medicine.

10.4.7 DRUG DISCOVERY

Traditionally, pharmaceutical companies were interested only in introducing new drugs when any well-known pharmaceutical company had been successful in developing them. However, in today's society, companies have invested heavily in approaches that can speed up the development process. The pressure of producing drugs in short period of time with a high standard of safety has resulted in extremely enhancing interest of the researchers in bioinformatics. Bioinformatics involves identification of biological candidate and could be the storage of information. Drugs only can be produced if the drug target is studied and identified. For example, human genome sequence information can be found in the system, which can help in drug-making process.

10.5 THE COMMERCIALIZATION OF BIOINFORMATICS

The success of development in combining both computer technology and biomedicine has helped scientists to become more efficient and productive with the ability to predict the upcoming trend of biosolution with the help of bioinformatics. With the bioinformatics technology being widely exploited, the opportunity for commercialization is unquestionable. However, many investors were reluctant to invest in bioinformatics sector due to its history of investment during the late 1990s which resulted in high losses. Despite the losses, in 2002, Philip Green, biologist in the University of Washington, wanted to decipher human genome with more accurate reading of DNA letters. Celera-made machine was the only tool he used which was supplied by Applied Biosystems. Due to the lack of functionality of this software, he then designed his own software to cater to the needs of his project. As a result, Green's innovation is categorized as industry standard and its source code is available for free.

The advancement of technology played a huge role in the success of bioinformatics, where computer tools are being used for managing biological information and computational biology is used to identify the molecular components of living things. The involvement of computer technology in biological area also introduces the studies of principle and operation of data manipulation and data analysis of biomolecules, structure, or composition of various materials, such as nucleic acids and, gene products, such as proteins. Research and data gathering in biological field involves laboratory experiments where mathematical operations and computation are used to obtain meaningful information from meaningless data, especially in genomics. Computers have been used as the backbone in bioinformatics and have eventually become one of the major tools in storing biological data and compare them with the existing data set. This provides important useful inputs, which biologists usually gather during their hands-on laboratory experiment, to computer-user/researcher and saves a lot of time and resources. Cost of laboratory has been cut down with the use of computer and software, where most of the operations and experiments are performed via computer with the help of special-purpose software.

The popularity of using open-source software has increased over the years, which has affected bioinformatics companies. Open-source software can be defined as computer software freely available to the public. In general, it is free software which can be freely copied, distributed, modified, and manufactured. Characteristics of open source are as follows: there should be no discrimination to people, groups, or endeavor; the license distribution should be costless and general to every product; and the license must be restriction-free with respect to other software and technology neutral. The Linux system is one of the great examples of open-source software with a major success. Bioinformatics firms find it hard to gain profit due to this open-source software movement.

10.6 ADVANTAGES

Bringing together large data sets of medical data and tools to analyze the data offers the potential to enhance the research capabilities of medical researchers, who could use this vast source of biological and clinical data to discover and develop new treatments and better understand illnesses. Pharmaceutical companies could use the biomedical data to create drugs targeted at specific populations. Furthermore, health care providers can use the data to better inform their treatments and diagnoses.[2]

Etheredge, as cited in a study by Castro,[2] claimed that applying informatics to health care creates the possibility of enabling "rapid learning" health applications to aid in biomedical research, effectiveness research, and drug safety studies. For example, using this technology, the side effects from drugs newly introduced to the market can be monitored in real time, and problems, such as those found with the recently withdrawn prescription drug Vioxx, can be identified more quickly. Moreover, the risks and benefits of drugs can be studied for specific populations yielding more effective and safer treatment regimens for patients.

Etheredge concluded that using rapid learning techniques not only can improve patient safety but also can lead to substantial improvements in the quality and cost of care by turning all of the raw digital data into knowledge where these rapid learning health networks can enable doctors and researchers to better practice evidence-based medicine. Evidence-based medicine is the use of treatments judged to be the best practice for a certain population on the basis of scientific evidence of expected benefits and risks.

Bioinformatics deals with computer management and analysis of biological information: genes, genomes, proteins, cells, ecological systems, medical information, robots, and artificial intelligence, as there are many applications of bioinformatics resulting from the combination of computer and biology. The evolution of technology helps in supporting bioinformatics in discovering diseases and applications in forensics using software packages and bioinformatics tools, for example, the evolution of technology such as Illumina NGS to provide accurate sequencing. NGS technology can provide valuable and useful information for a better understanding of health and diseases.

Bioinformatics can also be useful in determining the order of the four chemical building blocks called bases that make up the DNA molecule. This is because the sequence provides scientists data regarding what kind of genetic information is carried in a particular DNA segment. Moreover, the sequence data can highlight the changes in a gene that may cause disease.

10.6.1 FORENSIC DNA AND BIOINFORMATICS

Bioinformatics and forensic DNA analysis are fundamentally characterized by studies in both law and biology and draw their techniques from statistics and computer science, which facilitates solving problems in law and biology. It could be useful to identify a victim and suspect with personal relatedness to other individuals; these are the two major focuses of forensic

DNA analysis. It is a common event in forensic analysis, especially by crime and investigation unit or CSI, to look at close connections, for example, paternity disputes, suspected incest case, corpse identification, alimentary frauds (e.g., OGM, poisonous food, etc.), semen detection on underwear for suspected infidelity, insurance company fraud investigations when the actual driver in a vehicle accident is in question, criminal matters, and autopsies for human identification following accident investigations. All of these problems may be solved by using bioinformatics methods.

Also, genetic tests have been widely used for major catastrophic events such as terrorist attacks, airplane crash, and tsunami disaster. It can be used for mass fatality identification and forensic evidences. Personal identification relies on identifiable characteristics, as the human body has a personal identity that is unique biologically (such as blood, saliva, and DNA), physiological difference (such as fingerprints, eye irises, and retinas, hand palms, and geometry and facial geometry), behavioral difference (such as body posture, habits, signature, keystroke dynamics, and lip motion), and also on the combination of physiological and dynamical characteristics such as the voice.

Hence, genetic testing results are integrated with the information collected by multidisciplinary teams composed of medical examiners, forensic pathologists, anthropologists, forensic dentists, fingerprint specialists, radiologists, and experts in search and recovery of physical evidence. Officers could have access to the personal information, where biological data can be obtained from hospital records and behavioral data may be collected from banks or office document such as fingerprint or signature just by looking at the database.

Therefore, the application of genetic testing in large-scale tissue sampling and long-term DNA preservation plays an important role in mass fatalities which have recently occurred. Thus, DNA has become the most important personal identification characteristics because all genetic differences, whether being expressed regions of DNA (genes) or some segments of DNA, are characteristic of a person. DNA possesses coding pattern of inheritance that can be monitored and used as marker.

10.6.2 CANCER AND BIOINFORMATICS

According to the Cancer Research UK (2012), cancer is one of the leading causes of death worldwide with 14.1 million cases of cancer recorded and about 8.2 million deaths from cancer estimated in 2012. A leading cause of cancer is malignant growth or tumor which is caused by abnormal and uncontrolled division due to the changes in DNA of cell by mutation. Also, errors in the genes may cause this abnormal behavior to be cancerous. These changes occur when exposed to a certain type of cancer-causing substances.

In the post-genomics era, age holds phenomenal promise for identifying the mechanistic bases of organismal development, metabolic processes, and diseases. Bioinformatics research will lead to a wide understanding of the regulation of gene expression, protein structure determination, comparative evolution, and drug discovery. Presently, two-dimensional gel protein pattern can be easily analyzed using bioinformatics technology, where these software applications possess user-friendly interfaces that are incorporated with tools for linearization and merging of scanned images.

New techniques and new collaborations between computer scientists, biostatisticians, and biologists are required in today's research. There is a need to develop and integrate database repositories for the various types of data being collected, to develop tools for transforming raw primary data into forms suitable for public dissemination or formal data analysis, to obtain and develop user interfaces to store, retrieve, and visualize data from databases, and to develop efficient and valid methods of data analysis.

Cancer DNA sequencing using NGS provides better information and requires less time compared with normal gene sequencing using gel structure. With NGS, researchers can perform whole-genome studies, targeted gene profiling, and tumor-normal comparisons. Therefore, it is easy to detect tumor and DNA fragments with detailed quantitative measurements from the database.

Furthermore, the prediction of genes is likely to be linked to newly developed diseases or modified version of an old disease that evolves or mutates. The use of bioinformatics can easily recognize related genes that have any similar function or characteristic of the original gene, such as the similarity in percentage of DNA sequence. The biggest challenge is to identify enormous markers of DNA, as the application of molecular links to diseases will continue to face technological as well as biological and algorithm challenges. The human body consists of very complicated and diverse features because it is continually evolving and responding to changes.

As for using bioinformatics to replicate, the structure of new DNA structure provide a challenge in technology; this is because, unlike cells, other interrelationships may not be visible through the microscopic view. Thus, the already designed computer frameworks or databases may not cope with the expansion in network-level measurements and information.

10.7   FUTURE DIRECTION OF BIOINFORMATICS

Bioinformatics makes the information accessible and shareable in comparison with traditional biological records, with the help of developing tools that make it easy to send, receive, and share information. For example, electronic medical records (EMR) reduce the chances of error that are caused by obstruction and other researcher's conflicts during the manual data entry processing after paper-based data collection. Besides that, it also supports elimination of the manual task of extracting data from charts or filling out specified data sheets. The data stored can be obtained directly from the EMR. By referring to the EMR, the researchers do not need to examine or observe the task again. Bioinformatics has grown rapidly and diversified into subdisciplines, such as cheminformatics in chemistry; neuroinformatics, which is related to gathering data across all scales and neuroscience levels to understand the complex function of brain and work toward treatments for brain-related illness; and immunoinformatics, which uses informatics techniques to study molecules of the immune system.

Usually, the organizations, such as Antigen Discovery, Rasa Life Science Informatics, and LabCentrix, use bioinformatics to store large amount of data, or in other words, big data, which contain both structured and unstructured data that are hard to process if the organization uses traditional database and software techniques. Examples of big data are patient information, types of disease, and DNA. Hence, by applying bioinformatics, this can be done easily as it can store huge amount of data.

Bioinformatics interface seems to be unattractive, the designing of the tools has to be user-friendly, where people can easily understand it and use it. A user-friendly interface can make the researchers analyze the information more efficiently and effectively to convert the information into knowledge. Other than that, the users must be properly trained to use bioinformatics tools to prevent difficulties. This would be a tough situation for developing regions, as training and knowledge is required along with time and financial support.

Bioinformatics is important, yet the biggest challenge being faced by the molecular biology society nowadays is to make sense of the wealth of data that have been produced by the genome sequencing projects. With the advent of new tools and databases in molecular biology, researchers are now enabled to carry out research not only at genome level but at proteome, transcriptome, and metabolome levels. Therefore, incisive computer tools must be improved to accept the extraction of meaningful biological information.

10.8 CONCLUSION

In conclusion, the bioinformatics has wide applications from single traditional methods in handling genomic studies to more advanced methodologies toward the improvements in biological studies. Understanding the vast collection of bioinformatics tools and technologies that exist to serve different purposes is also vital to the advancement in IT. Bioinformatics helps in improving the ways of biological research and provides innovative ways, for instance, the use of databases to store any biological information with the help of more advanced software tools and computational algorithm. These databases, which are being stored in the computer memory, are mostly related to microorganisms. In business studies, the research on scientific fields gives opportunities for businesses to broaden their chances of making profits, for instance, the invention of new medicines, drug developments, and so forth.

In the world of biomedicine and microbiology, IT through bioinformatics has provided numerous resources and tools for new drug discovery, and personalized and preventive medicine. Moreover, the success in cancer treatment via gene therapy is also gradually increasing, owing to the NGS techniques and technologies. This increase in treatment success is observed not only for cancer but also for other known diseases such as HIV and AIDS.

On the other hand, databases are also being widely used in bioinformatics applications. The relational database concept of computer science and information retrieval concept are implemented to fully interpret biological database. The aim of the database is basically to be accessible and be useful in both single- and multiuser system environment. There are several types of databases provided for different purposes such as for protein development, human brain studies, diseases, drug discoveries and development, and also for agricultural studies, in both plants and animals. These data can be considered highly accessible and reliable for other researchers and scientists. Furthermore, it has data integration and even they can add new findings or even update, as it is most liked and easy to use. Scientists and researchers are also being trained in handling databases.

Finally, bioinformatics applications are widely being used for more advanced technological developments, there are pros and cons not only in biological manner but also in business manner. As stated, these are being handled properly and their usage is being improved throughout the years.

KEYWORDS

- physical chemistry
- bioinformatics
- computational methods
- biological investigations
- sequence database
- structural biology
- genomics

REFERENCES

1. Benson, D. A. GenBank. Nucleic Acids Research. Biological Database. 2009 (May 8, 2015).
2. Castro, D. *The Role of Information Technology in Medical Research*. The Information Technology and Innovation Foundation. 2009.
3. Chen, R. On Bioinformatic Resources. *Genom. Proteom. Bioinform.* **2015,** *13*(1), 1–3. http://dx.doi.org/10.1016/j.gpb.2015.02.002
4. Holford, M.; McCusker, J.; Cheung, K.; Krauthammer, M. A Semantic Web Framework to Integrate Cancer Omics Data with Biological Knowledge. *BMC Bioinform.* **2011,** *13*(Suppl. 1), S10. http://dx.doi.org/10.1186/1471-2105-13-s1-s10
5. Huang, Y. H.; Chun-Nan, H.; Peter, W. R. Citing a Data Repository: A Case Study of the Protein Data Bank. *PLoS One.* **2015,** *10*(8).
6. Iranbaksh, A.; Seyyedrezaei, S. H. The Impact of Information Technology in Biological Sciences. *Procedia Comput. Sci.* **2011,** *3*, 913–916.
7. Kumar, M. P. Information Technology: Roles, Advantages and Disadvantages. *Int. J. Adv. Res. Comput. Sci. Softw. Eng.* **2014,** *4*(6), 1020–1024.
8. Kanehisa, M. KEGG: Kyoto Encyclopedia of Genes and Genomes. *Nucleic Acids Res.* **2000,** *28*(1), 27–30.
9. Kijak, G.; Pham, P.; Sanders-Buell, E.; Harbolick, E.; Eller, L.; Robb, M., et al. Nautilus: A Bioinformatics Package for the Analysis of HIV Type 1 Targeted Deep Sequencing Data. *AIDS Res. Hum. Retrovir.* **2013,** *29*(10), 1361–1364. http://dx.doi.org/10.1089/aid.2013.0175
10. Leung, E.; Cao, Z.; Jiang, Z.; Zhou, H.; Liu, L. Network-based Drug Discovery by Integrating Systems Biology and Computational Technologies. *Brief. Bioinform.* **2012,** *14*(4), 491–505. http://dx.doi.org/10.1093/bib/bbs043
11. Luscombe, N.; Greenbaum, D.; Gerstein, M. *What Is Bioinformatics? An Introduction and Overview*; Department of Biophysics and Biochemistry, Yale University: New Haven, USA, 2001.
12. Ma, D.; Liu, F. Genome Editing and Its Applications in Model Organisms. *Genom. Proteom. Bioinform.* **2015,** *13*(6), 336–344. http://dx.doi.org/10.1016/j.gpb.2015.12.001

13. Morgan, R.; Chinnasamy, N.; Abate-Daga, D.; Gros, A.; Robbins, P.; Zheng, Z., et al. Cancer Regression and Neurological Toxicity Following Anti-MAGE-A3 TCR Gene Therapy. *J. Immunother.* **2013,** *36*(2), 133–151. http://dx.doi.org/10.1097/cji.0b013e3182829903

14. Mount, D. Using Bioinformatics and Genome Analysis for New Therapeutic Interventions. *Mol. Cancer Ther.* **2005,** *4*(10), 1636–1643. http://dx.doi.org/10.1158/1535-7163.mct-05-0150

15. Oliva, A.; Rendy, L.; Racheal, B.; Janet, W. Bioinformatics Modernization and the Critical Path to Improved Benefit-Risk Assessment of Drugs. *Drug Inf. J.* **2008,** *42*, 273–279.

16. Priyadarshi, M. B. *Sequence Databases.* Retrieved from Biotecharticles (2014). Obtained from: http://www.biotecharticles.com/Bioinformatics-Article/ (accessed Aug 8, 2017).

17. Toomula, N. Biological Databases-Integration of Life Science Data. *J. Comput. Sci. Syst. Bio.* **2011,** *4*(5), 87–92.

18. Wu, D.; Rice, C.; Wang, X. Cancer Bioinformatics: A New Approach to Systems Clinical Medicine. *BMC Bioinform.* **2012,** *13*(1), 71–80. http://dx.doi.org/10.1186/1471-2105-13-71

19. Yan, Q. Translational Bioinformatics Support for Personalized and Systems Medicine: Tasks and Challenges. *Transl. Med.* **2013,** *3*(2). http://dx.doi.org/10.4172/2161-1025.1000e120

CHAPTER 11

ADSORBENTS FOR DNA SEPARATIONS

SENEM YETGIN[1] and DEVRIM BALKOSE[2,*]

[1]*Food Engineering Department, Kastamonu University, Kastamonu, Turkey*

[2]*Chemical Engineering Department, Izmir Institute of Technology, Izmir, Turkey*

Corresponding author. E-mail: devrimbalkose@gmail.com

CONTENTS

ABSTRACT

DNA can easily be adsorbed to positively charged surfaces as well as to negatively charged surfaces by electrostatic bridges with the water of hydration of charged cations. The functional groups and the pore structure and surface charge are all the important parameters for DNA adsorption. DNA adsorption to alumina, clay, silica, single-walled carbon nanotubes (SWNT), metals, organic and bioorganic molecules, polypyrrole (PPy)–silica nanocomposite particles, polymers and hydrogels, poly-L-lysine-immobilized pHEMA, and HAP are reviewed in this chapter. The experimental results for preparation and characterization of silica, alumina, and HAP are reported in detail.

11.1 INTRODUCTION

Adsorption is a widely used DNA purification technique in biological analysis. The main parameters affecting the adsorption process are its size, shape, polarity, and chemical structure of the substance wanted to be removed. DNA is a biopolymer and it structurally depends on nucleotide sequence and has been known to bind many classes of substances—alumina, clay, and silica, including single-walled carbon nanotubes (SWNT) as biosensor,[72] metals, organic and bioorganic molecules such as drugs. Under most environmental conditions, DNA molecules are negatively charged, and they can easily be adsorbed to positively charged surfaces such as the edges of clay minerals[29] as well as to negatively charged surfaces such as the surfaces of clays by electrostatic bridges with the water of hydration of charged cations.[38,47] It should be noted that DNA double helix form has been affected by physical adsorption of the DNA on a surface, which also influences its denaturation.[2]

Some composites were also reported as DNA adsorbents such as polypyrrole (PPy)–silica monocomposite particles,[57,58] polymers, and hydrogels.[61] Polypyrrole is a conductive polymer, stable and biocompatible compared with other polymers.[33] Its good thermal and mechanical stability let its use in biological areas such as biosensor design. Composite form of it, with DNA absorbable material, makes them acceptable for DNA adsorption. Therefore, polypyrrole–silica nanocomposites (untreated and amine or carboxylic powder acid functionalized) were studied at neutral pH in sodium phosphate buffer by Saoudi and coworkers. [57,58] DNA adsorption was measured to be 32 and 22 mg/g for the aminated silica sol and the aminated PPy–silica particles, respectively, and 6.5 mg/g for the carboxylated particles. DNA adsorption is not applied only for purification of DNA but also for therapies of autoimmune diseases. DNA-immobilized supports, such as polymeric adsorbent, have been used. Amount of immobilized DNA on the material is directly related to antibody removal rate. For instance, poly-L-lysine-immobilized poly(2-hydroxyethyl methacrylate) (pHEMA) membrane was used for that purpose.[61] At 4°C from phosphate-buffered salt solution, maximum DNA adsorption was obtained as 5849 mg/m^2.

Different surfaces used in DNA adsorption are listed in Table 11.1. Glass, silica, polylactic acid, functionalized polystyrene, polypyrrole powders, soil, gold, montmorillonite, kaolinite, goethite, gibbsite, and allophane are some of the adsorbents used for DNA adsorption.

TABLE 11.1 DNA Adsorption onto Different Surfaces.

Adsorbent	DNA type	Reference
Glass	*Escherichia coli* DNA Human spleen DNA	Vogelstein and Gillespie[68]
Silica particles or diatoms	DNA and RNA from, e.g., human serum and urine	Boom et al.[6]
Polymerized poly-L-lactic acid film surface	Plasmid DNA	Jiang et al.[26]
Poly-L-lysine immobilized poly(2-hydroxyethyl methacrylate)	Calf thymus	Şenel et al.[61]
Polypyrrole (PPy)-silica monocomposite	Calf thymus DNA	Saoudi et al.[58]
Positively charged amidine functionalized polystyrene microspheres	Double stranded DNA extracted from herring	Hodrien et al.[25]
Chemically synthesized polypyrrole powders	Calf thymus	Saoudi et al.[58]
Soil	Calf thymus DNA	Ogram et al.[43]
N-methylpyridinium terminated gold	Type XIV from herring testes	Aslanoğlu et al.[3]
Silica beads were packed into glass microchips	PCR-amplifiable DNA	Breadmore et al.[7]
Montmorillinite	Salmon sperm DNA	Saeki et al.[56]
Kaolinite	Salmon sperm DNA	Saeki et al.[56]
Silica	Salmon sperm DNA	Saeki et al.[56]
Goethite	Salmon sperm DNA	Saeki et al.[56]
Gibbsite	Salmon sperm DNA	Saeki et al.[56]
Synthetic allophane	Salmon sperm DNA	Saeki et al.[56]
Naturel allophane	Salmon sperm DNA	Saeki et al.[56]

A detailed review about preparation and characterization of adsorbents namely silica, alumina, and hydroxyapatite (HAP) will be made in the following section. Experimental results will be reported for representative commercial samples.

11.2   SILICA, ALUMINA, AND HAP

11.2.1   SILICA

Nucleic acids are readily adsorbed on silica and on a glass surface under chaotropic solution conditions.[6,9] DNA adsorption by silica was extensively investigated.[20,39,40] Silica offers a quick and less chemical purification. Therefore, silica, the most commonly used matrix for solid-phase extraction, was shown to effectively bind DNA when it was used to purify DNA from agarose gels.[68] Silica is extensively used in commercial DNA purification. Silica-based kits are the most common method of nucleic acid purification for PCR. A typical spin column comprises a nearly 5-mm-diameter and 1-mm-thick glass fiber disc (membrane), or other form of silica, filled in a polypropylene tube.[30]Spin column was designed to allow centrifugation or vacuum suction to facilitate flow through the porous silica membrane. Silica-based DNA extraction kits are widespread, produced by companies such as Clontech (Nucleo Spin), Mo Bio Laboratories (UltraClean BloodSpin), Qiagen (QiaAmp), Promega (Wizard), Epoch Biolabs (EconoSpin), and Sigma-Aldrich (GenElute). Promega products are available for plasmid, genomic, and fragment/PCR product purification. Promega has sold and supported silica-based DNA purification systems for nearly two decades.

According to some studies, DNA molecules bind to OH groups on the edge of phyllosilicates such as montmorillonite[37,46] as seen in Figure 11.1(a) and (b). DNA adsorption in soils using montmorillonite as adsorptive particle has been investigated by several researchers.[28,50] Adsorption mechanism in Figure 11.1(c) represents DNA molecules associated with the surface, which negatively charged materials via a bridging of cations.[28,46] Divalent cations such as Mg^{2+} and Ca^{2+} support binding about 100 times greater than monovalent cations such as Na^+, K^+, and NH^{+4}.[19]According to Figure 11.1(d), DNA molecules directly bind to solid organic surface. On the contrary, Saeki and Sakai[55] proved that organic matter addition decreased the adsorption capacity of DNA on the soil material.

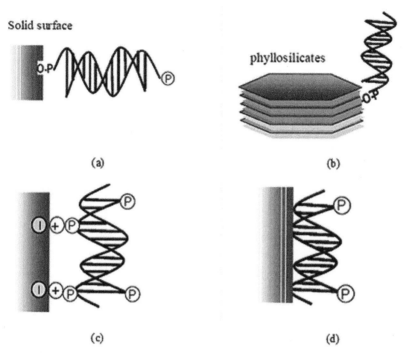

FIGURE 11.1   Conceptual figures of DNA adsorptions.

DNA adsorption by silica was controlled by three effects (1) weak electrostatic repulsion forces, (2) dehydration, and (3) hydrogen bond formation.[40] The presence of a monovalent cation such as Na^+ neutralizes the negative charges on the phosphate backbone of DNA, reducing the electrostatic barrier between DNA and silica. Therefore, DNA adsorption capacity is increased.

Mao et al.[37] have also indicated that it is possible to form hydrogen bonds between the phosphate groups in DNA and the silanol groups on the surface of silica (Fig. 11.2). This hydrogen bond formation has a weak and additive effect on adsorption. Melzak et al.[40] reported that intermolecular electrostatic forces between DNA and silica surface strongly disfavors adsorption at low ionic strength. The adsorption capacity as a function of ionic strength can be explained by electrostatic double-layer repulsion between the DNA and the natural organic matter (NOM) layer.[42] Silica adsorption capacity of DNA was found to be greater than that of montmorillonite.[56] The solution chemistry and the type of surface will influence the amount of DNA adsorbed.

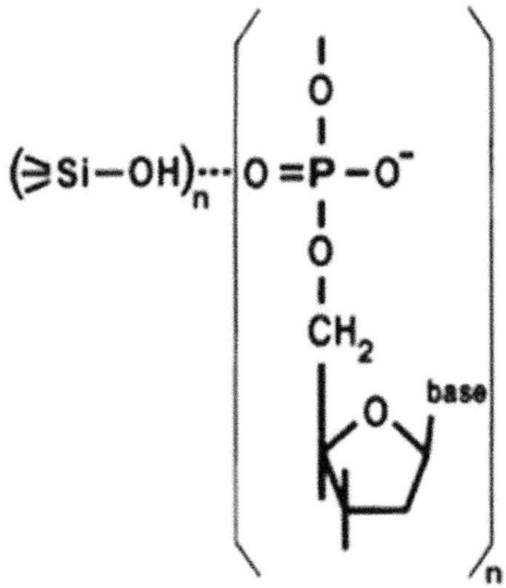

FIGURE 11.2   Proposed binding of DNA to silanol groups on the silica surface. (Reproduced from Environmental Health Perspectives, 1994; 102(suppl 10):165–171).

Consequently, there are other important situations where both adsorption and denaturation of DNA occur simultaneously. For example, adsorption of plasmid and chromosomal DNA on microcrystalline silica surface and the effect of ionic strength, temperature, pH, DNA size, and conformation on the adsorption phenomenon were reported by Melzak et al.[40] Adsorption isotherms pointed out those three effects: (1) shielded intermolecular electrostatic forces, (2) dehydration of the DNA and silica surfaces, and (3) intermolecular hydrogen bond formation in the DNA–silica contact layer. These three effects mentioned above make the dominant contributions to the overall driving force for adsorption. For instance, adsorption of DNA from *Bacillus subtilis*, calf thymus, and salmon sperm on montmorillonite, kaolinite, and silica increased with an increase of ionic strength or a decrease of pH.[40]

Adsorption of DNA onto bare silica under high ionic strength and chaotic solution conditions is supported.[7] High ionic strength serves to protect the negative surface, reducing the electrostatic repulsion between the negative DNA and the surface of the silica, while the mono- and divalent salt dehydrates the silica surface and DNA, thus promoting hydrogen bonding between the DNA molecules and the protonated silanol groups.

Solberg and Landry [63] focused on mesoporous silica material to find its gene therapy potential. Ammonium ion containing chemicals was found to be two to three times more effective in cross-link between DNA and meso-porous silica surface than mono- or divalent cations

Adsorbed DNA was 100 times more resistant against DNase I than was DNA free in solution. Khanna and Stotzky [28] reported that the protective effect of clays—montmorillonite (M) and kaolinite (K)—against the activity of nucleases did not eliminate the transforming ability of bound DNA. In order to determine clay–DNA complexes x-ray diffractometry (X-RD) and transmission (TEM) and scanning electron microscopy (SEM) were used. The results figured out possible place of the DNA bound and the adsorption process. SEM images showed that the binding of this DNA was mainly on the edges of M and K, although some binding was also apparent on the planar surfaces.[29] Extension from the edges of the clays enables the unbound end of DNA to interact with receptor sites on competent cells and result in their transformation; and binding on clays alters the electron distribution and/or conformation of DNA, which reduces its hydrolysis by nucleases.

Recently, there has been an increasing interest in methodology of DNA extraction for the synthesis of microchips and biosensors. Silica solid phase is also most commonly and easily adaptable to microdevices. Since standard procedures are time-consuming, microdevices are preferable. Price et al.[52] reviewed DNA microchips. They concluded that DNA purification by microchips is related to the availability of equipment, reagents, and sample, as well as the balance between speed, extraction efficiency, and quality.

Approximately 5×10^{-7} dm^3 bed volume filled silica resin was used in a microchip DNA purification device by Tian et al.[65] Sol–gel-derived silica adsorption was used in DNA purification on microchips under chaotropic conditions and was described in many studies. Breadmore et al.[7] used a tetraethoxyorthosilicate-based sol–gel to immobilize the silica beads in a microdevice, and also showed effective extraction of DNA from only 0.2 μL of whole blood sample. Figure 11.3 shows the microchip channel filled with silica beads which was immobilized with sol–gel method. They improved DNA purification from blood in less than 15 min.

Different oxide forms of iron are used in the studies in biological applications that focus on magnetic bioseparation, biological labeling, and diagnostics contrast enhancement agents for magnetic resonance imaging, tumor hyperthermia, and drug carrier design.[64] Magnetic particles are also used in DNA purification method. Magnetic separation is also a good way to shorten both adsorption and separation steps. In DNA purification method, magnetic silica microspheres are used. They consist of magnetic nanoparticles uniformly dispersed in a silica matrix where there are large amounts of hydroxyl groups on their surfaces. These groups interaction efficiency can be enhanced with silanol, epoxide, diol, and carboxyl groups. Chao et al.[10] tested over 100 ng/μL concentrated calf thymus DNA adsorption behavior onto silica-coated magnetic particle with modified surfaces. As a result, surface-modified magnetic particles are promising materials to DNA purification.

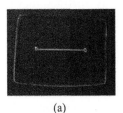 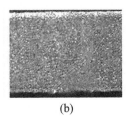 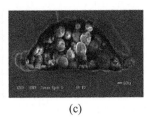

(a)   (b)   (c)

FIGURE 11.3 Microchip packed with silica particles (a) 1× magnification, (b) 10× magnification, (c) cross section of packed channel at 500× magnification. (Reprinted with permission from Breadmore, M. C.; Wolfe, K. A.; Arcibal, I. G.; Leung, W. K.; Dickson, D.; Giordano, B. C.; Power, M. E.; Ferrance, J. P.; Feldman, S. H.; Norris, P. M.; Landers, J. P. Microchip-Based Purification of DNA from Biological Samples. *Anal. Chem.* **2003**, *75*, 1880–1886. © 2016) American Chemical Society.)

Silica is one of the commonly used adsorbents and support materials for scientific research. It is mainly found in soil material.[41] Adsorption of DNA onto soil, variable-charged minerals, humic substance, and organomineral complexes are the most frequently used materials understanding the DNA extraction from soil as model solid surfaces. DNA of various organisms has an advantage to adjust its activity and susceptibility to biodegradation. Saeki et al.[54] also emphasized that the presence of oxide minerals in soil enhance DNA adsorption. Silica is one of them. Higher number of cross-linking sites on the surface of silica enables it higher attachment rate occurred by hydroxyl groups that easily form hydrogen bonding with amine groups.

11.2.1.1 SILICA AEROGELS

Aerogels have been used in a wide range of applications as listed in Table 11.2, such as catalytic supports, thermal insulation in solar window systems, acoustic barriers, supercapacitors, refrigerators, hydrophobic adsorbents for nonpolar compounds, Cerenkov radiation detector media in high energy physics, or inertial confinement fusion targets for thermonuclear fusion reactions.[14,49] Ru et al.[53] studied drying of silica aerogels and xerogels by conventional and CO_2 and ethanol supercritical drying (SCD) methods.

TABLE 11.2   Identification of Aerogel Properties and Features with Their Applications.

Property	Features	Applications
Thermal conductivity	Best insulating solid Transparent High temperature Lightweight	Architectural and appliance insulation, portable coolers, transport Vehicles, pipes, cryogenic, skylights, space vehicles and probes, casting molds
Density/porosity	Lightest synthetic solid Homogeneous High specific surf. area Multiple compositions	Catalysts, sorbers, sensors, fuel storage, ion exchange targets for ICF, X-ray lasers
Optical	Low refractive index solid Transparent Multiple compositions	Cherenkov detectors, lightweight optics, light guides, special effect optics
Acoustic	Lowest sound speed	Impedance matchers for transducers, range finders, speakers
Mechanical	Elastic Lightweight	Energy absorber, hypervelocity particle trap
Electrical	Lowest dielectric constant High dielectric strength High surface area	Dielectrics for ICs, spacers for vacuum electrodes, vacuum display spacers, capacitors

Silica aerogels, which are synthesized through the association of a chemical step, called as "sol–gel" chemistry. Physical step is a particular way of drying the wet gel at supercritical conditions.

11.2.1.1.1   Sol–Gel Process

The sol–gel process is a wet-chemical technique widely used for producing promising materials in scientific areas such as material science and ceramic engineering. Special property of sol–gel method is its capability to convert the molecular precursor to the product. It makes possible to control the process and structural composition. This method is used primarily for the fabrication of materials (typically metal oxides) starting from a colloidal solution (sol). Sol is a colloidal suspension which consists of particles having sizes less than 1000 nm. Thus, the gravitational force become negligible and short-time forces, such as wander walls and surface charge forces, are the main forces acting on the particles. A gel is described as a solid molecular three-dimensional network that included a liquid network of the same size and shape. Chemical step has three general reversible reactions:

hydrolysis–esterification, alcohol condensation–alcoholysis, and water condensation–hydrolysis as indicated in Figure 11.4. In the first reversible reaction, alkoxide groups (OR) are replaced with hydroxyl groups (OH). Then condensation reaction occurs rather than hydrolysis or alcoholysis.

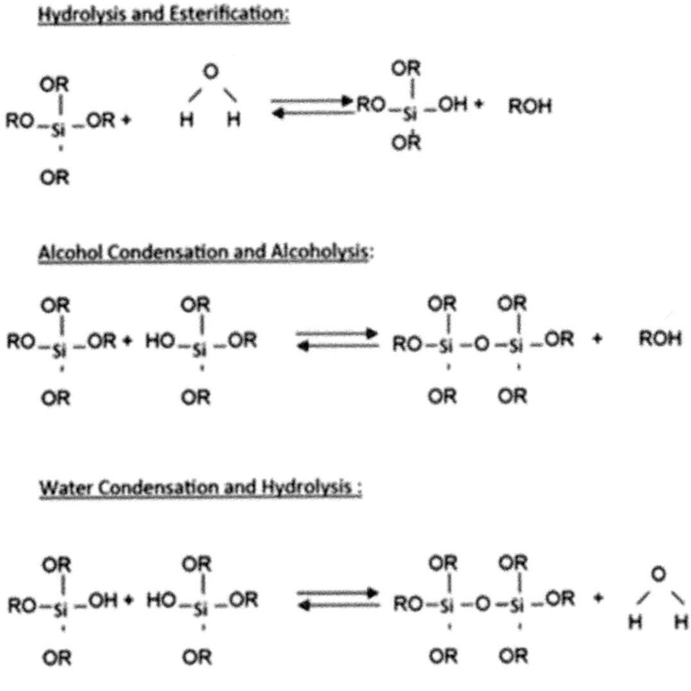

FIGURE 11.4   Hydrolysis and condensation for silicon alkoxides.

These reversible reaction rates define the generally assumed product structures. According to this, relative slow hydrolysis and fast condensation rates are resulted as controlled precipitation. Expected products from different hydrolysis and condensation reaction rates are reported by Legrand.[35] Colloids and sols were obtained if both rates are slow for hydrolysis and condensation. On the other hand, if both reactions are fast, colloidal gels and precipitates are obtained.

Hydrolysis starts with water. However water and alkoxysilanes such as tetraethoxysilane (TEOS) and tetramethoxysilane (TMOS) are partially immiscible. Additional solvent is needed to homogenize the mixture. Therefore, alcohols, acetone, dioxane, and tetrahydrofuran are the solvents added to this mixture.

In the sol–gel process, a sol is first formed by mechanically mixing a liquid alkoxysilane precursor, such as TEOS, deionized water, a solvent, and an acid or alkaline catalyst at ambient conditions. During this step, the alkoxysilane groups are transformed to silanol groups by acid catalysis.

Several important papers have investigated the significance of the chemical parameters such as the nature of the starting alkoxide,[1,13] the hydrolysis ratio (water:alkoxide molar ratio), the cosolvent,[8] and the catalysis conditions that determine the kinetics and mechanisms of the hydrolysis and condensation reactions.[16] Some attention has also been devoted to the influence of the processing factors such as the reaction temperature, the ageing period and conditions, the washing solvent, and the drying conditions.

11.2.1.1.2 Gel and Gelation

A gel is composed of solid and liquid phases which are independent of each other. A wet gel was prepared by two-step process which was developed by Brinker et al..[8] The first step involves acid (hydrochloric) catalysis of the hydrolysis of tetraethyl orthosilicate (TEOS). The second step involves condensation polymerization of the silanol groups resulting from the hydrolysis reaction. The condensation reaction is reserved by the acid added in the first step; thus, the second step includes the addition of ammonium hydroxide to neutralize acid.

11.2.1.1.3 Drying of the Gels

Three main routes are commonly used for drying[5]:

1. Freeze-drying (which needs to avoid the triple point)
2. Evaporation (which implies crossing the liquid–gas equilibrium curve)
3. SCD (which needed to avoid the critical point)

Depending on the drying process, gel is named differently. Drying of the gel is a critical step. A xerogel is the result of that liquid evaporation at ambient temperature and pressure (it often retains the same form as the original gel with up to 90% shrinkage).

An aerogel is the result of the removal of the liquid part without damaging the solid structure (often by supercritical removal). During SCD, especially under high pressure and high temperature, not only drying but also super-critical extraction and the chemical reaction of alcogel component occurs. This drying process, discovered by Kistler (1931),[31] involves removal of the solvent from wet gel.[34] This process preserves the texture of the dry material. In practice, it strongly reduces the pore collapse. SCD is consisted in heating a gel in a device such as autoclave, until the pressure and temperature exceeded the critical temperature T_c and critical pressure P_c of the liquid entrapped in the gel pores. Aerogel shows very high surface area and important nanoporosity. This is provided by the absence of capillary forces. The surface forces attracting the liquid and solid phases were not present, since only the gas phase and the solid existed above critical temperature. Shrinkage of the gels during drying is driven by the capillary pressure, P, which can be represented by eq 11.1. Capillary force can also be measured by contact angle as represented in eq 11.2.

$$P = \frac{\gamma_{lv}}{(r_p - \delta)} \qquad (11.1)$$

$$P = \frac{2\gamma x Cos\theta}{r_p}, \qquad (11.2)$$

where γ_{lv} is the surface tension of the pore liquid, r_p is the pore radius, which can be represented by eq 11.3, and δ is the thickness of a surface adsorbed layer.[8]

$$r_p = \frac{2V_p}{S_p}, \qquad (11.3)$$

where V_p and S_p are pore volume and surface area, respectively. When liquid evaporates from the pores of a gel, a concave (meniscus) forms. At the moment that the liquid/vapor meniscus enters a pore, the walls will be covered with a film of adsorbed liquid and then capillary pressure will increase according to eq 11.1 attracting the walls of the capillary closer to each other.

By transforming the solvent into a supercritical fluid (SCF), the surface tension disappears along with the capillary pressure gradient built up in the pore walls, avoiding the potential collapse of the pore volume due to the capillary force. SCD in organic solvents, which are usually alcohols, leaves the pores without damaging and the resulting materials are generally hydrophobic since their surfaces are covered with alkoxy groups.[49]

It is important that during SCD, mass transport and heat transfer occur simultaneously. Therefore, homogenizer solvent should be selected carefully. The most commonly used SCFs and their properties are given in Table 11.3. SCFs are highly compressed gases which combine properties of gases and liquids in same conditions. Figure 11.5 is the phase diagram of a pure substance in a two coordinate system: pressure (P) and temperature (T). For every substance, there is a value of T and P where liquid phase and gaseous phase have the same density. They are the coordinates of the so called critical point, which ends the liquid–vapor coexistence curve. The area corresponding to temperature and pressure beyond the critical point coordinated is the supercritical region. The fluid in this region is called as a SCF. Subra and Justin have reported that SCF can transfer heat and mass better than gases do.

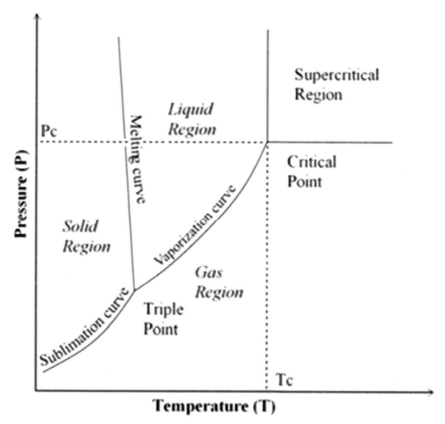

FIGURE 11.5   P-T plane phase diagram of a pure substance.

TABLE 11.3 Properties of Some Supercritical Fluids.

Fluid	Formula	Critical constant	
		Pressure (MPa)	Temperature (°C)
Carbon dioxide	CO_2	7.36	31.1
Freon 116	CF_3CF_3	2.97	19.7
Ethanol	C_2H_5OH	6.36	243
Methanol	CH_3OH	7.93	240
Nitrous oxide	N_2O	7.24	36.4
Acetone	$(CH_3)_2O$	4.66	235
Water	H_2O	22	374

(Reprinted with permission from Pierre, A. C.; Pajonk, G. M. Chemistry of Aerogels and Their Applications. Chem. Rev. **2002,** *102,* 4243–4265. © 2002 American Chemical Society.)

Adsorption of calf thymus DNA to a silica aerogel, a mesoporous silica gel, and a silica wafer was investigated by Yetgin and Balkose (2015).[71] Silica aerogel was synthesized from TEOS by supercritical ethanol drying process. The DNA adsorption capacity of silica aerogel was nearly two times of that of the mesoporous silica gel due to its macroporous structure and its higher silanol content. Silica aerogel was found to be a very promising material for DNA adsorption.

11.2.2 ALUMINA

Alumina, also called aluminum oxide, is the only solid oxide form of aluminum. It has the chemical formula Al_2O_3. It is commonly referred to as alumina (α-alumina) or corundum which is the naturally occurring alumina. Alumina materials have been technologically significant ceramic materials throughout human history. Type of α-alumina has hexagonal close-packed structure while γ-alumina type signifies cubic close-packed crystalline form. In a hexagonal unit cell, the sides that provide depth and width to the cell are equal, and are at an angle of 120° to each other. c is at angles of 90° to sides "a" and "b." The hexagonal parameters for α-Al_2O_3 are $c = 1.297$ nm and $a = 0.475$ nm, with $c/a = 2.73$.[36]

α-phase alumina, also called corundum, is the only thermodynamically stable oxide of aluminum and is the final product of the calcination process. Calcination order can be defined as follows: gibbsite → boehmite (γ-AlOOH) → γ-alumina (γ-Al_2O_3) → δ-alumina (δ-Al_2O_3) → θ-alumina (θ-Al_2O_3) → α-alumina. Metastable Al_2O_3 phases transform toward stable α-alumina by calcination procedure.

Zeta potential measurement has identified that the α-alumina has isoelectric point at nearly pH 9.1.[51]

Alumina has a melting point nearly 2000°C. Therefore, it is used as insulator or refractory material. It is used in orthopedic and dental implants surgery.[24] Furthermore, it is used as catalytic support material in reaction engineering.

Alumina is also classified as nonbioactive material.[44] It has been used in orthopedic surgery because of its high biocompatibility.[27] Its adsorption capacity of DNA was studied and already published by Chattoraj and Upadhyay in 1968 and reported as a review by Chattoraj and Mitra.[11] Alumina has been also tested on DNA and RNA purification. DNA and RNA were also denaturized with heat and acid presence. Adsorption isotherm was found to fit Langmuir equation in linear form. Therefore, authors concluded that denaturized DNA and RNA were adsorbed by alumina more than their native form. Alumina adsorbed 0.25 and 0.75 mg amount of DNA and RNA, respectively at pH 6.5 with maximum 10 mg/cm^3 concentration.

11.2.3 HYDROXYAPATITE

HAP's chemical formula is $Ca_{10}(PO_4)_6(OH)_2$ which is the main mineral element of teeth and bones and belongs to calcium phosphates family. Different techniques are used for HAP powder production. HAP's morphology, stoichiometry, and level of crystallinity change with preparation technique, which define biomaterial-phase thermal stability, mechanical stability, and dissolution behavior. HAP has lattice parameters $a = b = 0.934$ nm, $c = 0.687$ nm, $\alpha = \beta = 90°$, $\gamma = 120°$, and hexagonal crystal structure with the space group P63/m.[48] Ca/P ratio is an important ratio which defines properties of calcium phosphate mineral such as acidity and solubility.[66] Lower Ca/P ratio represents larger acidity and solubility of the mixture, for example, Ca/P < 1. These properties change while increasing the Ca/P ratio to the value of 1.67. The Ca/P ratios of tetracalcium phosphate $(Ca_4P_2O_9)$, HAP $(Ca_{10}(PO_4)_6(OH)_2)$, α-tricalcium phosphate $(Ca_3(PO_4)_2)$, and β-tricalcium phosphate $(Ca_3(PO_4)_2$ are 2, 1.67, 1.5, and 1.5, respectively.[66]

HAP has been used clinically for many years. HAP has good biocompatibility in bone contact as its chemical composition is similar to that of bone material. HAP shows excellent biocompatibility with hard tissues and also with skin and muscle tissues.[23] Synthetic HAP is a very important biomaterial used for several applications in medicine as a bulk ceramic, implant coating materials, ceramic coating, or as one of the components of composites.

Porous HAP was used for cell loading. Ohgushi and Caplan [44] applied HAP for cell loading, drug carrier for controlled drug release based on adsorption/desorption properties, [30] most extensively for hard tissue scaffolds, affinity chromatography analysis for DNA binding for protein purification, [19] and the separation of ssDNA and dsDNA by chromatography [12] or both of them such as plasmid DNA and protein adsorption. [59] This can be explained by positively charged pairs of calcium ions and six negatively charged oxygen atoms which are associated with triplets of crystalline phosphates placed onto the surface. It is possible that amino and guanidinyl groups of proteins can attach to phosphate ions of HAP, and DNA's phosphate part can attach to calcium ions.

HAP has osteogenic cell proliferation capacity onto implant surface, therefore should be investigated of surface interaction on coated material or smooth surface. Therefore, coating is main usage area of HAP. In practice, HAP surface can be formed on either HAP thin film or HAP pellet forms. [62] Sol–gel deposition technique is a well-known film formation procedure. Sol or particulate sol can be used. The final form can be obtained by dip or spin coating. In dipping process, substrate such as inert glass film contacts with the dip solution for a few seconds. HAP pellet can be prepared from powder form. For instance, Şimsek [62] compacted HAP powder in a stainless steel die with 1-cm internal diameter under 160 MPa pressure. All pellets were sintered different temperatures between 800°C and 1300°C for 2 h. The sol–gel approach affords conditions for the synthesis of HAP films. HAP film can also be prepared directly from sol or indirectly from particulate sol solutions. Ozcan and Çiftçioğlu prepared particulate sol route HAP thin film. [45] Yetgin [70] prepared HAP pellets by sintering and examined the AFM image of calf thymus DNA which was placed on its surface by dropping and drying of a DNA aqueous solution.

11.3 MATERIALS AND THE METHOD

11.3.1 MATERIALS

Silica (SiO_2), alumina (Al_2O_3), and HAP [$Ca_{10}(PO_4)_6(OH)_2$] were examined as adsorbents for calf thymus DNA. Silica (Sigma-Aldrich-Silicagel, Grade 7744 pore size 60 Å, 70–230 mesh), alumina from Seydişehir, Turkey, and HAP supplied by Sigma-Aldrich (23093-6) were characterized for this purpose.

11.3.2   CHARACTERIZATION OF SILICA, ALUMINA, AND HAP

The characterization of the adsorbents included the determination of the functional groups, pore size, surface area, and chemical composition. The adsorbents used in this study were characterized by x-ray fluorescence (XRF) spectroscopy, Fourier transform infrared (FTIR) spectroscopy, x-ray diffraction (XRD), and physical adsorption of nitrogen.

Elemental analyses were made by XRF method. XRF analyses were carried out by using Spectro IQ II on the opaque surfaces of the pellets prepared by heating the mixture of lithium tetraborate and sample at 1100°C and cooling to room temperature.

FTIR spectroscopy is used to obtain information about functional groups present in a material using KBr disc method. Shimadzu 8601 FTIR spectro-photometer was used to obtain the spectra.

Philips X'Pert Pro diffractometer was used to investigate phase structure of the adsorbents and crystalline form of the component present in the adsorbent samples. The operating conditions were 45 kV and 40 mA, Cu K$_\alpha$ radiation, $\lambda = 0.154$ nm. The registrations were performed in the range over 2θ values of 5–70° with a scan speed of 0.06°/s. The data were collected with X'Pert data collector. The collected data were analyzed by X'Pert Graphics & Identify software.

Specific surface area, pore diameter, and pore volume of the adsorbents were obtained by using the physical adsorption of nitrogen technique at 77 K, using Micromeritics ASAP 2010.

11.4   RESULTS AND DISCUSSION

The elemental composition of silica, alumina, and HAP which was determined by XRF analysis is given in Table 11.4. While the silica gel contained 80.77% silica, alumina contained 82.38% alumina. Both samples contained free and bound water that was lost during heating for sample preparation. HAP's theoretical Ca/P value of 1.67 was obtained in freshly prepared sample, but during heating at 1000°C for sample preparation for XRF analysis, volatile components such as water and phosphoric acid were removed decreasing P content. Thus the molar ratio of Ca/P was found as 2.0 from the analysis indicating HAP was transformed to tetracalcium phosphate during sample preparation for XRF analysis.

TABLE 11.4 Composition (Mass, %) of Representative Adsorbents by XRF.

	Na	Mg	Al	Si	P	Ca	Fe	Cu
Silica	0.02	0.55	0.46	37.75	0.02	0.13	0.09	0.28
Alumina	1.11	1.16	43.61	0.001	0.01	0.06	0.09	0.25
HAP	0.21	0.27	0.66	0.68	9.84	26.69	0.08	0.23

Table 11.5 shows the main vibration bands of silica related to FTIR spectrum. The peak at 3400 cm^{-1} is related to the hydrogen-bonded v O–H mode of residual silanol (Si–OH) groups and of adsorbed water.[15] Low-intensity peak at 1630 cm^{-1} is related to bending vibration of water.[14] The weak band near 970 cm^{-1} was assigned to the oscillating oxygen atoms in the silica network, including silanol groups and broken Si–O–Si bridges.[14] The peak at 1080 cm^{-1} wavenumber is related to Si–O–Si bond tensile strain.

TABLE 11.5 Assignments of the Main Infrared Bands in the FTIR Spectrum of Silica Gel.

Wavenumber (cm^{-1})	Assignment	Reference
~460	ρSi–O–Si	Fidalgo and Ilharco[15]
800	v_aSi–O–Si	Fidalgo and Ilharco[15]
950	Si–OH stretching vibration	Fidalgo and Ilharco[15]
~1080	Asymmetric stretching vibration of the Si–O bond	Fidalgo and Ilharco[15] Fidalgo and Ilharco[17]
1600	Bending vibration of H$_2$O δH–O–H mode	Balkose et al.[4] Fidalgo and Ilharco[17]
3400	Hydrogen bonded –OH stretching	Balkose et al.[4]

However, it is reported that perfect silica surface such as quartz has no free surface hydroxyl groups.[65] On the other hand, acid treatment of silica hydrolyzes the surface, and increases the concentration of surface silanol groups.[41] In Figure 11.6, FTIR spectrum of silica gel is seen. It clearly shows that the silica gel contains OH groups, H$_2$O, and SiO$_2$.

The XRD diagram of silica gel shown in Figure 11.7 has no crystalline diffraction peaks indicating that it was totally amorphous and not crystalline.

The FTIR spectrum of alumina shown in Figure 11.8 had peaks related with water (at 1600 cm^{-1}) and hydrogen-bonded OH groups (3464 cm^{-1}) and Al–O stretching vibrations (1014 cm^{-1}).

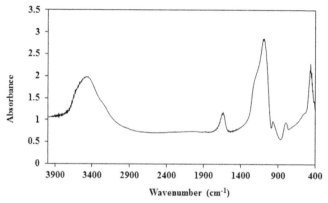

FIGURE 11.6   FTIR spectrum of silica gel.

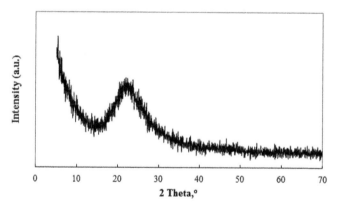

FIGURE 11.7   X-ray diffraction diagram of silica gel.

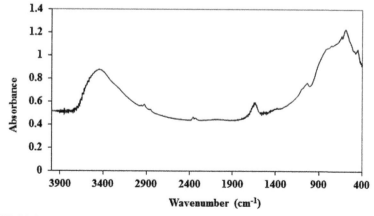

FIGURE 11.8   FTIR spectrum of Alumina from Seydişehir, Turkey.

The XRD diagram of the alumina from Seydişehir, Turkey is seen in Figure 11.9. X'Pert Graphics & Identify software of XRD pattern and obtained data for α-alumina were listed in Table 11.6. The observed peaks and the peaks of α-alumina had the same 2θ values.

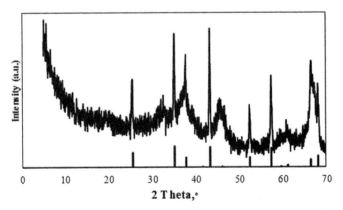

FIGURE 11.9   XRD pattern of Alumina from Seydişehir, Turkey (Cu K$_\alpha$ radiation was used).

TABLE 11.6   The Observed and the X-ray Diffraction Peaks of α-Alumina.

Observed 2θ, °	JCPDS # (75-1865) 2θ, °
25.6	25.546
35.14	35.105
37.74	37.734
41.65	41.619
43.36	43.302
52.64	52.486
57.45	57.423
59.73	59.670
61.35	61.056
66.40	66.434
68.19	68.126
76.70	70.756

FTIR spectrum of HAP is shown in Figure 11.10. Bands at 3400 and 1600 cm^{-1} belong to hydrogen-bonded OH stretching vibrations and bending vibrations of water, respectively. The bands at 1092 and about 1040 cm^{-1} were assigned to the components of the triply degenerated asymmetric P–O stretching mode.[60] The bands at 601 and 571 cm^{-1} are related to components of the triply degenerate O–P–O bending mode and the bands in the range 462–474 cm^{-1} are also connected to the components of the doubly degenerate O–P–O bending mode.[17]

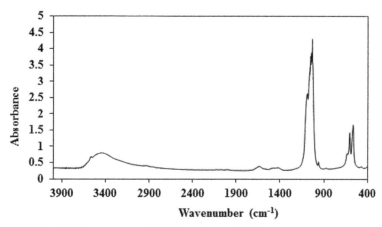

FIGURE 11.10   FTIR spectrum of HAP from Sigma Aldrich.

The XRD diagram of HAP is shown in Figure 11.11. The observed peaks were analyzed by X'Pert Graphics & Identify software and the obtained data are listed in Table 11.7. Joint Committee on Powder Diffraction Standards (JCPDS) numbers which are known to identity of the x-ray powder diffraction patterns were also obtained by this software. The HAP sample had all the peaks related to HAP which was reported in JCPDS 09-0432.

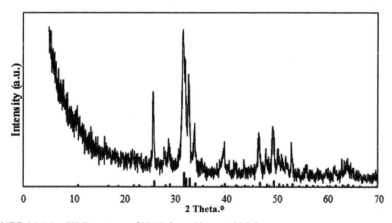

FIGURE 11.11   XRD pattern of HAP from Sigma Aldrich.

TABLE 11.7 The Observed and Expected X-ray Diffraction Peaks of HAP. (Source: JCPDS number 09-0432 by Wu et al.[69])

Observed 2θ	JCPDS # (09-0432) 2θ
10.79	10.820
16.90	16.741
21.70	21.819
22.77	22.902
25.77	25.879
28.05	28.126
28.79	28.966
31.64	31.773
32.05	32.196
32.78	32.902
39.05	39.204
40.27	40.452
41.82	42.029
46.54	46.711
49.30	49.467
50.29	50.490
63.15	63.011
63.88	64.078

11.4.1 SURFACE CHARACTERISTICS OF SILICA, ALUMINA, AND HAP

The surface characteristics of the adsorbents can be determined by nitrogen gas adsorption at 77 K. In Figures 11.12–11.14, the nitrogen gas adsorption isotherms of silica, alumina, and HAP are seen. Silica, alumina, and HAP isotherms were all of the Type IV of the UIPAC classification. Hysteresis loop is formed by adsorption/desorption isotherms due to condensation and evaporation of liquid nitrogen from mesopores and depends upon the shape and size of pores.[22] Hysteresis cycle was commonly observed for materials with interrelated pore networks with different size and shape. The isotherms of silica, alumina, and HAP have hysteresis loops, which are consistent with the presence of mesopores. Relative pressure of (P/P_0) greater than 0.6 is associated with capillary condensation in mesopores, which is characteristic of type IV isotherms.[53] Single-point, BET, and Langmuir surface areas, average pore diameter, total pore volume, and micropore volume of the adsorbents are listed in Table 11.8. Silica had the higher surface area, total pore volume, micropore volume, and pore diameter than alumina and HAP.

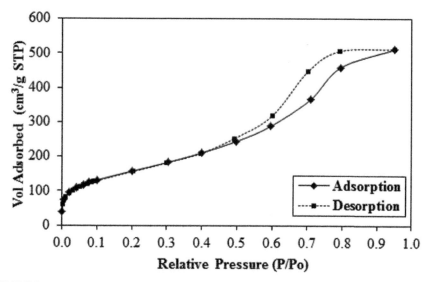

FIGURE 11.12   Nitrogen adsorption and desorption isotherms of Silica at 77 K.

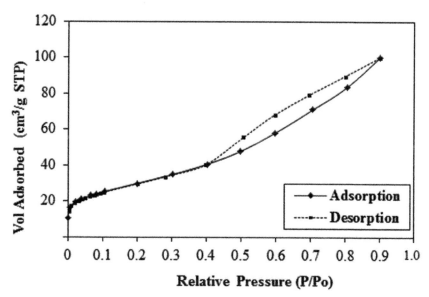

FIGURE 11.13   Nitrogen adsorption and desorption isotherms of Alumina at 77 K.

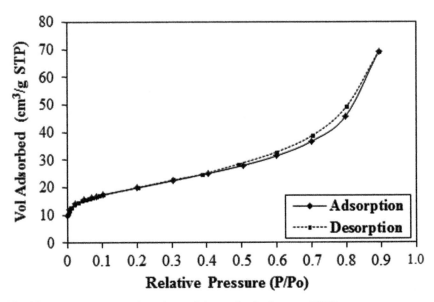

FIGURE 11.14   Nitrogen adsorption and desorption isotherms of HAP.

TABLE 11.8   Surface Characteristics of the Adsorbents.

Properties	Silica	Alumina	HAP
Single point surface area (m^2/g)	556	106	68
BET surface area (m^2/g)	571	108	69
Langmuir surface area (m^2/g)	787	148	96
Average pore diameter (nm) (4V/A by BET)	5.5	5.7	6.2
Single point total pore volume (cm^3/g)	0.79	0.16	0.11
Max micropore volume (cm^3/g)	0.20	0.04	0.03

11.5 CONCLUSIONS

Adsorption of DNA was made for DNA purification, biosensor development, and gene transfer purposes. DNA can easily be adsorbed to positively
charged surfaces as well as to negatively charged surfaces by electrostatic
bridges with the water of hydration of charged cations. DNA adsorption
to alumina, clay, silica, SWNT, metals, organic and bioorganic molecules,
polypyrrole (PPy)–silica nanocomposite particles, polymers and hydrogels,
poly-L-lysine-immobilized pHEMA, and HAP are reviewed in this chapter.
The preparation and characterization of silica, alumina, and HAP were

reviewed in detail. Experimental results are reported for the silica, alumina, and HAP samples which were used in calf thymus DNA adsorption in our further studies. OH functional groups that are potential bonding sites for DNA were present in all three adsorbents. The silica gel sample had highest surface area, total pore volume and micropore volume, and the pore diameter among the three adsorbents under investigation. The functional groups and the pore structure and surface charge are all the important parameters for DNA adsorption.

KEYWORDS

- adsorbents
- DNA purification
- silica
- gelation
- polypyrrole

REFERENCES

1. Alie, C.; Pirard, R.; Pirard, J. P. The Role of the Main Silica Precursor and the Additive in the Preparation of Low-density Xerogels. *J. Non-Cryst. Solids* **2002**, *311*, 304–313.
2. Allahverdyan, A. E.; Gevorkian, Z. S.; Hu, C. K.; Nieuwenhuizen, T. M. How Adsorption Influences DNA Denaturation. *Phys. Rev. E* **2009**, *79*, 031903.
3. Aslanoglu, M.; Houlton, A., Horrocks, B. R. Functionalised Monolayer for Nucleic Acid Immobilisation on Gold Surfaces and Metal Complex Binding Studies. *Analyst* **1998**, *123*, 753–757.
4. Balköse, D.; Alp, B.; Ülkü, S. Water Vapour Adsorption on DNA. *J. Therm. Anal. Calorim.* **2008**, *94*, 695–698.
5. Bisson, A.; Rigacci, A.; Lecomte, D.; Rodier, E.; Achard, P. Drying of Silica Gels to Obtain Aerogels: Phenomenology and Basic Techniques. *Prog. Dry. Technol.* **2003**, *21*, 593–628.
6. Boom, R.; Sol, C. J. A.; Salimans, M. M. M.; Jansen, C. L.; Wertheim-van Dillen, P. M. E.; van der Noordaa, J. Rapid and Simple Method for Purification of Nucleic Acids. *J. Clin. Microbiol.* **1990**, *28*, 495–503.
7. Breadmore, M. C.; Wolfe, K. A.; Arcibal, I. G.; Leung, W. K.; Dickson, D.; Giordano, B. C.; Power, M. E.; Ferrance, J. P.; Feldman, S. H.; Norris, P. M.; Landers, J. P. Microchip-Based Purification of DNA from Biological Samples. *Anal. Chem.* **2003**, *75*, 1880–1886.

8. Brinker, C. J.; Scherer, G.W. *Sol–Gel Science: The Physics and Chemistry of Sol–Gel Processing*; Academic Press: Boston, 1990.
9. Buffone, G. J.; Demmter, G. J.; Schimbor, C. M.; Greer, J. Improved Amplification of Cytomegalovirus DNA from Urine after Purification of DNA with Glass Beads. *Clin. Chem.* **1991**, *37*(11), 1945.
10. Chao, Z. Z.; Cui, Y.; Hong, W. Q. Surface Modification of Magnetic Silica Microspheres and its Application to the Isolation of Plant Genomic Nucleic Acids. *Chin. J. Anal. Chem.* **2007**, *35*(1), 31–36.
11. Chattoraj, D. K.; Mitra, A. Adsorption of DNA at Solid–water Interfaces and DNA–surfactant Binding Interaction in Aqueous Media. *Curr. Sci.* **2009**, *97*(10), 1430–1438.
12. Chen, W. Y.; Lin, S. M.; Lin, P. S.; Tasi, S. P.; Chang, Y.; Yamamoto, S. Studies of the Interaction Mechanism Between Single Strand and Double-strand DNA with Hydroxyapatite by Microcalorimetry and Isotherm Measurements. *Colloids Surf. A Physicochem. Eng. Aspects* **2007**, *295*, 274–283.
13. Dorcheh, A. S.; Abbasi, M. H. Silica Aerogel; Synthesis, Properties and Characterization. *J. Mater. Process. Technol.* **2008**, *199*, 10–26.
14. Estella, J.; Echeverria, J. C.; Julian, M. L.; Garrido, J. Effect of Supercritical Drying Conditions in Ethanol on the Structural and Textural Properties of Silica Aerogels. *J. Porous Mater.* **2008**, *15*, 705–713.
15. Fidalgo, A.; Ilharco, L. M. The Defect Structure of Sol–Gel Derived Silica/Polytetrahydrofuran Hybrid Films by FTIR. *J. Non-Cryst. Solids* **2001**, *283*, 144.
16. Fidalgo, A.; Rosa. M.E.; Ilharco M. Chemical Control of Highly Porous Silica Xerogels: Physical Properties and Morphology. *Chem. Mater.* **2003**, 15, 2186-2192
17. Fidalgo, A.; Farinha, J. P. S.; Martinho, J. M. G.; Rosa, M. E.; Ilharco, L. M. The Influence of The Wet Gels Processing on The Structure and Properties of Silica Xerogels. *Microporous Mesoporous Mater.* **2005**, *84*, 229–235.
18. Fowler, B. O. Infrared Studies of Apatites. I. Vibrational Assignments for Calcium, Strontium, and Barium Hydroxyapatites Utilizing Isotopic Substitution. *Inorg. Chem.* **1974**, *13*, 194.
19. Franchi, M.; Ferris, J.P.; Gallori, E. Cations as Mediators of The Adsorption of Nucleic Acids on Clay Surfaces in Prebiotic Environments. *Origins of Life and Evolution of the Biosphere*, **2003**, 33, 1–16
20. Fujiwara, M.; Yamamoto, F.; Okamoto, K.; Shiokawa, K.; Nomura, R. Adsorption of Duplex DNA on Mesoporous Silicas: Possibility of Inclusion of DNA into Their Mesopores. *Anal. Chem.* **2005**, *77*, 8138–8145.
21. Gadgil, H.; Oak, S. A.; Jarrett, H. W. Affinity Purification of DNA-Binding Proteins. *J. Biochem. Biophys. Methods* **2001**, *49*, 607–624.
22. Gregg, S. J.; Sing, K. S. W. *Adsorption, Surface Area and Porosity*; Academic Press: London, 1982.
23. Hench, L. L. Bioceramics: From Concept to Clinic. *J. Am. Ceram. Soc.* **1991**, *74*, 1487–1510.
24. Hench, L. L. Bioceramics. *J. Am. Ceram. Soc.* **1998**, *81*, 1705–1728.
25. Hodrien, A. J.; Waigh, T. A.; Voice, A. M.; Blair, G. E.; Clarke, S. M. Adsorption of DNA onto Positively Charged Amidine Colloidal Spheres and the Resultant Bridging Interaction. *Int. J. Biol. Macromol.* **2007**, *41*, 146–153.
26. Jiang, T.; Chang, J.; Wang, C.; Ding, Z.; Chen, J.; Zhang, J.; Kang, E. T. Adsorption of Plasmid DNA onto N,N'-(dimethylamino)ethyl-methacrylate Graft-polymerized

Poly-L-lactic Acid Film Surface for Promotion of In-situ Gene Delivery. Biomacromolecules **2007**, *8*, 1951–1957.

27. Karlsson, M.; Palsg, E.; Wilshaw, P. R.; Silvio, L. D. Initial in Vitro Interaction of Osteoblasts with Nano-Porous Alumina. *Biomaterials* **2003**, *24*, 3039–3046.

28. Khanna, M.; Stotzky, G. Transformation of *Bacillus subtilis* by DNA Bound on Montmorillonite and Effect of DNase on the Transforming Ability of Bound DNA. *Appl. Environ. Microbiol.* **1992**, *58*, 1930–1939.

29. Khanna, M.; Yoder, M.; Calamai, L.; Stotzky, G. X-ray Diffractometry and Electron Microscopy of DNA from *Bacillus subtilis* Bound on Clay Minerals. *Sci. Soils* **1998**, *3*, 1–10.

30. Kim, J.; Mauk, M.; Chen, D.; Qiu, X.; Kim, J.; Galeb, B.; Baua, H. H. A PCR Reactor with an Integrated Alumina Membrane for Nucleic Acid Isolation. *Analyst* **2010**, *135*, 2408–2414.

31. Kistler, S. S. Coherent Expanded Aerogels and Jellies. *Nature* **1931**, *127*, 741.

32. Komlev, V. S.; Barinov, S. M.; Koplik, E. V. Method To Fabricate Porous Spherical Hydroxyapatite Granules Intended For Time-controlled Drug Release. *Biomaterials* **2002**, *23*, 3449.

33. Kumar, M. A.; Jung, S.; Ji, T. Protein Biosensors Based on Polymer Nanowires, Carbon Nanotubes and Zinc Oxide Nanorods. *Sensors* **2011**, *11*, 5087–5111.

34. Land, V. D.; Harris, T. M.; Teeters, D. C. Processing of Low Density Silica Gel by Critical Point Drying or Ambient Pressure Drying. *J. Non-Cryst. Solids* **2001**, *283*(1–3), 11–17.

35. Legrand, A. P. (Ed.). *The Surface Properties of Silicas*; Wiley: London, 1998.

36. Levin, I.; Brandon D. Metastable Alumina Polymorphs: Crystal Structures and Transition Sequences. *J. Am. Ceram. Soc.* **1998**, *81*(8), 1995–2012.

37. Lorenz, M. G.; Wackermagel, W. Adsorption of DNA to Sand and Variable Degradation Rates of Adsorbed DNA. *Appl. Environ. Microbiol.* **1987**, *53*, 2948–2952.

38. Lorenz, M. G.; Wackermagel, W. Bacterial Gene Transfer by Natural Genetic Transformation in the Environment. *Microbiol. Rev.* **1994**, *58*(3), 563–602.

39. Mao, Y.; Daniel, L. N.; Whittaker, N.; Saffiotti, U. DNA Binding to Crystalline Silica Characterized by Fourier-Transform Infrared Spectroscopy. *Environ. Health Perspect.* **1994**, *102*, 165–171.

40. Melzak, K. A.; Sherwood, C. S.; Turner, R. B. F.; Haynes, C. A. Driving Forces for DNA Adsorption to Silica in Perchlorate Solutions. *J. Colloid Interface Sci.* **1996**, *181*, 635.

41. Nawrocki, J. The Silanol Group and its Role in Liquid Chromatography. *J. Chromatogr. A* **1997**, *779*(1–2), 29–71.

42. Nguyen, T. H.; Elimelech, M. Adsorption of Plasmid DNA to a Natural Organic Matter-Coated Silica Surface: Kinetics, Conformation, and Reversibility. *Biomacromolecules* **2007**, *8*, 24–32.

43. Ogram, A.; Sayler, G. S.; Gustin, D.; Lewis, R. J. DNA Adsorption to Soils and Sediments. *Environ. Sci. Technol.* **1988**, *22*, 982–984.

44. Ohgushi, H.; Caplan, A. I. Stem Cell Technology and Bioceramics: From Cell to Gene Engineering. *J. Biomed. Mater. Res.* **1999**, *48*, 913.

45. Ozcan, S.; Çiftçioğlu, M. Particulate Sol Route Hydroxyapatite Thin Film–Silk Protein Interface Interactions. *G. U. J. Sci.* **2010**, *23*(4), 475–485.

46. Paget, E.; Monroziera, L. J.; Simonet, P. Adsorption of DNA on Clay Minerals: Protection Against DNaseI and Influence on Gene Transfer. *FEMS Microbiol. Lett.* **1992**, *97*, 31–39.

47. Paget, E.; Simonet, P. On The Track of Natural Transformation in Soil. *FEMS Microbiol. Ecol.* **1994**, *15*, 109–118.

48. Peroos, S.; Du, Z.; de Leeuw, N. H. A Computer Modeling Study of the Uptake, Structure and Distribution of Carbonate Defects in Hydroxyapatite. *Biomaterials* **2006**, *27*, 2150–2161.

49. Pierre, A. C.; Pajonk, G. M. Chemistry of Aerogels and Their Applications. *Chem. Rev.* **2002**, *102*, 4243–4265.

50. Pietramellara, G.; Franchi, M.; Gallori, E.; Nannipieri, P. Effect of Molecular Characteristics of DNA on its Adsorption and Binding on Homoionic Montmorillonite and Kaolinite. *Biol. Fertil. Soils* **2001**, *33*, 402–409.

51. Polat, M.; Sato, K.; Nagaoka, T.; Watari, K. Effect of pH and Hydration on The Normal and Lateral Interaction Forces Between Alumina Surfaces. *J. Colloid Interface Sci.* **2006**, *304*, 378–387.

52. Price, C. W.; Leslie, D. C.; Landers, J. P. Nucleic Acid Extraction Techniques and Application to the Microchip. *Lab Chip* **2009**, *9*, 2484–2494.

53. Ru, Y.; Guoqiang, L.; Min, L. Analysis of The Effect of Drying Conditions on the Structural and Surface Heterogeneity of Silica Aerogels and Xerogel by Using Cryogenic Nitrogen Adsorption Characterization. *Microporous Mesoporous Mater.* **2010**, *129*, 1–10.

54. Saeki, K.; Morisaki, M.; Sakai, M. The Contribution of Soil Constituents to Absorption of Extracellular DNA by Soils. *Microbes Environ.* **2008**, *23*(4), 353–355.

55. Saeki, K.; Sakai, M. The Influence of Soil Organic Matter on DNA Adsorption on Andosols. *Microbes Environ.* **2009**, *24*(2), 175–179.

56. Saeki, K.; Morisaki, M.; Sakai, M.; Wada, S. I. DNA Adsorption on Synthetic and Natural Allophanes. *Appl. Clay Sci.* **2010**, *50*, 493–497.

57. Saoudi, B.; Jammul, N.; Abel, M. L.; Chehimi, M. M.; Dodin, G. DNA Adsorption onto Conducting Polypyrrole. *Synth. Metals* **1997**, *87*, 97–103.

58. Saoudi, B.; Jammul, N.; Chehimi, M. M.; McCarthy, G. P.; Armes, S. P. Adsorption of DNA onto Polypyrrole–Silica Nanocomposites. *J. Colloid Interface Sci.* **1997**, *192*, 269–273.

59. Schmoeger, E.; Paril, C.; Tscheliessnig, R.; Jungbauer, A. Research Article Adsorption of Plasmid DNA on Ceramic Hydroxyapatite Chromatographic Materials. *J. Sep. Sci.* **2010**, *33*, 3125–3136.

60. Slosarczyk, A.; Paluszkiewiczb, C.; Gawlicki, M.; Paszkiewicf, Z. The FTIR Spectroscopy and QXRD Studies of Calcium Phosphate Based Materials Produced from the Powder Precursors with Different Ca/P Ratios. *Ceram. Int.* **1996**, *23*, 297–304.

61. Şenel, S.; Bayramoglu, G.; Arıca, M. Y. DNA Adsorption on a Poly-L-lysine-Immobilized Poly(2-hydroxyethyl methacrylate) Membrane. *Polym. Int.* **2003**, *52*, 1169–1174.

62. Şimşek, D. Preparation and Characterization of HA Powders—Dense and Porous HA Based Composite Materials. Master Thesis Izmir Institute of Technology, Materials Science and Engineering, Izmir, 2002.

63. Solberg, S. M.; Landry, C. C. Adsorption of DNA into Mesoporous Silica. *J. Phys. Chem. B* **2006**, *110*, 15261–15268.

64. Souza, K. C.; Mohallem, N. D. S.; Sousa, E. B. M. Mesoporous Silica-Magnetite Nanocomposite: Facile Synthesis Route for Application in Hyperthermia. *J. Sol–Gel Sci. Technol.* **2010**, *53*, 418–427.

65. Tian, H.; Huhmer, A. F. R.; Landers, J. P. Evaluation of Silica Resins for Direct and Efficient Extraction of DNA from Complex Biological Matrices in a Miniaturized Format. *Anal. Biochem.* **2000**, *283*, 175–191.

66. Vallet-Regi, M.; Gonzalez-Calbet, J. M. Calcium Phosphates as Substitution of Bone Tissues. *Prog. Solid State Chem.* **2004,** *32*(1–2), 1–31.
67. Vandeventer, P. E.; Lin, J. S.; Zwang, T. J.; Nadim, A.; Johal, M. S.; Niemz, A. Multiphasic DNA Adsorption to Silica Surfaces Under Varying Buffer, pH, and Ionic Strength Conditions. *J. Phys. Chem. B* **2012,** *116*(19), 5661–5670.
68. Vogelstein, B.; Gillespie, D. Preparative and Analytical Purification of DNA from Agarose. *Proc. Natl. Acad. Sci. U. S. A.* **1979,** *76*, 615–619.
69. Wu, C. H.; Wang, T. W.; Sun, J. S.; Wang, W. H.; Lin, F. H. A Novel Biomagnetic Nanoparticle Based on Hydroxyapatite. *Nanotechnology* **2007,** *18*, 165601.
70. Yetgin, S. DNA Adsorption on Silica, Alumina and Hydroxyapatite and Imaging of DNA by Atomic Force Microscopy. PhD Dissertation, Izmir Institute of Technology, Izmir, 2013.
71. Yetgin, S.; Balkose, D. Calf Thymus DNA Characterization and its Adsorption on Different Silica Surfaces. *RSC Adv.* **2015,** *5*, 57950–57959.
72. Zhao, X.; Johnson, J. Karl Simulation of Adsorption of DNA on Carbon Nanotubes. *J. Am. Chem. Soc.* **2007,** *129*, 10438–10445.

CHAPTER 12

ADSORPTION OF DNA ON SILICA, ALUMINA, AND HYDROXYLAPATITE

SENEM YETGIN[1] and DEVRIM BALKOSE[2,*]

[1]Food Engineering Department, Kastomonu University, Kastomonu, Turkey

[2]Chemical Engineering Department, Izmir Institute of Technology, Izmir, Turkey

*Corresponding author. E-mail:devrimbalkose@gmail.com

CONTENTS

ABSTRACT

The equilibrium and kinetics of DNA adsorption on silica, alumina, and hydroxylapatite (HAP) powder as a function of pH and $MgCl_2$ concentration were investigated. The adsorption isotherms of DNA on silica, alumina, and HAP were fitted to Langmuir model better than Freundlich model and Dubinin–Radushkevich model. Higher correlation coefficients were obtained with pseudo-second-order model than the pseudo-first-order reaction model in adsorption kinetics of DNA. The DNA adsorption to the external surface of the particles was very fast kinetically and accomplished within about 15 min. After that, intraparticle diffusion mechanism becomes dominant. Intraparticle diffusion coefficient decreased significantly in the order of alumina, silica, and HAP. The kinetic models, reaction, and diffusion models showed that adsorption at the external surface was dominant at initial stages of the DNA adsorption and it was followed by slower intraparticle diffusion.

12.1 INTRODUCTION

Adsorption is one of the most widely used techniques for removal of pollutants from contaminated media or desired substance in solution. Solid-phase extraction method is one of the useful extraction techniques for DNA purification based on adsorption. Immobilization of DNA onto solid particles, beads, or column is known as solid-phase extraction method. DNA is a negatively charged biopolymer. This charge defines interaction possibilities with surface. DNA can be attached to positively charged surfaces easily.

The forces between the surface and DNA might be repulsive or attractive. The adsorption phenomenon depends on the interaction between the surface of the adsorbent and the adsorbed species. Attractive interactions lead to adsorption. The interaction may be due to: chemical bonding, hydrogen bonding, hydrophobic bonding, Van der Waals force, and electrostatic interaction.[17] DNA adsorbed at the silica surface depends on solution pH, ionic strength, electrolyte type, and valency and conformation of DNA (linear, plasmid, supercoiled).[1,11,14,19]

Surface charge is one of the important parameters to understand electrostatic interaction between DNA and solid surface. In the double helix structure, the bases exist in a highly hydrophobic environment inside the helix, while the outer, negatively charged backbone allows the dsDNA molecule to interact freely with the hydrophilic environment. General idea is about DNA that is not adsorbed by same charged surface due to the electrostatic repulsion. It is reported that based on the physicochemical circumstance DNA has superior adsorption onto different hydrophilic and hydrophobic, inorganic and organic solid surface because of its surface active characteristic.[3] DNA is surface active due to its preferential adsorption from solution to different hydrophilic and hydrophobic, inorganic and organic solid surfaces depending upon physicochemical conditions.

Adsorption process is defined by isotherm. The adsorption isotherm is the relationship that shows the distribution of adsorbate between the adsorbed phase and the solution phase at equilibrium. Adsorption isotherms are essential for the description of how adsorbate concentration interacts with adsorbents and are useful in optimizing their use. Therefore, empirical equations are important for adsorption data interpretation and predictions. The Langmuir and Freundlich isotherms are the most frequently used models used in data evaluation.

Possible electrostatic and hydrophobic mechanisms for the adsorption of DNA in solutions containing monovalent salt are discussed and compared with the observations in divalent salt.[10] Also, other parameters that effect DNA adsorption have been classified as; base composition, ionic strength, pH, DNA chain length, and DNA concentration.[4]

12.1.1 SORPTION ISOTHERM METHOD: ADSORPTION EQUILIBRIA

The adsorption isotherm is the relationship that shows the distribution of adsorbate between the adsorbed phase and the solution phase at equilibrium. Adsorption isotherms are essential for the description of how adsorbate concentration interacts with adsorbents and are useful in optimizing their use. Therefore, empirical equations are important for the adsorption data, interpretation, and predictions. A wide variety of other adsorption models have been formulated over the years. Freundlich, Langmuir, and Brunauer–Emmett–Teller (BET) are the most useful models; they figure out adsorption profile despite the fact that they are old adsorption models. These fundamental adsorption equation's general equilibrium constants were analyzed on the basis of equilibrium concentration. Langmuir and Freundlich differs in that while Langmuir expression explains monolayer adsorption process. Freundlich can explain multilayer adsorption process. BET equation is a theoretical expression of adsorption and is commonly used in the gas-phase adsorption process for finding the monolayer adsorption capacity of an adsorbent.

12.1.1.1 LANGMUIR MODEL

The Langmuir equation is based on a kinetic approach and assumes a uniform surface, a single layer of adsorbed material at constant temperature. The model is useful when there is no strong specific interaction between the surface and the adsorbate. Thus, single adsorbed layer forms and no multi-layer adsorption occurs. It also assumes that the surface is homogeneous and adsorption energy is constant over all sites. Adsorption occurs at specific homogeneous sites within the adsorbents and there is no interaction between the sorbate molecules that are bound to the next active sites. The Langmuir equation has the following form:

$$q_e = \frac{Q°bC_e}{1+bC_e} \tag{12.1}$$

where, q_e is the amount adsorbed at equilibrium, C_e is the equilibrium concentration, b and $Q°$ are Langmuir coefficients related to the energy of adsorption and the maximum adsorption capacity, respectively. Langmuir equation can be described by the linearized form as follows:

$$\frac{1}{q_e} = \frac{1}{Q°} + \frac{1}{bQ°C_e} \tag{12.2}$$

The linear Langmuir plot can be obtained by plotting $1/q_e$ versus $1/C_e$. The coefficients $Q°$ and b can be evaluated from the intercept and slope, respectively.

12.1.1.2 FREUNDLICH MODEL

The Freundlich model is an empirical equation, which assumes that the adsorbent has a heterogeneous surface composed of adsorption sites with different adsorption potentials and energy. The model equation is as follows:

$$q_e = K_f C_e^{1/n} \tag{12.3}$$

where q_e is amount adsorbed at equilibrium and C_e is the equilibrium concentration. K_f and n are equilibrium constants (temperature dependent) related to adsorption capacity and intensity, respectively. Graphically, a plot of q_e versus C_e gives the adsorption isotherm. The linearized form of Freundlich sorption isotherm is:

$$log\, q_e = log\, K_f + \frac{1}{n} log\, C_e \tag{12.4}$$

A plot of $log\, q_e$ versus $log\, C_e$ gives a linear graph. The coefficients K_f and n can be calculated from the intercept and slope, respectively.

The dimension of K_f depends on the value C_e while the exponent "n" is dimensionless. Adsorbent total adsorption capacity increases resulting in K_f parameter increase. The n value might vary along adsorption process and is related to the adsorption efficiency and also to the energy of adsorption.

12.1.1.3   DUBININ–RADUSHKEVICH MODEL

Dubinin–Radushkevich (D–R) equation is widely used for description of adsorption in microporous materials, where adsorption process follows a pore-filling mechanism. Instead of surface layering, in pore filling mechanism the chemical potential is a function of adsorbed amount. Dubinin–Radushkevich equation can be represented in eq 12.5.[5] This model is based on the assumptions of a change in the potential energy between the adsorbate and adsorbent phases and a characteristic energy of a given solid. Adsorption energy value has a clue about the type of sorption, whether it is physical or chemical.

The linear plot of ln qe versus ε^2 are used to define sorption energy constant (K) and maximum adsorption capacity based on D–R isotherm (qm), respectively. Then adsorption energy (E) is obtained by the use of sorption energy constant with eq 12.7. Units of the isotherm formula parameters are Ce (ng/μL), qm is the maximum adsorption capacity based on D–R isotherm (μg/g), and K is the constant related to the sorption energy (mol²/kJ²), therefore, E has kJ/mol unit. If the value is lower than 8 kJ/mol, adsorption is defined physical. Between the values of 8 and 16 kJ/mol, sorption is chemically controlled.[6]

$$q_e = q_m e^{\left(-K\varepsilon^2\right)} \tag{12.5}$$

$$\varepsilon = RT\left(1 + \frac{1}{C_e}\right) \tag{12.6}$$

$$E = \frac{1}{\sqrt{2K}} \tag{12.7}$$

where, ε is Polanyi potential, R is the gas constant (8.314×10^{-3} kJ/mol K), and T is the temperature (K), where E is the mean adsorption energy (kJ/mol).

12.1.2 ADSORPTION KINETICS

Adsorption is one of the most widely used technique to remove pollutants from contaminated media or desired substance in solution. Several system variables should be considered. For instance, initial concentration, sorbent particle size, solution temperature, solution pH, and agitation have a greater effect on the sorption of solutes in reaction controlled sorption processes. As a result, the correlation coefficients between experimental and theoretical data will provide the 'best fit' model. Some kinetic models are involved in adsorption reaction models; the pseudo-first-order equation, the pseudo-second-order equation, or diffusion model; the Elovich equation and intra-particle diffusion model.

Sorption rate is described by three main series of resistances due to external mass transfer, intraparticle diffusion, and reaction. The first one occurs through transport of adsorbate molecules from the bulk solution to the adsorbent external surface. The second is the diffusion of the adsorbate from the external surface into the pores of the adsorbent.

When adsorption is performed, thermodynamic and kinetic parts should be involved to get more details about its performance and mechanisms.[19] To determine the kinetic performance of a given adsorbent has great significance for the small application as much as adsorption capacity of used adsorbent. The cost and performance of a product or the mode of application are the concerns in controlling the efficiency of this process. Therefore, the sorption capacity and required contact time are the two most important parameters.

Adsorption is not only to remove substance from solution but also is a fundamental parts of a reaction mechanism. Reaction mechanism requires the adsorption of each reactant on the catalysis surface. Therefore, adsorption kinetic is important. Chemical kinetics is a useful way of describing the reaction pathway and time needed to reach equilibrium. In other words in kinetics analysis, the solute uptake rate, which determines the residence time required for completion of adsorption reaction, may be established. The dependence of the sorption kinetics on the physical and chemical characteristics of the adsorbent is obvious and affects the mechanism. Numerous studies were carried out in formulating a general expression to analyze the kinetics of adsorption on solid surfaces for the liquid–solid adsorption system. Adsorption rate covers a series of resistance resulted from diffusion and reaction steps.

12.1.2.1 REACTION MODELS

12.1.2.1.1 The Pseudo-First-Order Equation

Scientists have proposed mathematical models to describe adsorption data, which can generally be classified as adsorption reaction models and adsorption diffusion models. In 1898, Lagergren presented the first-order rate equation for the adsorption of oxalic acid and malonic acid onto charcoal.[10] Lagergren kinetics equation may have been the first one in describing the adsorption of liquid–solid systems based on solid capacity. Lagergren rate equation is one of the most widely used sorption rate equations for the sorption of a solute from a liquid solution. It is represented in the following formula.[6]

$$\frac{dq_t}{dt} = k\left(q_e - q_t\right) \tag{12.8}$$

where: q_t is the adsorption capacity at time t, q_e is the adsorption capacity at equilibrium, and k_l is the rate constant of pseudo-first-order adsorption.

It can be linearized after taking integration and applying boundary conditions for $t = 0$ to $t = t$. The integrated form becomes:

$$log(q_e - q_t) = log(q_e) - \frac{k_l}{2.303}t \tag{12.9}$$

When the log $(q_e - q_t)$ versus t plot is drawn, the slope gives the $k_l/2.303$ and the intercept gives the log q_e. This model assumes that the rate of change of adsorbate uptake with time is related to the difference in saturation concentration and the amount of solid uptake with time.[5]

12.1.2.1.2 The Pseudo-Second-Order Equation

A pseudo-second-order model describing the adsorption kinetic may be expressed in the following form:

$$\frac{dq}{dt} = K_2\left(q_e - q\right)^2 \tag{12.10}$$

where, q_e: adsorbate amount per unit weight of adsorbent amount at equilibrium, K_2: the rate constant of pseudo-second-order adsorption. It can be linearized by taking integration and applying the initial conditions.

$$\frac{t}{q} = \frac{1}{K_2 q_e^2} + \frac{1}{q_e} t \tag{12.11}$$

If t/q versus t is plotted, slope gives the $1/q_e$ and the intercept gives the $1/K_2 q_e^2$. This model assumes that the sorption process is a pseudo-chemical reaction process. The driving force is difference between the average solid concentration and the equilibrium concentration. Also, the overall sorption rate is proportional to the square of the driving force.[5]

12.1.2.2   DIFFUSION MODELS

Adsorption and desorption rate in porous material is generally defined transport within the pore structure rather than surface kinetic.[20] Diffusion models derived from Fick's law of mass transport as represented in eq 12.12. It is important that in this equation, mass transport not only depends on concentration. Because it is known that for any transport mechanism, true driving force is the gradient of the chemical potential (μ). The adsorption process only be independent of the adsorbate concentration in case of thermodynamically ideal system.[21]

$$F = -D\frac{\partial C}{\partial x} \tag{12.12}$$

Adsorption process includes fluid film diffusion, intraparticular diffusion, and mass action. Mass action can be neglected because of high rate for kinetic study in physisorption processes. Consequently, adsorption process is predominantly controlled by film diffusion or intraparticular diffusion.

Adsorption kinetics is described by many models: The two-resistance models such as the film-solid model, the film-pore model, and the branched pore model give detailed analysis of the adsorption dynamics.

Distinction of kinetic and diffusion control mechanism generally is not simple. It is a general idea that if system reaches equilibrium within 3 h, the process is usually kinetic controlled. If process takes place above 24 h it is diffusion controlled. Either diffusion or kinetic control adsorption equilibria when the duration of the process between 3 and 24 h. A more suitable quantitative approach to distinguish between kinetic and diffusion rate control can be performed by the square root of contact time analysis according to equation.

However, these models include complex partial differential equations solution and their solution needs computer programs aid and extensive computer time.[12] Mathews and Webber [13] intraparticle diffusion models are the two most widely used models for studying the mechanism of the adsorption without complex solution and computer program.

12.1.2.2.1 *External Mass Transfer: Mathew–Weber Model*

This model assumes that only external film diffusion is the main driving force during the initial sorption period and controls the sorption process. In other words, the film diffusion controls the adsorbate uptake rate in the initial stages, followed by particle diffusion in the latter stages. The model formula is given in eq 12.13.

$$ln \frac{C_t}{C_0} = -k_f St \tag{12.13}$$

where, C_t is the concentration of adsorbate in the solution at time t, C_0 is the initial concentration of adsorbed in the solution, k_f is the external mass transfer coefficient (m/s), t is the time, and S is the surface area for mass transfer (m^{-1}). The initial slope of the linear plot of ln C_t/C_0 versus t is used in the determination of the external mass transfer coefficient (k_f).

12.1.2.2.2 *Intraparticle Diffusion: Weber–Morris Model*

This model was derived from the Fick's second law. According to this, the effects of the external mass transfer resistance can be ignored, the direction of the diffusion is radial, and the intraparticle diffusivity is constant. This model was used to calculate the intraparticle diffusion rate constant. The model equation is given below.

$$q_t = k_d t^{0.5} \tag{12.14}$$

A plot of the amount of adsorbed, q_t, against the square root of time, $t^{0.5}$, gives a straight line plot of slope k_d, a diffusional rate parameter. This straight line, passing through the origin, indicates intraparticle diffusion control.[8] It can be seen from eq 12.14 that if intraparticle diffusion is the rate limiting step, then a plot of q versus $t^{0.5}$ will give a straight line with a slope that equals k_d and an intercept equal to zero.

The approximate solution of the diffusion equation for $q/q_0 < 0.6$ is shown in eq 12.15.

$$\frac{q_t}{q_0} = \frac{6}{\sqrt{\pi}} \left(\frac{Dt}{r^2} \right)^{1/2} \tag{12.15}$$

12.1.2.2.3 Micropore Diffusion

For a spherical particle at constant surface concentration for short time and long time periods can be written in eqs 12.16 and 12.18, respectively by integrating eq 12.17.

$$\frac{m_t}{m_\infty} = 6 \left(\frac{D_t}{r^2} \right)^{1/2} \left[\frac{1}{\sqrt{\pi}} + 2 \sum_{n=1}^{\infty} ierfc \left(\frac{nr_c}{\sqrt{D_c t}} \right) - 3 \frac{D_c t}{r_c^2} \right] \tag{12.16}$$

$$\frac{\partial_q}{\partial_t} = D \left[\frac{\partial^2 q}{\partial r^2} + \frac{2}{r} \frac{\partial q}{\partial r} \right] \tag{12.17}$$

$$\frac{m_t}{m_\infty} = 1 - \frac{6}{\pi^2} exp \left[-\frac{\pi^2 D_c t}{r_c^2} \right] \tag{12.18}$$

In literature, a variety of studies focused on developing a formula to be used in analyzing the kinetics of adsorption on solid surface. To identify the exact mechanism, it is necessary to carry out experiments. Therefore, determination of DNA adsorption kinetics requires adsorption experiments.

DNA adsorption onto silica coated natural organic matter (NOM) surface kinetic was found 120 mg/dm³ supercoiled and the linear plasmid DNA by QCM-D by Nyugen and Chen [15]. Their result also revealed that adsorption took place within 20 min for Ca^{2+} and 40 min for Mg^{2+} cations. Just, 7 mM Na^+ ion positively affects DNA adsorption capacity when both divalent cations are present. They concluded that adsorption was mainly controlled by specific bridging between the DNA phosphate groups and NOM carboxyl groups. Additionally, same group reported that kinetics of DNA adsorption to NOM-coated silica surfaces depended on ionic strength, and that DNA adsorption was significant at moderately high ionic strength.[16] Adsorption kinetic measurements were performed by quartz crystal microbalance device which is sensitive for any little change in the amount mass.

Kinetics model of DNA adsorption onto silica surface was reported by Vandeventer et al.[22] Their model schematic representation can be seen in Figure 12.1. According to the study, DNA can loosely or tightly bind to available silica surface sites.

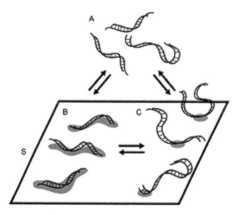

FIGURE 12.1 Kinetic model representation of DNA adsorption to silica. (Reprinted with permission from Vandeventer, P. E.; Lin, J. S.; Zwang, T. J.; Nadim, A.; Johal, M. S.; Niemz, A. Multiphasic DNA Adsorption to Silica Surfaces Under Varying Buffer, pH, and Ionic Strength Conditions. *J. Phys. Chem. B* **2012**, *116*(19), 5661–5670). © 2016 American Chemical Society.)

"A" represents bulk DNA in solution, and "S" denotes the available surface binding sites. "B" and "C" represent DNA tightly and loosely bound to the surface through multiple binding sites or a single binding site, respectively. Assumed adsorption pathway is given in eqs 12.19–12.21.

$$A + nS \xrightleftharpoons[k_2]{k_1} B \tag{12.19}$$

$$B \xrightleftharpoons[k_4]{k_3} C + (n-1)S \tag{12.20}$$

$$B \xrightleftharpoons[k_4]{k_3} C + (n-1)S \tag{12.21}$$

Experiments were performed with Sigma Aldrich silica particles and MagPrep silica particles (MagPrep silica particles (70912) from Merck) as a function of time by using different buffer solution at pH 5 such as in 6 M sodium perchlorate (SP), acetic acid containing 400 mM K$^+$ (AA), glycine containing 400 mM KCl (GL), or sodium citrate containing 400 mM KCl. MagPrep beads and Sigma Aldrich silica particle were used in different amounts. Tubes loaded with 250 µg magnetic beads were filled with 100 ng/µL salmon sperm DNA in 150 µL of the buffer under investigation. Aldrich silica was used 3–6 mg amount 200 ng/µL salmon sperm DNA for experiments using sodium perchlorate buffer or 80 ng/µL for all other buffers. Sodium perchlorate buffer solution has great effect on DNA adsorption on both types of silica material as seen in Figure 12.2. Buffer ions play a role in the DNA adsorption process further than simply adjusting the pH.

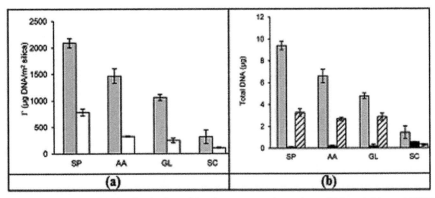

FIGURE 12.2 (a) DNA adsorbed to silica (mass per unit area) at pH 5 and (b) total DNA adsorbed to 250 µg of MagPrep silica particles as a function of buffer (gray). (Reprinted with permission from Vandeventer, P. E.; Lin, J. S.; Zwang, T. J.; Nadim, A.; Johal, M. S.; Niemz, A. Multiphasic DNA Adsorption to Silica Surfaces Under Varying Buffer, pH, and Ionic Strength Conditions. *J. Phys. Chem. B* **2012**, *116*(19), 5661–5670). © 2016 American Chemical Society.)

In the present study, adsorption equilibrium and kinetics of calf thymus DNA on silica, alumina, and hydroxyapatite from aqueous solutions at different pH values is aimed to be investigated. Calf thymus DNA adsorption was determined by using Nano Drop method for this purpose. The fit of experimental data to different adsorption and kinetic models was investigated by linear regression analysis.

12.2 MATERIALS AND METHODS

12.2.1 MATERIALS

Calf thymus DNA was used and supplied by Sigma Aldrich (D1501) and it is used in adsorption experiments. Zeta potential measurements by Yetgin and Balkose [24] showed that calf thymus DNA in aqueous solutions had isoelectric point near pH 2 value and their particle size ranged from 300 to 800 nm and it was 490 nm on the average. As adsorbent, alumina (Al_2O_3), silica (SiO_2), and hydroxyapatite (HA) ($Ca_{10}(PO_4)_6(OH)_2$) were used. HAP had Ca/P mass ratio of 2.71.[23]. Silica (Sigma Aldrich-Silicagel, Grade 7744 pore size 60 Å, 70–230 mesh) and alumina was obtained from Seydişehir in Turkey and hydroxyapatite was supplied by Sigma Aldrich (23093-6). The average particle sizes of silica, alumina, and hydroxyapatite was 162 and 59 μm, and 102 nm, respectively.[23] Their pore sizes are 5.7, 5.5, and 6.2 nm for alumina, silica, and hydroxyapatite, respectively. The BET surface area values were 108, 571, and 69 m²/g for alumina, silica, and hydroxyapatite, respectively.[23]

12.2.2 METHODS

12.2.2.1 ADSORPTION EXPERIMENTS

Before the adsorption experiment, all adsorbents were dried at 120°C for 18 h under vacuum. In order to the control the adsorption of buffer solution, adsorbents were controlled free of DNA. Blank control of UV analysis is done as in the following to adsorbent surface and release of ingredients from the adsorbent to buffer solution 1.5 cm³ buffer was equilibrated with 0.0075 g of adsorbent. UV spectrum of the supernatant was taken. The DNA concentration has been estimated at 260 nm using a standard curve obtained by measuring absorbance of a series of solutions of known DNA concentrations by spectrophotometric method. This method is applied by Nano-Drop 2000 (Thermo Scientific ND 1000). DNA solutions were prepared in 10–20–40–60–80–100 ng/μL range for obtaining the calibration curve. The following buffer solutions in Table 12.1 were prepared to obtain solutions with pH values 2–9.

Stock solutions of DNA were diluted with pH buffer solutions to obtain the 10–100 ng/μL concentration range. Subsequently, 0.0075 mg of adsorbent (SiO_2, Al_2O_3 and HA) was added to the 1.5 mL DNA solutions to get a solid to liquid ratio 5 g/dm³, and DNA concentration of the supernatant was measured after mixing at 100–200 rpm by Fine PCR mixer incubator at 25°C for 24 h. Cation addition effect was tested by adding 0.5 and 20 mM $MgCl_2$ into the DNA stock solution at pH 5.

TABLE 12.1   The Buffer Solutions with Ionic Strength 0.001 M.

pH	Buffer type
2	Glycin-HCl
3	Potassium hydrogen phthalate-HCl
4	Potassium hydrogen phthalate-HCl
5	Acetate
6	Acetate
7	Phosphate buffer
8	Phosphate buffer
9	Phosphate buffer

12.2.2.2   ADSORPTION KINETICS

In total, 0.008 mg of adsorbent (SiO_2, Al_2O_3, or HAP) was added to the 1.6 mL 100 ng/µL DNA concentration solutions to keep a solid to liquid ratio 5 g/dm^3. Kinetic measurements were carried out at 25°C under constant stirred conditions for maximum 3 h by Fine PCR mixer incubator. About 2–4 µL supernatant sample was taken at each time interval. Separated samples were stored in clean Eppendorf tubes until analyses. Data were collected for each pH with respect to time.

For the control of adsorption of buffer solution to adsorbent surface and release of ingredients from the adsorbent to buffer solution, 1.5 cm^3 buffer was equilibrated with 0.0075 g of adsorbent and blank measurement was done by taking the UV spectrum of the supernatant. The DNA concentration has been estimated at 260 nm using a standard curve obtained by measuring absorbance of a series of solutions of known DNA concentrations by spectrophotometric method. This method was applied again by Nano Drop 2000 (Thermo Scientific ND 1000).

12.3   RESULTS

12.3.1   ADSORPTION EQUILIBRIUM

Charging of solid surfaces in aqueous environments is change with the pH of the aqueous solutions. DNA molecule is negatively charged due to phosphate groups when the pH is above the isoelectric point. Therefore, surface charge inevitably influences adsorption phenomena considering the role played in electronic interaction. Therefore, pH 5 is the optimum value that figures out influence of other parameters on adsorption process, which is away from DNA isoelectric point (pH 2) and thus, makes the DNA stable. Structural or chemical properties might impact on DNA purification method. Adsorption isotherms of DNA on silica alumina and hydroxyapatite are shown at different pH values.

The isoelectric point of DNA used in the present study was at pH 2.[24] The isoelectric points of silica, alumina, HAP were reported to be 2,[22] 9,[18] and 8[9], respectively. DNA is negatively charged above pH 2. Since silica surface is also negatively charged above pH 2, the adsorption of DNA should be due to hydrogen bonding, Van der Waals attractions between silica and DNA. Alumina and DNA had opposite surface charges between pH 2 and 9, thus ionic attractions play an important role on adsorption on alumina. The surface charge of HAP is positive up to pH 8. Thus, ionic attractions between DNA and hydroxyapatite had a great contribution to the adsorption on HAP. DNA adsorption isotherms for silica, alumina and hydroxyapatite at different pH values are seen in Figures 12.3–12.5.

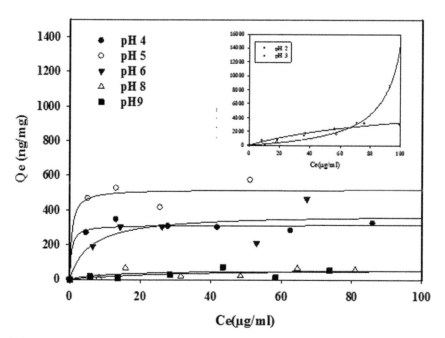

FIGURE 12.3   DNA adsorption on Silica at different pH values.

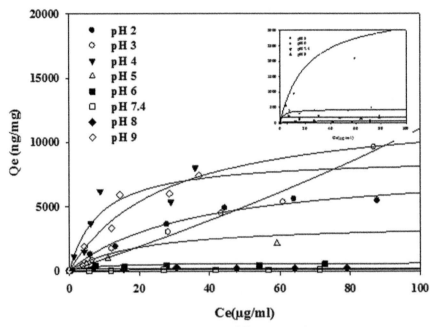

FIGURE 12.4   DNA adsorption on Alumina at different pH values.

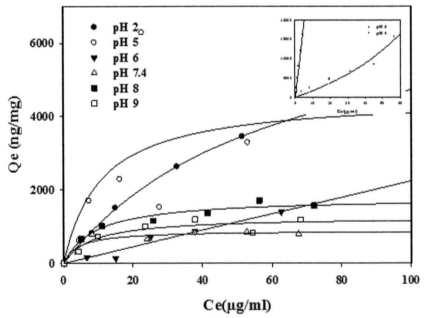

FIGURE 12.5   DNA adsorption on HAP at different pH values.

Cation addition was also tested by adding $MgCl_2$ into the DNA stock solution. Also, 0.5 and 20 mM $MgCl_2$ were tested. Adsorption isotherms about this situation are shown in Figures 12.6 and 12.7, respectively. Consequently, adsorbents adsorption capacities were enhanced. Mg^{2+} ions help the DNA adsorption onto the surface as a bridge. Approximately 3 mg/g adsorption capacity of HAP is greater than fine organic clay minerals and montmorillonite (1 mg/g) reported in Cai et al.[2] at 0.5 mM $MgCl_2$ concentration and with 25 g/dm^3 solid to liquid ratio. However, at 20 mM $MgCl_2$ concentration clay minerals, DNA adsorption capacities are greater than HAP. It should be considered that their initial concentration is 200 µg salmon sperm DNA in 400 µL tris buffer, which is higher than this study's initial concentration.

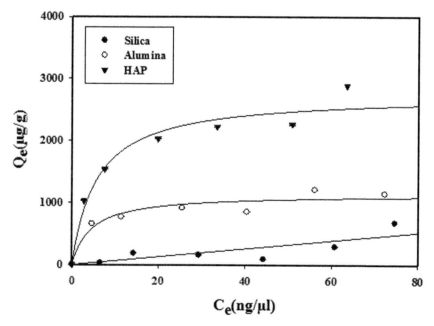

FIGURE 12.6   Adsorption on Silica, Alumina, and HAP at pH 5 with 0.5mM $MgCl_2$.

As a summary, Figure 12.8 shows the adsorption capacity of all adsorbents at pH 2–9. HAP has always higher adsorption capacity than silica and alumina. The results indicate that there is significant potential for the removal of DNA from aqueous solution using HAP as an adsorbent. Nano sized HAP which has Ca/P mol ratio of 2.0 is promising adsorbent for DNA purification. DNA's phosphate part can easily attract calcium ions.

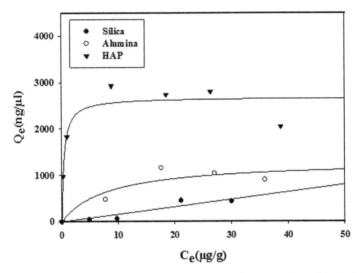

FIGURE 12.7   DNA adsorption on Silica, Alumina, and HAP at pH 5 with 20mM $MgCl_2$.

$MgCl_2$ addition enhanced the adsorption capacity as shown in Figure 12.9. As the divalent cation molarity increase from 0.5 to 20 mM, adsorption capacity was enhanced but not proportionally.

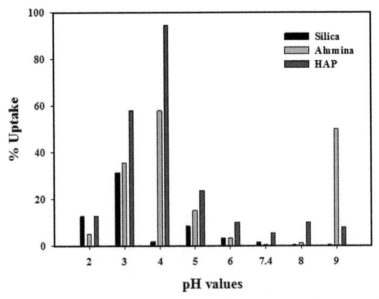

FIGURE 12.8   Calf thymus DNA removal % by Silica Alumina and HAP versus pH values.

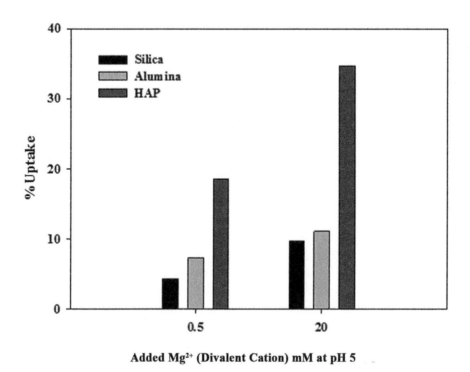

FIGURE 12.9   Calf thymus DNA removal % by Silica Alumina and HAP versus divalent cation addition at pH 5.

12.3.1.1   ADSORPTION ISOTHERMS

Sorption isotherms data shown in Figures 12.3- 12.5 can be subject to the Langmuir (eq 12.2) and Freundlich (eq 12.3) plots. The Langmuir plots yielded two important parameters: Langmuir monolayer capacity, Q^o, giving the amount of DNA required to occupy all the available sites in unit mass of the adsorbent, and the Langmuir equilibrium parameter, b, related to the equilibrium constant. The Freundlich model gives parameters K_f and n. Parameters of Langmuir and Freundlich models for adsorption of DNA at different values are shown in Table 12.2. The regression coefficients were not different for Langmuir and Freundlich models. Negative Q^o and b values were found for silica at pH 3 and pH 5 with $MgCl_2$ and alumina at pH 9 with high regression coefficients indicated, the adsorption did not fit to Langmuir model. The negative n value found for Alumina at pH 8 shows the Freundlich model was not followed. Freundlich model regression coefficients were lowered when $MgCl_2$ was added.

TABLE 12.2 Parameters of Langmuir and Freundlich Models for Calf Thymus DNA Adsorption Isotherms.

pH	Adsorbent	Langmuir			Freundlich		
		R^2	Q^o	b	R^2	K_f	n
2	Silica	0.992	4347.8	0.0022	0.932	127.7	1.41
	Alumina	0.967	7142.8	0.0344	0.964	437.03	1.66
	HAP	0.997	9090.9	0.0125	0.984	188.67	1.35
3	Silica	0.982	−3333	−0.0058	0.878	0.064	0.404
	Alumina	0.996	40000	0.0019	0.993	95.37	1.083
	HAP	0.997	30703	0.0045	0.992	157.3	1.096
4	Silica	0.909	76.16	0.3730	0.942	23.98	14.81
	Alumina	0.909	16826	0.0093	0.997	16.74	1.06
	HAP	0.999	70338	0.0711	0.72	305.8	0.89
5	Silica	0.995	584.79	0.747	0.959	408.9	11.21
	Alumina	0.861	172.4	0.988	0.88	258.8	2.07
	HAP	0.772	8333	0.0201	0.79	364.1	1.68
6	Silica	0.912	524.1	0.094	0.934	104.3	2.83
	Alumina	0.912	524.1	0.094	0.883	142.9	3.59
	HAP	0.928	3333	0.011	0.986	27.7	1.04
7.4	Silica	–	–	–			
	Alumina	0.785	6.48	0.072	0.776	1.288	3.32
	HAP	0.811	101	0.125	0.947	19.15	2.668
8	Silica	0.914	6.39	0.201	0.835	1.47	1.232
	Alumina	0.991	20.37	0.093	0.769	250.1	−15.52
	HAP	0.968	169.49	0.129	0.947	420.8	3.087
9	Silica	0.857	60.79	0.068	0.764	7.0	1.973
	Alumina	0.624	−5000	−0.029	0.893	813.8	1.612
	HAP	0.936	1449.2	0.072	0.875	204.8	2.158
pH 5 and 0.5 mM MgCl$_2$	Silica	0.9	−357.1	−0.012	0.651	6.86	1.07
	Alumina	0.86	1111	0.281	0.849	464	4.89
	HAP	0.959	2500	0.222	0.962	796.3	3.40
pH 5 and 20 mM MgCl$_2$	Silica	0.915	−714.2	−0.011	0.909	464.1	4.87
	Alumina	0.817	1666.7	0.055	0.571	228.14	2.24
	HAP	0.712	5000	0.042	0.823	2335	7.8

Dubinin–Radushkevich(D-R) model eq 12.5 was also applied to obtained experimental data. Parameters of q_m, K, and energy (E) were determined and listed in Tables 12.3–12.5. Regression coefficients of D–R isotherm models for silica alumina and HAP were lower than 0.92, 0.97, and 0.95, respectively. Model parameter could not be obtained at some pH values such as for silica at pH 8, alumina at pH 5, and HAP at pH 3. Furthermore, no reliable data were obtained in the presence and absence of MgCl2 at pH 5 for silica. It should be noted that D–R adsorption isotherm is formulated for microporous materials. Silica has the highest micropore volume among to all adsorbents. All energy values were lower than 8 kJ/mol therefore, DNA adsorption on silica, alumina, and HAP were all physical adsorption.

TABLE 12.3 Dubinin–Radushkevich Parameters for Silica at 25°C.

pH value	q_m (ng/mg)	K (mol²/kJ²)	R^2	E (kJ/mol)
2	1.3×10^{10}	2.440	0.92	0.453
4	4.49×10^2	0.057	0.86	2.972
5	5.1×10^3	0.304	0.23	1.282
6	6.06×10^3	0.432	0.74	1.076
9	1.2×10^7	1.929	0.65	0.509
20 mM MgCl$_2$	9.1×10^6	1.504	0.91	0.557

TABLE 12.4 Dubinin–Radushkevich Parameters of Calf Thymus DNA for Alumina at 25°C.

pH value	q_m (ng/mg)	K (mol²/kJ²)	R^2	E (kJ/mol)
2	2.3×10^8	1.741	0.97	0.536
4	1.31×10^5	0.455	0.80	1.048
6	2.44×10^4	0.631	0.87	0.890
7	6.32×10^3	0.780	0.88	0.801
8	8.39×10^3	0.611	0.97	0.904
9	4.12×10^5	0.631	0.91	0.890
0.5 mM MgCl$_2$	4.8×10^3	0.234	0.79	1.463
20 mM MgCl$_2$	1.91×10^4	0.444	0.60	1.061

TABLE 12.5 Dubinin–Radushkevch Parameters of Calf Thymus DNA for HAP at 25°C.

pH value	q_m (ng/mg)	K (mol²/kJ²)	R^2	E (kJ/mol)
2	4.3×10^6	1.075	0.95	0.682
4	5.45×10^4	0.135	0.94	1.925
5	5.2×10^3	0.138	0.95	1.903
6	1.08×10^{10}	2.509	0.95	0.446
7	3.39×10^3	0.222	0.73	1.499
8	1.65×10^4	0.377	0.86	1.151
9	1.17×10^4	0.374	0.82	1.156
0.5 mM MgCl$_2$	9.12×10^3	0.210	0.83	1.544

It is known that DNA adsorption is possible because DNA has negative charge due to the phosphate backbone above its isoelectric point, it can make hydrogen bond with the solid surfaces and there are Van der Waals attractions between DNA and the surfaces. Calf thymus DNA used in the adsorption experiment had an isoelectric point less than pH 2.2. Zeta potential measurement showed that the α-alumina has isoelectric point at nearly pH 9.1.[18] The isoelectric points of silica, alumina, and HAP were reported to be 2, 9, and 8, respectively. DNA was negatively charged above pH 2.2. Since the silica surface was also negatively charged above pH 2, the adsorption of DNA on silica should be due to hydrogen bonding and Van der Waals attractions. Alumina and DNA had opposite surface charges between pH 2 and 9, thus ionic attractions play an important role on adsorption of DNA on alumina. The surface charge of HAP was positive up to pH 8. Thus, ionic attractions between DNA and hydroxyapatite had a great contribution to the adsorption of DNA on HAP.

12.3.2 ADSORPTION KINETICS

12.3.2.1 PSEUDO-FIRST- AND SECOND-ORDER MODELS

Experimental data show that those adsorptions occurred in short time. Equilibrium was generally achieved within 15 min as seen in Figures 12.10–12.18 that show the amount adsorbed DNA versus time curves for silica, alumina, and hydroxyapatite at pH 2, pH 5, and pH 8. Pseudo-first-order (eq 12.8)

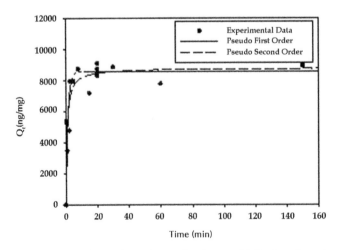

FIGURE 12.10   Pseoudo first order and second orders model fits for adsorption of DNA to silica from pH 2 solution.

and pseudo-second-order (eq 12.11) model constants were determined by nonlinear regression analysis of the experimental adsorbed amount (q) versus time (t) data. In Figures 12.10–12.18 first-order model fits by nonlinear regression are shown. The first- and second-order model constants are reported in Table 12.6, Table 12.7, and Table 12.8 for silica, alumina, and hydroxyapatite, respectively. High correlation coefficients were obtained with pseudo-second-order reaction than the pseudo-first-order reaction model.

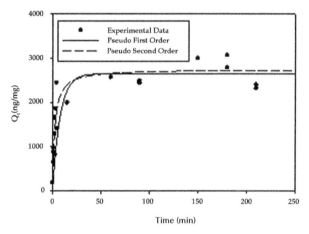

FIGURE 12.11   Pseoudo first order and second orders model fits for adsorption of DNA to silica from pH 5 solution.

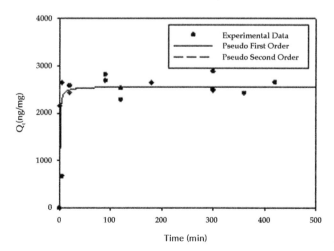

FIGURE 12.12   Pseoudo first order and second orders model fits for adsorption of DNA to silica from pH 8 solution.

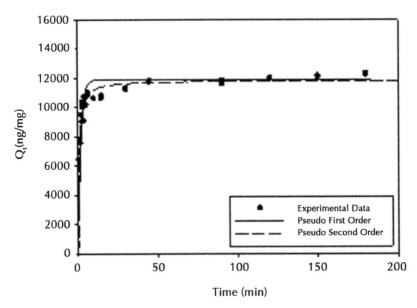

FIGURE 12.13   Pseoudo first order and second orders model fits for adsorption of DNA to alumina from pH 2 solution.

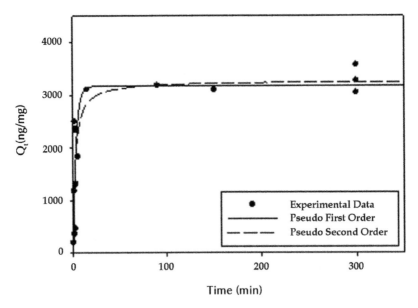

FIGURE 12.14   Pseudo first order and second orders model fits for adsorption of DNA to alumina from pH 5 solution.

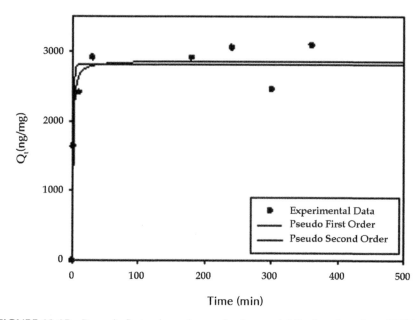

FIGURE 12.15 Pseoudo first order and second orders model fits for adsorption of DNA to alumina from pH 8 solution.

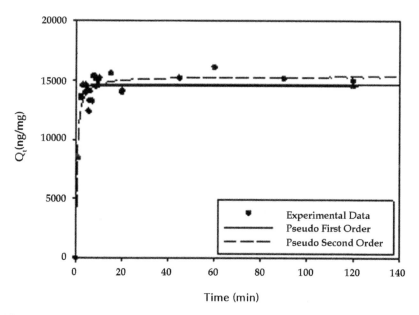

FIGURE 12.16 Pseoudo first order and second orders model fits for adsorption of DNA to HAP from pH 2 solution.

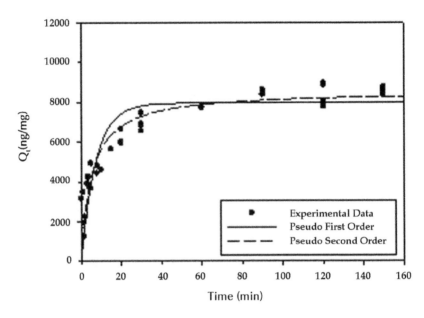

FIGURE 12.17   Pseoudo first order and second orders model fits for adsorption of DNA to HAP from pH 5 solution.

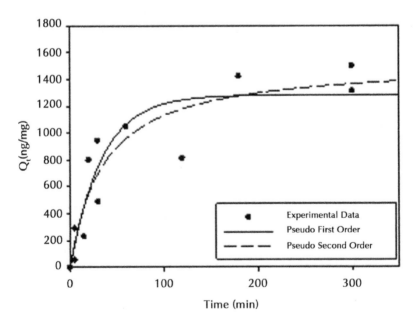

FIGURE 12.18   Pseoudo first order and second orders model fits for adsorption of DNA to HAP from pH 8 solution.

TABLE 12.6 The Reaction Model Parameters of Calf Thymus DNA on Silica at 25°C.

pH	Pseudo first-order			Pseudo second-order		
	q_e (ng/mg)	k_1 1/min	R^2	q_e, ng/mg	k_2, g/(mg.min)	R^2
2	8615	0.731	0.78	8821	1.46×10^{-4}	0.83
3	899	0.529	0.74	932	7.65×10^{-4}	0.72
4	5605	0.010	0.94	8270	8.95×10^{-7}	0.94
5	2647	0.130	0.48	2750	1.66×10^{-4}	0.86
7.4	8449	0.495	0.89	8713	8.5×10^{-5}	0.93
8	2483	1.992	0.59	2561	7.1×10^{-4}	0.63

TABLE 12.7 The Reaction Model Parameters of Calf Thymus DNA on Alumina at 25°C.

pH	Pseudo first-order			Pseudo second-order		
	q_e (ng/mg)	k_1, 1/min	R^2	q_e (ng/mg)	k_2, g/(mg.min)	R^2
2	11420	0.876	0.87	11839	1.47×10^{-4}	0.96
3	2356	0.043	0.99	2635	2.28×10^{-5}	0.99
4	15973	0.026	0.98	19185	1.44×10^{-6}	0.96
5	3175	0.297	0.77	3271	1.31×10^{-4}	0.78
7.4	6461	0.102	0.88	6672	3.27×10^{-5}	0.91
8	2810	0.876	0.94	2872	4.3×10^{-4}	0.96

TABLE 12.8 The Reaction Model Parameters of Calf Thymus DNA on HAP 25°C.

pH	Pseudo first-order			Pseudo second-order		
	q_e (ng/mg)	k_1, 1/min	R^2	q_e (ng/mg)	k_2, g/(mg.min)	R^2
2	14592	1.04	0.92	15333.9	1.42×10^{-4}	0.91
3	8594	3.67	0.89	8664.7	1.18×10^{-3}	0.91
4	14157	11.05	0.82	14442.1	1.35×10^{-3}	0.91
5	7961	0.133	0.84	8574.7	2.1×10^{-5}	0.91
7.4	4779	0.197	0.88	5132.4	5.4×10^{-5}	0.83
8	1283	0.029	0.72	1524.7	1.8×10^{-5}	0.86

12.3.2.2 INTRAPARTICLE DIFFUSION

The initial stage of DNA adsorption was defined as kinetically very fast and completed within about 15 min. After that, mechanism of intraparticle diffusion becomes dominant. Diffusivity of DNA in the adsorbents was calculated by using the solution of the diffusion equation for the long time period and is shown in Table 12.9. R^2 values range obtained between 0.62 and 0.93 for Silica, 0.47 and 0.84 for Alumina, and 0.15 and 0.84 for HAP, as reported in Table 12.8. Long-time diffusion coefficient—in other word intraparticle diffusion rate was decreased significantly in the order of alumina, silica, and HAP. The average pore diameter of alumina (5.7 nm) was greater than that of silica (5.5 nm). Diffusion coefficient of DNA in alumina was determined to be nearly 10 times than in silica. HAP has the highest average pore diameter value (6.2 nm) and on the contrary the smallest particle size (100 nm) in all adsorbents. Therefore, diffusion is not expected in HAP, thus the diffusivity 10^{-20} m²/s is very low.

TABLE 12.9 Diffusivity of DNA in Silica, Alumina, and HAP at 25°C.

pH	Silica		Alumina		HAP	
	D (m²/s)	R^2	D (m²/s)	R^2	D (m²/s)	R^2
2	7.8×10^{-15}	0.75	2.5×10^{-14}	0.86	5.2×10^{-19}	0.87
3	5.3×10^{-16}	0.45	–	–	9.6×10^{-21}	0.88
4	1.5×10^{-15}	0.82	1.5×10^{-14}	0.56	7.34×10^{-20}	0.90
5	7.3×10^{-16}	0.73	1.1×10^{-14}	0.66	8.2×10^{-20}	0.89
7.4	2.5×10^{-15}	0.75	1.5×10^{-14}	00.87	6.2×10^{-20}	0.94
8	–	–	1.6×10^{-14}	0.74	6.2×10^{-20}	0.57

12.4 CONCLUSIONS

DNA is adsorbed to the surfaces by ionic bonds, by Van der Waals attractions and by hydrogen bonds. Calf thymus DNA used in the adsorption experiment had an isoelectric point less than pH 2.2. The isoelectric points of silica, alumina, and HAP were reported to be 2, 9, and 8, respectively. DNA was negatively charged above pH 2. Since the silica surface was also negatively charged above pH 2, the adsorption of DNA on silica should be due to hydrogen bonding and Van der Waals attractions. Alumina and DNA had opposite surface charges between pH 2 and 9, thus ionic attractions play an important role on adsorption of DNA on alumina. The surface charge of HAP was positive up to pH 8. Thus, ionic attractions between DNA and hydroxyapatite had a great contribution to the adsorption of DNA on HAP.

Enhancements of the adsorption capacities of adsorbents were obtained with the addition of $MgCl_2$. The adsorption capacity of DNA on adsorbents increased when divalent cation molarity was changed from 0.5 to 20 mM.

The adsorption isotherms of DNA on silica, alumina and HAP were fitted to Langmuir model better than Freundlich model and Dubinin–Radushkevich model. Higher correlation coefficients were obtained with pseudo-second-order model than the pseudo-first-order reaction model in adsorption kinetics of DNA. The DNA adsorption to the external surface of the particles was very fast kinetically and accomplished within about 15 min. After that, intraparticle diffusion mechanism becomes dominant. Intraparticle diffusion coefficient decreased significantly in the order of alumina, silica, and HAP. The kinetic models, reaction, and diffusion models showed that adsorption at the external surface was dominant at initial stages of the DNA adsorption and it was followed by slower intraparticle diffusion.

KEYWORDS

- adsorption
- DNA
- silica
- alumina
- hydroxylapatite

REFERENCES

1. Allemand, J. F.; Bensimon, D.; Jullien, L., Bensimon, A.; Croquette, V. pH-Dependent Specific Binding and Combing of DNA. *Biophys. J.* **1997**, *73*, 2064–2070.
2. Cai, P.; Huang, Q.; Jiang, D.; Rong, X.; Liang, W. Microcalorimetric Studies on the Adsorption of DNA by Soil Colloidal Particles. *Colloids Surf. B Biointerfaces* **2006**, *49*, 49–54.
3. Chattoraj, D. K.; Mitra, A. Adsorption of DNA at Solid–Water Interfaces and DNA–Surfactant Binding Interaction in Aqueous Media. *Curr. Sci.* **2009**, *97*(10), 1430–1438.
4. Dias, R. S.; Lindman, B. *DNA Interactions with Polymers and Surfactants*; John Wiley & Sons: New Jersey, 2008.
5. ......Do, D.D. *Adsorption Analysis: Equilibria and Kinetics*. Imperial College Press, London, 1998
6. Erdoğan, B. C.; Ülkü, S. Cr(VI) Sorption By Using Clinoptilolite and Bacteria Loaded Clinoptilolite Rich Mineral. *Microporous Mesoporous Mater.* **2012**, *152*, 253–261.
7. Ho, Y. S.; Mckay, G. Research Note the Sorption of Lead (Ii) Ions on Peat. *Water Res.* **1999**, *33*(2), 578–584.

8. Ho, Y. S.; Ng, J. C. Y.; Mckay, G. Kinetics of Pollutant Sorption by Biosorbents Review. *Sep. Purif. Methods* **2000**, *29*(2), 189–232.

9. Komulski, M. pH Dependent Surface Charging and Points of Zero Charge IV. Update and New Approach. *J. Colloid Interface Sci.* **2009**, *337*, 439–449.

10. Lagergren, S. About the Theory of So-Called Adsorption of Soluble Substances. *Kungliga Svenska Vetenskapsakademiens. Handlingar* **1898**, *24*, 1–39.

11. Lorenz, M. G.; Wackernagel, W. Bacterial Gene Transfer by Natural Genetic Transformation in the Environment. *Microbiol. Rev.* **1994**, *58*(3), 563–602.

12. Malash, G. F.; El-Khaiary, M. I. Piecewise Linear Regression: A Statistical Method for the Analysis of Experimental Adsorption Data by the Intraparticle-Diffusion Models *Chem. Eng. J.* **2010**, *163*, 256–263.

13. Mathews, A. P.; Weber, W. J. Effects of External Mass Transfer and Inter-particle Diffusion on Adsorption. *Phys. Chem. Waste Water Treat. AIChE Symp. Ser. 7* **1976**, *3*, 91–98.

14. Melzak, K. A.; Sherwood, C. S.; Turner, R. B. F.; Haynes, C. A. Driving Forces for DNA Adsorption to Silica in Perchlorate Solutions. *J. Colloid Interface Sci.* **1996**, *181*, 635–644.

15. Nguyen, T. H.; Chen, K. Role of Divalent Cations in Plasmid DNA Adsorption to Natural Organic Matter-Coated Silica Surface. *Environ. Sci. Technol.* **2007**, *41*, 5370–5375.

16. Nguyen, T. H.; Elimelech, M. Adsorption of Plasmid DNA to a Natural Organic Matter-Coated Silica Surface: Kinetics, Conformation, and Reversibility. *Biomacromolecules* **2007**, *8*, 24–32.

17. Parida, S. K.; Dash, S.; Patel, S.; Mishra, B. K. Adsorption of Organic Molecules on Silica Surface. *Adv. Colloid Interface Sci.* **2006**, *121*, 77–110.

18. Polat, M; Sato, K.; Nagaoka, T.; Watari, K. Effect of pH and Hydration on The Normal and Lateral Interaction Forces between Alumina Surfaces. *J. Colloid Interface Sci.* **2006**, *304*, 378–387.

19. Qiu, H.; Lv, L.; Pan, B.; Zhang, Q. J.; Zhang, W.; Zhang, Q. Critical Review in Adsorption Kinetic Models. *J. Zhejiang Univ. Sci. A.* **2009**, *10*(5), 716–724.

20. Romanowski, G.; Lorenz, M. G.; Wackernagel, W. Adsorption of Plasmid DNA to Mineral Surfaces and Protection against DNase I. *Appl. Environ. Microbiol.* **1991**, *7*, 1057–1061.

21. Ruthven, D. *Principles of Adsorption and Adsorption Processes*; John Wiley and Sons: New York, 1984.

22. Vandeventer, P. E.; Lin, J. S.; Zwang, T. J.; Nadim, A.; Johal, M. S.; Niemz, A. Multiphasic DNA Adsorption to Silica Surfaces under Varying Buffer, pH, and Ionic Strength Conditions. *J. Phys. Chem. B* **2012**, *116*(19), 5661–5670.

23. Yetgin, S. DNA adsorption on Silica, Alumina and Hydroxyapatite and Imaging of DNA by Atomic Force Microscopy. PhD Dissertation, Izmir Institute of Technology, Izmir, 2013.

24. Yetgin, S.; Balkose, D. Calf Thymus DNA Characterization and Its Adsorption on Different Silica Surfaces. *RSC Adv.* **2015**, *5*, 57950–57959.

CHAPTER 13

THERMAL BEHAVIOR OF SODIUM SALT OF CALF THYMUS DNA

AYSUN TOPALOĞLU[1], GÜLER NARIN[2,*], and DEVRIM BALKÖSE[1]

[1]*Department of Chemical Engineering, Izmir Institute of Technology, Gulbahce 35430, Urla Izmir, Turkey*

[2]*Department of Chemical Engineering, Uşak University, Uşak, Turkey*

[*]*Corresponding author. E-mail: gulernarin@gmail.com*

CONTENTS

ABSTRACT

Dehydration, melting, and thermal degradation of sodium salt of calf thymus DNA in dry state were investigated. For investigation of kinetics of reactions taking place during the nonisothermal heating, thermal gravimetric analysis (TGA) data recorded at different heating rates were analyzed using various model-free isoconversional methods. Differential scanning calorimetry (DSC) analysis was performed both in open and hermetically sealed cells. Melting temperature of the DNA was higher in the sealed cells (110°C) than in the open cells (80°C). The residues formed after the DNA was repeatedly heated and cooled in sealed DSC cells were analyzed by transmittance Fourier transform infrared (FTIR) spectroscopy. These residues were also analyzed by scanning electron microscope (SEM) for the morphological changes. The original fibrous form of the DNA turned to a black residue during heating to 175°C in the hermetically sealed cells with upper pressure limit of 0.3 MPa. Microbubbles indicating the intumescence characteristic of the DNA were observed on the surface of the residues. In the transmission FTIR spectrum of the residue obtained by heating the DNA to 600°C under nitrogen flow, bands characteristics for inorganic phosphates were detected. The DNA was also analyzed in situ by diffuse reflectance infrared Fourier transform (DRIFT) spectroscopy during heating/cooling under vacuum. The results were discussed based on the effects of heating rate and reaction atmosphere on the intumescence behavior of the DNA.

13.1 INTRODUCTION

Deoxyribose nucleic acid (DNA) molecule is the storage site for genetic information and fulfils the replication of that information during the cell division.[1] DNA is a biopolymer composed of repeating units of nucleotides. A nucleotide unit consists of the segment of the backbone of the molecule which holds the chain together and also a nucleobase (purines and pyrimidines) which interacts with the other DNA strand in the helix. Alternating phosphate and sugar (deoxyribose) molecules form the backbone of the DNA strand. The sugars are linked together by phosphodiester bonds between the third and fifth carbon atoms of the adjacent sugar rings. The helix of the DNA is held by two forces: hydrogen bonds between nucleotides and base-pairing interaction of the nucleobases. DNA has been investigated for various applications like DNA-based drugs, recombinant DNA for the production of industrial microorganisms, biosensor development, environmental monitoring, and so on.[2-5]

Since there is a close correlation between the biological function of DNA and its structure, understanding the relation of water to DNA is of both practical and fundamental importance. The stability and conformational integrity of DNA are largely controlled by the interactions with the surrounding water molecules.[6–8] Two factors are mainly responsible for the stability of the DNA double helix: base-pairing between complementary strands and stacking between adjacent bases.[9] Water is essential to the stabilization of the secondary and tertiary structures, thereby biologically active structures of DNA. DNA has been shown to exist in a variety of secondary structures or conformations such as right-handed double-helical A, B, C, and D forms and the left-handed double-helical Z form.[8] One of the important factors in determining the conformation of DNA is the degree of hydration. At high relative humidity (≥90%), DNA fibers or films adopt the B conformation at room temperature. As the relative humidity is reduced to a certain threshold, the decrease in hydration leads to the transformation of B-DNA into other conformations, depending on the DNA sequence and the nature of the counterions involved.[10] There are two kinds of hydration shells around DNA: the water molecules which hydrate the DNA helix up to a relative humidity of about 80% form the tightly bound primary hydration shell (about 20 water molecules per nucleotide that hydrate the phosphate, sugar, and base) while those which adsorbed at higher humidities form the more loosely bound secondary hydration shell.[11] In the primary hydration shell, at least 11–12 water molecules per nucleotide (wpn) that are directly bound to the DNA are incapable of crystallization to ice upon cooling to temperatures well below 0°C.[12] These tightly bound water molecules in the primary shell constitute an integral part of the stability of DNA-ordered structures. These water molecules are difficult to remove and their removal leads to the collapse of the double helical structure of DNA. In the absence of water, the ordered conformation is lost and base stacking is destroyed.[13–18] All three molecular subgroups in DNA (the heterocyclic bases, deoxyribose, and the diesterified phosphate groups) provide sites where water molecules can be adsorbed.[11] Presumably, specific water–DNA interactions arise primarily from hydrogen bonding to the groups on the edges of the bases facing the grooves of the double helix and to a lesser extent from hydrogen bonding to the sugars and phosphate groups.[8]

The structural changes during the melting of DNA include disruption of base pairs, unstacking of the bases, and disordering of the deoxyribose–phosphate backbone.[19] It is known from infrared (IR) spectroscopic studies that the structural changes of DNA induced by dehydration are reversible upon rehydration.[14,15] Duguid and coworkers used differential scanning calorimetry (DSC) and laser Raman spectroscopy to probe the energetic and structural changes that accompany melting of the B form of DNA of 160 base pair fragments of calf thymus DNA in solution. Raman bands diagnostic of purine and pyrimidine unstacking, conformational rearrangements in the deoxyribose–phosphate moieties, and changes in environment of phosphate groups have been identified. Melting behaviors of the separate Raman markers (834, 1240, and 1668 cm^{-1}) have been correlated with one another over the temperature interval 11–93°C. From the DSC results, the standard thermodynamic parameters (vant Hoff enthalpy and effective number of base pairs in a cooperative melting) were obtained and correlated with the localized structural changes monitored by Raman spectroscopy. The changes observed in specific Raman band frequencies and intensities as a function of temperature revealed that thermal denaturation was accompanied by disruption of Watson–Crick base pairs, unstacking of the bases, and disordering of the B-form backbone. These three types of structural change were highly correlated in the temperature range of 20–93°C.[19]

There are relatively few published works on the thermal stability of nucleic acids at low hydration.[20–25] Falk and coworkers investigated hydration of solid calf thymus DNA (NaDNA) gravimetrically as a function of relative humidity at 21°C. The Brunauer–Emmett–Teller (BET) model fitted the data between 0% and 80% relative humidity. It was found that two molecules of water are strongly bound to each phosphate group with energy about 8.4 kJ, higher than that for the adsorption of water molecules. Above 75% relative humidity, the water molecules hydrate the C=O groups and ring nitrogen atoms fill the grooves of the DNA helix. This process is completed by about 80% relative humidity and further hydration of DNA is accompanied by swelling. All exposed hydration sites are probably filled at this point.[13]

From the frequency and intensity changes of infrared bands of solid films of calf thymus LiDNA and NaDNA as a function of relative humidity, Falk and colleagues defined five preferred hydration sites. These sites have different binding affinities for the water of hydration. The $PO_2^-Na^+$ groups (phosphate groups of the phosphodiester backbone) become hydrated at relative humidity below 65% (about six water molecules are adsorbed completing the hydration of the phosphate group), while the hydration of the bases begins at higher relative humidities. Above 65% relative humidity, the C=O groups and the ring nitrogen atoms became hydrated. The strength of hydrogen bonding is greatest for the first water molecules adsorbed and decreases thereafter, approaching the strength of hydrogen bonding of liquid water.[23]

Falk and coworkers investigated polarized infrared and ultraviolet spectra of oriented films of the NaDNA as a function of relative humidity. It was concluded that DNA films were stable in the B configuration at relative humidity values as low as 75% and that at still lower humidities a reversible transition occurs to a disordered form in which the bases are no longer stacked one above another and are no longer perpendicular to the axis of the helix. The loss of base stacking upon drying suggested that the B configuration of DNA was stabilized by the stacking of the bases in the presence of water.[14]

In another study, Falk and colleagues measured the width of OH stretching infrared absorption band (3400 cm^{-1}) of water bound to solid calf thymus DNA as a function of temperature. It was found that the first 10 water molecules per base pair in the primary hydration shell do not freeze into an ice-like state at temperatures down to $-150°C$. An additional intermediate layer of about 3 wpn crystallizes with difficulty and tends to supercool if the cooling rate is high.[24]

Lee and coworkers investigated thermal denaturation of salmon testes NaDNA at low water contents (1–12 wpn) as a function of temperature using DSC and IR spectroscopy in the temperature range from $-10°C$ to 160°C. The samples with different moisture contents (1–12 wpn) were analyzed by DSC in closed cells. The DSC curves demonstrated a sharp endothermic maximum near 90°C in the first DSC scan for the sample with 12 wpn. This peak became broader, less intense, shifted to lower temperatures as the initial wpn of the DNA decreased (up to 6.3 wpn), and eventually vanished at 3.3 wpn due to the loss of ordered structure upon dehydration, as confirmed by the IR data. During the second scan, this endothermic maximum was replaced by a step increase in the heat flow curve for the samples with 3.3–4.7 wpn. DNA thermal denaturation in the 3.3–12 wpn range was irreversible but structural disruptions caused by dehydration were fully reversible. The calorimetric manifestation of the glass transition was established for denatured DNA at low hydration.[22]

Abdurakhman and coworkers studied effects of hydration on the structure of poly(dG)–poly(dC) DNA using infrared spectroscopy. The water content of the samples reduced as temperature increased and reached a very low value of about 0.1 wpn above 120°C. As humidity was lowered to 0%, the water content decreased to 1 wpn. The molecular vibrations, which characterize the backbone structure and base stacking and pairing, indicated that the samples maintained an A-form double helical structure at all the values of water content. The disorder in the base stacking was observed as a result of the decrease in the water content. Denaturation at high temperatures above

100°C was reversible with the decreasing temperature. At high temperature, the disorder was accompanied by denaturation, where the changes in the backbone occurred.[25]

Whitson and coworkers measured the enthalpy of desorption of the water of primary hydration bound to wet-spun films of CsDNA at 59% relative humidity by DSC as 30.88 kJ/mol H_2O. The net activation energy for desorption of this water from CsDNA was calculated as 60.8 kJ/mol H_2O from the DSC data by the Kissinger method.[26]

Hoyer and coworkers performed DSC experiments on desorption of water from poly(A) or DNA in the region of less than 10 water molecules per base pair. A broad endothermic transition centered near 70°C in the DSC thermogram was interpreted as being due to desorption of the water of hydration. For hydrations above 10 water molecules per base pair, the enthalpy of the desorption was the same as that for bulk water (about 41.9 kJ/mol) and increased to about 126 kJ/mol at two water molecules per base pair.[27]

Marlowe and colleagues performed DSC experiments on hydrated wet-spun films of calf thymus Cs, K, and NaDNA (with about eight water molecules per base pair) to study effect of counterions on binding of the water of primary hydration. Two prominent features were revealed by DSC: a broad endothermic transition near 75°C due to desorption of the primary water of hydration and a sharp exothermic transition near 240°C due to pyrolysis of the sample. The temperatures of both transitions increased with the increasing heating rate. The DNA films which were heated at a scan rate of 10°C/min to 180°C once yielded crystalline diffraction patterns indicating that A-DNA-like helices were preserved (even at very low relative humidity). But heating to 180°C thrice destroyed the double helical structure of DNA as well as the crystallinity of the sample.[18]

Lavalle and coworkers measured the water content and swelling of wet-spun films of calf thymus Na, K, Rb, and CsDNA as a function of relative humidity. The water content of the films was independent of counterion species. The A to B transition occurred at different relative humidity values implying that the intermolecular interactions were quite different in the DNAs with different counterions. The swelling experiments revealed that the intermolecular bonds changed in a very similar manner up to 84% relative humidity. Dramatic and irreversible changes were observed in the dimensions of the films above 84% relative humidity.[10]

Understanding of DNA degradation is important for studies that use or target DNA, as well as for biomedical applications such as thermal targeting of DNA in cancer cells. Such targeting can be achieved, for example, by using localized heating with gold nanoparticles.[28] Knowledge on DNA degradation can also be used in obtaining purines and pyrimidines. The sublimation of adenine, cytosine, guanine, and thymine from λ-DNA was tested under reduced pressure (0.13 kPa) at temperatures above 150°C. With the exception of guanine, ~60–75% of each base was sublimed directly (not acid hydrolyzed prior to heating) from the λ-DNA and recovered on a cold finger of the sublimation apparatus after heating to 450°C.[29]

Thermal degradation of pUC19 and pGY1 plasmids,[28] λ-DNA,[290] herring sperm DNA,[30,31] and NaDNA from salmon testes[22] were previously investigated. Karni and coworkers found that under dry conditions, degradation of pGY1 and pUC19 plasmids occurs in a linear manner, starts at about 130°C with complete degradation at around 190°C. Upon heating of the plasmids to 170–200°C, aggregates of short DNA fragments were formed resulting from incomplete degradation. After heating at 210°C, these fragments were destroyed (complete degradation). Degradation of DNA in solution was studied using a pressure system that prevents evaporation of the water. DNA degradation occurred between 100°C and 110°C, much lower than under dry conditions. This degradation temperature was lower than the boiling temperature of water, a temperature range in which DNA does not degrade when pressure is not applied. This result indicates that applying pressure (1 or 1.3 MPa) to a DNA solution greatly increases the sensitivity of DNA to heating. It is possible that the pressure weakens the chemical bonds between the atoms within the DNA molecule. Another possibility is that the exposure to oxygen within the compressed air that was used to create the pressure damaged the DNA and made it more sensitive to heat-induced degradation. They could not determine the degradation of DNA in aqueous solution using this method, since applying pressure on the DNA solution caused the DNA to be more sensitive to heat and to degrade already above 90°C.[28]

Recently DNA was investigated as a novel green natural alternative to organohalogen flame retardants. DNA's fire retardant activity is related to its intumescent-type thermal behavior. When an intumescent material is subjected to a heat flow, it develops a multicellular foamed carbonaceous shield (char) on its surface. This shield acts as a physical barrier able to limit the heat, fuel, and oxygen transfer between flame and polymer, leading to flame extinguishment.[32,33]

The intumescent behavior of DNA is due to its structure containing phosphate groups (produce phosphoric acid), deoxyribose units (act as a carbon source, dehydrates to form char and release blowing gases), and nitrogen-containing bases (release ammonia). Supramolecular assembled structure of DNA in which the three components are molecularly organized is also important for its "all-in-one" intumescent fire retardant behavior.[30]

Thermal degradation of DNA from herring sperm occurred through a number of steps following the removal of adsorbed water (completed until 150°C). In the first degradation step (150–200°C), two sharp weight losses were detected and a small amount of combustible volatiles were produced. The chemical modifications involved in thermal degradation of DNA were followed by the attenuated total reflectance (ATR) spectra of the residues after heating at selected temperatures in the range of 180–600°C. In the attenuated total reflectance Fourier-transform infrared (ATR-FTIR) spectra of the residue heated to 180°C, changes in intensities/frequencies of the bands at 3000 cm^{-1} (OH stretching vibrations), 890 cm^{-1} (sugar–phosphate vibrations), 1260 cm^{-1} (PO_2^- asymmetric stretching), 1430–1475 cm^{-1} (base-sugar vibrations), and 1660 cm^{-1} were observed. Water, CO_2, and isocyanic acid evolved in the 50–180°C range under air while OH groups appeared in the residue. The DNA underwent a significant physical modification at this stage: the original white DNA powder grains were blown to hard, blackish/red closed cells which is the typical behavior of intumescent fire retardant materials. At higher temperatures (180–230°C), the closed cells formed at 180°C disrupted likely due to increasing pressure of the gases trapped in the cells. DNA from herring sperm transformed into a multicellular, foamed, thermally insulating material at a relatively low temperature (160–200°C) as compared with an intumescent model compound, pentaerythritol diphosphate (300–350°C). Furthermore, on heat exposure, the intumescent residue was converted into a highly thermally stable ceramic-like material which is resistant to thermal oxidation unlike organically derived intumescent chars and hence preserved the intumescent fire barrier properties.[30] It was also shown that DNA can promote char formation either when coated on cotton fabrics[31–34] or on thick films of ethylene vinyl acetate copolymers.[35]

Very recently DNA was investigated as a flame retardant additive for melt-compounded formulations with low-density polyethylene (LDPE) and compared with LDPE compounded with melamine polyphosphate (MPP), one of the industry standard intumescent FR additives for plastics. DNA showed a much greater compatibility with the LDPE matrix than MPP. Biochemical characterization of heat-treated DNA revealed that DNA undergoes denaturation, fragmentation, and oxidation at 145°C.[36]

Using as a flame retardant agent requires bulk amounts of DNA, thus the major drawback is the cost. However, advances in DNA extraction from large-scale sources as well as its "all-in-one" intumescent formulation make DNA competitive intumescent fire retardant material.

In the present study, dehydration, melting, and thermal degradation kinetics of sodium salt of calf thymus DNA in dry state was investigated.

Kinetics of the reactions causing mass loss during the nonisothermal heating in the 15–200°C range was investigated using various model-free isoconversional methods. DSC analyses were performed both in open and hermetically sealed cells at different heating rates. The residues of the sealed cell DSC analysis, repeatedly heated/cooled in the temperature range of 10–175°C, were analyzed by transmission IR spectroscopy. The morphological changes in these residues were examined by scanning electron microscopy. Evolution of the IR bands during heating of the DNA under vacuum was followed by in situ diffuse reflectance infrared Fourier transform (DRIFT) spectroscopy.

13.2 MATERIALS AND METHODS

13.2.1 MATERIALS

Sodium salt of calf thymus DNA in fiber form from Sigma Aldrich (D1501) was used in the experiments as received. It is referred to as DNA further in the chapter for simplicity. It was stored at 4°C in refrigerator till it was used. The DNA from calf thymus is 41.9 mole% G–C and 58.1 mole% A–T with 2.94 wpn. The water content as received was measured as 11 wt% by thermal gravimetric analysis (TGA).

13.2.2 METHODS

13.2.2.1 THERMAL GRAVIMETRIC ANALYSIS

Thermal gravimetric analysis of DNA was made by heating the samples (5–10 mg) at 5, 10, and 15°C/min rates from 20°C to 600°C in Setaram Labsys TGA under nitrogen gas flow.

13.2.2.2 DSC ANALYSIS

DSC analyses of DNA under nitrogen flow in open aluminum pans and in hermetically sealed aluminum cells were performed using Shimadzu DSC 50 and TA instrument DSC Q10, respectively. The DNA samples were analyzed by heating at 5, 10, and 15°C/min rates in open pans from room temperature up to 600°C.

The DNA in hermetically sealed cells was heated at 5, 10, and 15°C/min rates from 10°C up to 175°C, kept at this temperature for 10 min, then cooled to room temperature at 5, 10, and 15°C/min rates and heated to 175°C again at the same rate.

13.2.2.3 TRANSMISSION FOURIER-TRANSFORM INFRARED (FTIR) ANALYSIS

FTIR spectra of the DNA as received, heated up to 600°C in open DSC cells and heated–cooled–reheated in hermetically sealed cells, were taken with KBr pellet technique using Shimadzu FTIR-8400. Two milligrams of DNA was mixed with 200 mg KBr and the mixture was pressed into 1-cm diameter pellets under 6 ton forces.

13.2.2.4 DIFFUSE REFLECTANCE INFRARED FOURIER TRANSFORM (DRIFT) SPECTROSCOPY ANALYSIS

DRIFT spectra were recorded using a praying mantis diffuse reflection attachment (Harrick Scientific Products Inc.) equipped with a high-temperature, low-pressure reaction chamber fitted with CaF_2 windows (HVC-DRP, Harrick Scientific Products Inc.), and FTS 3000 MX spectrophotometer (Digilab Excalibur Series). We placed 0.1 g DNA in the sample holder and after taking its spectrum at room temperature and 101.3 kPa pressure, the sample chamber was evacuated down to 0.1 Pa pressure and heated up to 175°C at 2°C/min rate under vacuum. The DRIFT spectrum of the sample was obtained at different temperatures during its dynamic heating. The sample chamber was cooled to room temperature under vacuum and opened to room atmosphere (25°C, 60% relative humidity). The DRIFT spectrum of the sample after exposure to the atmosphere was also recorded.

13.2.2.5 SCANNING ELECTRON MICROSCOPY

Microphotographs of DNA samples before and after the DSC analysis in hermetically sealed cells were taken by scanning electron microscope (Quanta FEG-250, FEI). The samples were coated with gold before the analysis.

13.3 RESULTS AND DISCUSSION

13.3.1 *THERMAL CHARACTERIZATION OF CALF THYMUS DNA*

13.3.1.1 *THERMAL GRAVIMETRIC ANALYSIS*

A polymer undergoes physical changes (such as glass transition and melting) and chemical changes (such as thermal degradation to smaller molecular mass fragments) on heating. While there are no mass losses for the glass transition and melting, the mass decreases as temperature increases if volatile molecules are formed on heating the polymer.

A general reaction can be written for dehydration and thermal degradation of a polymer even if it is much more complicated:

$$A_{(s)} \rightarrow B_{(s)} + C_{(g)},$$

for which the rate can be expressed as

$$\frac{d(1-\alpha)}{dt} = -k(T)(1-\alpha)^n, \tag{13.1}$$

where α is the conversion (extent of reaction), t is the time, $k(T)$ is the reaction rate constant, T is the absolute temperature, n is the reaction order, and $(1-\alpha)^n$ represents the reaction model describing the dependence of the rate on the fraction of solid remaining unreacted as the reaction proceeds and referred as $f(\alpha)$. The conversion is defined as

$$\alpha = \frac{m_o - m_T}{m_o - m_f}, \tag{13.2}$$

where m_0 is the initial mass of the sample, m_T is the mass of the sample at time t, and m_f is the final mass of the sample. For a nonisothermal thermogravimetric analysis under constant heating rate, that is, $\beta = dot/dt$, eq 13.1 becomes

$$\frac{d\alpha}{dT} = \frac{k(T)}{\beta} f(\alpha). \tag{13.3}$$

Substituting $k(T)$ by the Arrhenius expression, eq 13.3 yields

$$\frac{d\alpha}{dT} = \frac{A}{\beta} \exp\left(-\frac{E_a}{RT}\right) f(\alpha), \tag{13.4}$$

where A is the pre-exponential factor, E_a is the apparent activation energy, and R is the universal gas constant. The activation energy is denoted as apparent activation energy since the activation energy derived from thermogravimetric data is the sum of activation energies of chemical reactions and physical processes. Integrating eq 13.4 up to conversion α at the temperature T gives

$$\int_0^\alpha \frac{d\alpha}{f(\alpha)} = g(\alpha) = \frac{A}{\beta}\int_{T_o}^{T} \exp\left(-\frac{E_a}{RT}\right)dT \approx \frac{A}{\beta}\int_0^{T} \exp\left(-\frac{E_a}{RT}\right)dT$$

$$= \frac{AE_a}{R\beta}\int_0^x \frac{\exp(-x)}{x^2}dx = \frac{AE_a}{R\beta}p(x),$$

(13.5)

where T_0 is the initial temperature, $g(\alpha)$ is integral form of the reaction model, and $p(x)$ is the temperature integral for $x = E_a/RT$ which does not have analytical solution. If T_0 is low, the lower limit of the integral on the right-hand side of eq 13.5 can be approximated to be zero.

The Arrhenius parameters (A and E_a) together with the reaction model, $f(1 - \alpha)$, are called the kinetic triplet. These parameters can be estimated by model-free and model-fitting analysis methods. The isoconversional model-free methods allow the apparent activation energy to be determined as a function of the conversion without presumption of the reaction model. These methods assume that the reaction rate at a given conversion is only a function of the temperature. The most common isoconversional model-free kinetic methods used in evaluating activation energy are integral Flynn–Wall–Ozawa (FWO) and Kissinger–Akahira–Sunose (KAS) methods and differential Friedman method.

13.3.1.1.1 FWO Method

This method, suggested independently by Flynn and Wall,[37] and Ozawa,[38] uses Doyle's[39] approximation of $p(x)$ yielding

$$\ln\beta = \ln\left[\frac{AE_a}{Rg(\alpha)}\right] - 5.331 - 1.052\frac{E_a}{RT}.$$

(13.6)

For a constant conversion, the plot $\ln\beta$ versus $1/T$ from the data at different heating rates gives a straight line from the slope of which the apparent activation energy can be obtained.

13.3.1.1.2 KAS Method

KAS method[40,41] is based on the Coats–Redfern[42] approximation of $p(x)$ for $20 < x < 50$,

$$p(x) \cong \frac{\exp(-x)}{x^2}. \tag{13.7}$$

From eqs 13.5 and 13.7 it follows that

$$\ln\left(\frac{\beta}{T^2}\right) = \ln\left[\frac{AR}{E_a g(\alpha)}\right] - \frac{E_a}{RT}. \tag{13.8}$$

Thus, for constant conversion, the plot $\ln(\beta/T^2)$ versus $(1/T)$ gives a straight line whose slope can be used to calculate the apparent activation energy.

13.3.1.1.3 Friedman Method

The differential isoconversional method suggested by Friedman[43] is based on eq 13.4 in logarithmic form and leads to

$$\ln\left[\beta_i\left(\frac{d\alpha}{dT}\right)_{\alpha,i}\right] = \ln\left[Af(\alpha)\right] - \frac{E_{a,\alpha}}{RT_{\alpha,i}}, \tag{13.9}$$

where subscripts i and α designate a given value of heating rate and extent of conversion, respectively. The subscript α,i designates values related to a given conversion, and i is an ordinal number of the experiment conducted at the heating rates, β_i. This method leads to the activation energy for a given value of α by plotting $\ln(\beta_i (d\alpha/dT))$ against $1/T$.

13.3.1.1.4 Kissinger Method

Another model-free method is the Kissinger method, which is based on determination of the kinetic parameters by the shift of the peak maximum[40]:

$$\ln\left(\frac{\beta}{T_p^2}\right) = \ln\left(\frac{AR}{E_a}\right) - \frac{E_a}{RT_p}, \tag{13.10}$$

where T_p is the peak maximum temperature.

Figure 13.1 shows the TG curves recorded at different heating rates. The integrated and partial mass losses are given in Table 13.1. The mass loss in the 15–200°C range corresponds to elimination of water from the structure since it was heated under nitrogen gas flow. When DNA is heated in air, CO_2 and isocyanic acid are released, besides water, in this temperature range.[31] The mass loss in the 200–400°C and 400–600°C ranges correspond to pyrolysis of DNA and volatilization of DNA bases, respectively.[30]

TABLE 13.1   Partial and Integrated Mass Losses for DNA.

Heating rate (°C/min)	Mass loss (%)			
	15–200°C	200–400°C	400–600°C	Total
5	11.1	28.8	8.7	48.6
10	11.1	29.4	8.2	48.7
15	12.8	30.5	7.3	50.6

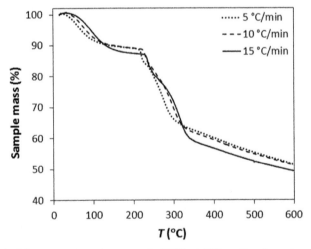

FIGURE 13.1   Thermogravimetric curves for DNA at different heating rates.

The derivative TG (DTG) curves (Fig. 13.2) showed mainly three peaks: a broad peak in the 25–200°C range, sharp second (200–250°C), and third (250–400°C) peaks. The maximum rate of mass loss increased with the increasing heating rate and the peak maxima shifted gradually to higher temperatures as the heating rate increased from 5°C to 15°C/min (from 51.4°C to 92.7°C, from 219.3°C to 235.6°C, and from 270.0°C to 318.3°C for the first, second, and third peaks, respectively).

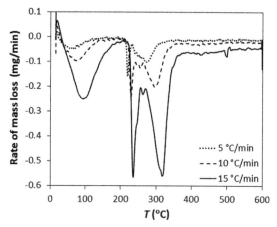

FIGURE 13.2   DTG curves for DNA at different heating rates under nitrogen flow.

The calf thymus DNA contains 11.1 wt% water initially corresponding to 2.94 wpn. This initial water content corresponds to the tightly bound water molecules in the primary shell and constitutes an integral part of the DNA-ordered structure. Variation of the water content of the DNA with temperature during heating up to 220°C under nitrogen flow is shown in Figure 13.3. The water content of the DNA at the temperature of the maximum mass loss rate was calculated as 2 wpn for all the heating rates. This wpn corresponds to about 21% relative humidity according to the relationship given by Lavalle and coworkers.[10]

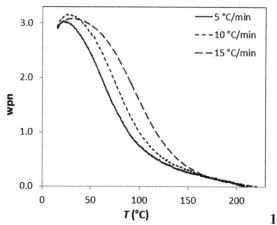

FIGURE 13.3   Variation of the water content of the DNA with temperature during the TG analysis.

The first peak in TG curves was analyzed using different kinetic models by assuming that this peak is due to desorption of the water from the DNA.

13.3.1.1.5   FWO Method

The FWO plots obtained for the temperature range of 15–200°C are presented in Figure 13.4. The lines have fairly equal slopes at low conversions but the slope has changed at high conversions ($\alpha > 0.70$). The apparent activation energy values obtained by FWO method were in the range of 37.15–99.38 kJ/mol with the average value of 47.26 kJ/mol as reported in Table 13.2.

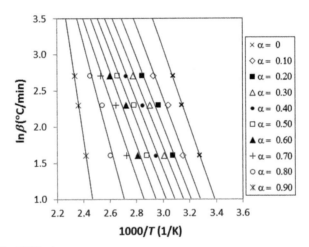

FIGURE 13.4   FWO plots.

TABLE 13.2   Average Activation Energies for the Individual Conversion Ranges.

	FWO		KAS		Freidman	
	$\alpha \leq 0.7$	$\alpha > 0.7$	$\alpha \leq 0.7$	$\alpha > 0.7$	$\alpha \leq 0.7$	$\alpha > 0.7$
Average E_a (kJ/mol)	39.70	66.06	35.96	62.85	41.67	70.58
Standard deviation (kJ/mol)	2.29	29.53	2.41	30.74	3.05	29.20

13.3.1.1.6   KAS Method

In accordance with the KAS method (eq 13.8), the $\ln(\beta/T^2)$ versus $1/T$ plots for the temperature range of 15–200°C gave straight lines (Fig. 13.5) from which the apparent activation energies were determined in the range of

33.32–97.56 kJ/mol with the average value of 43.72 kJ/mol as reported in Table 13.2.

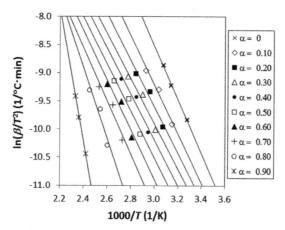

FIGURE 13.5 ln(β/T^2) versus $1/T$ plots (KAS method).

13.3.1.1.7 Friedman Method

Plotting of ln($\beta_i(d\alpha/dT)$) against $1/T$ (Fig. 13.6) gives a series of lines with high correlation coefficients ($R^2 > 0.879$) yielding the apparent activation energies in the range of 38.41–102.36 kJ/mol as reported in Table 13.2.

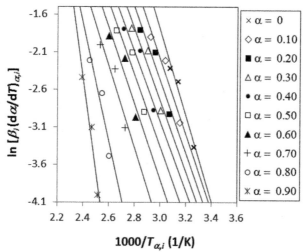

FIGURE 13.6 Friedman plots.

The dependence of activation energy, obtained from TG data in the 15–200°C range, on conversion for the three isoconversional methods (FWO, KAS, and Friedman) is shown in Figure 13.7. The apparent activation energies calculated by the different methods are compatible with each other and followed the similar trend with the conversion. The slight discrepancies in the apparent activation energies determined by the different methods are attributed to the assumptions in the derivation of the model equations.[44]

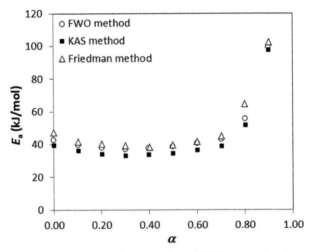

FIGURE 13.7 Dehydration activation energies of DNA as a function of conversion calculated from the isoconversional model-free methods (15–200°C).

E_a is found to vary with the extent of conversion indicating complexity of the dehydration reaction and reflects the changing mechanism (multistep mechanism) during the course of the reaction. The variation of activation energy as the reaction proceeds in homogenous state can be attributed to control of the reaction rate by several elementary steps, each having unique activation energy.[45] On the other hand, in an elementary solid-state reaction (heterogeneous reaction), the activation energy might vary with conversion due to heterogeneous nature of the solid sample, which might cause a systematic change in the reaction kinetics due to product formation, crystal defect formation, intracrystalline strain, or other similar effects. Kinetic complexities can also result from physical processes (e.g., sublimation, localized melting, adsorption–desorption, diffusion of a gaseous product, melting, particle size, and morphology effects, etc.) that have different activation energies.[46] In the present study, the initial water content of the DNA was 2.94 wpn as shown in Figure 13.3 and it is expected that removal of the water will result in a fully denaturized DNA at 175°C. These transformations might result in formation of solid products on the DNA surface having a different energy barrier to the vaporizing water molecules. Thus, the rate of dehydration might be controlled by the rate of diffusion of structural water molecules across this barrier.

In the observed E_a versus α curves, two regions exist: for α ≤ 0.70, where E_a is almost constant and for > 0.70, where E_a increases with the conversion. The averaged activation energies over each conversion range are reported in Table 13.2 together with the standard deviation values. For α ≤ 0.70 (corresponding to the final temperatures of 93.5°C, 104.4°C, and 121.8°C for 5°C, 10°C, and 15°C/min heating rates, respectively) with fairly constant activation energy with conversion, the volatilization of the structural water in DNA double helix can be assumed to follow a single-step reaction. This conversion range corresponds to 70 wt% of the total mass loss in the 15–200°C range and the final water content of 0.88 wpn.

The experimental values of E_a for the conversion below 0.70 are expected to reflect the energy barriers to volatilization of primary shell hydration water from DNA, thus E_a is dependent on the strength of interaction between water and DNA. The averaged activation energies over the conversion range of α ≤ 0.7 are slightly lower than the activation energy of bulk water vaporization (43.20 kJ/mol). This implied that vaporization of the strongly bound water molecules is at least partially limited by diffusion. The difference between the activation energies for vaporization of water from DNA and of bulk water (2–8 kJ/mol) is comparable to the hydrogen bond energy (5–10 kJ/mol) which would be the typical mechanism of interaction between water and DNA.[47] The increasing dependence of E_a on α observed for α > 0.7 might reflect concurrent (parallel, simultaneous) reactions.[48]

13.3.1.2 DSC ANALYSIS OF DNA IN OPEN AND HERMETICALLY SEALED CELLS

DSC thermograms recorded using open aluminum cells during heating of DNA up to 600°C under nitrogen flow are shown in Figure 13.8. There are mainly two peaks in these curves: an endothermic peak (with maximum at T_m) belongs to melting and dehydration of DNA, and an exothermic second peak (with maximum at T_d) due to pyrolysis of DNA. For 5°C/min heating rate, another exothermic peak was detected at 341.3°C (with enthalpy of 325.5 J/g).

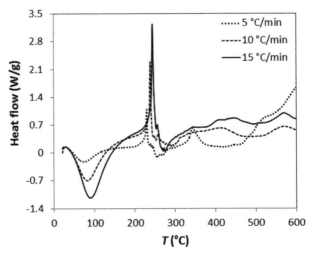

FIGURE 13.8   DSC curves for heating DNA in open cells under nitrogen flow.

One broad endothermic peak was detected in the DSC thermograms acquired in the hermetically sealed cells during the first heating to 175°C (Fig. 13.9). The transition has started at room temperature but has not completed at 175°C. The temperatures corresponding to the peak maxima (T_m) are 113.6°C, 112.9°C, and 114.3°C for 5°C, 10°C, and 15°C/min heating rates, respectively. These T_m values are higher than those detected in the DSC curves recorded using the open cells. At the end of the first heating, DNA was cooled at the same rate to 10°C and then reheated up to 175°C (second heating). The endothermic peak observed in the first scan is absent in the second scan thermogram indicating irreversibility of the transition occurred during the first scan.

The DSC thermograms presented here are similar to those obtained by Lee and coworkers for salmon testes NaDNA with initial water content of 12 wpn. They detected a sharp endothermic transition between 70°C and 100°C with maximum near 90°C in the first DSC scan in hermetically sealed pans during heating up to 140°C. Complete absence of the endothermic peak in the second scan thermogram indicated irreversibility of the thermal denaturation of DNA in solid state. In contrast, the endothermic peak reappeared in the second scan when the DNA solution was heated, implying reversibility of the denaturation in solution.[22]

In another study, reappearance of the endothermic peak during the second scan was observed by Marlowe and coworkers for wet-spun film of NaDNA after the sample was exposed to 75% relative humidity after the first heating at 180°C. If the sample was kept under dry atmosphere after the first scan, no transition was observed during the subsequent scan. The broad peak was assigned to dehydration of the DNA. Heating to 180°C once yielded crystalline diffraction patterns but thrice destroyed the double helical structure and crystallinity of the DNA.[18]

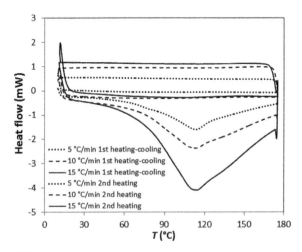

FIGURE 13.9 DSC curves for the analysis in hermetically sealed cell with heating up to 175°C, cooling to 10°C, and then reheating (second heating) to 175°C at different rates. The arrows indicate the direction of temperature change.

The extent of reactions taking place in the open and sealed cells during the DSC analyses in the temperature range of 15–200°C as a function of temperature was calculated from the ratio of the integrated area until the corresponding temperature to the total area of the peaks was detected in the DSC thermograms. Since the transition observed for the sealed cell during the first heating had not been completed at the final temperature (175°C), the upper end of the peaks were simulated numerically and the total area was calculated by numerical integration. According to this simulation, the endothermic transition in the sealed cell completed at around 200°C independent of the heating rate.

The extent of reactions taking place during the TG, open cell, and sealed cell DSC analyses as a function of temperature is shown in Figure 13.10. As seen in these plots, the extent of reactions followed the same trend during the TG and open cell DSC analyses up to the extent of reaction of about 0.7 for 5 and 10°C/min heating rates (Fig. 13.10a and b). After this limit, the reactions in the open DSC cells proceed faster with the increasing temperature.

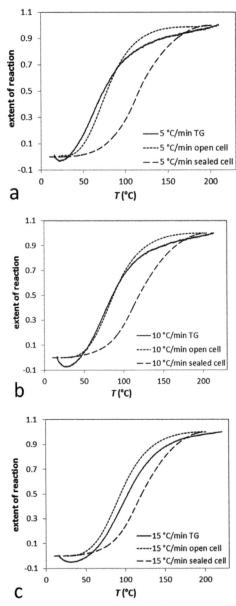

FIGURE 13.10 Extent of reactions as a function of temperature for TG, open cell, and sealed cell DSC analyses at the heating rate of (a) 5°C/min, (b) 10°C /min, and (c) 15°C/min.

Effect of heating rate on the temperature sensitivity of the reaction rates for the TG, open cell, and sealed cell DSC analysis can be seen in Figure 13.11. The extent of reaction in the sealed DSC cell is slightly sensitive to temperature while those occurring during the open cell DSC and TG analyses are highly temperature dependent.

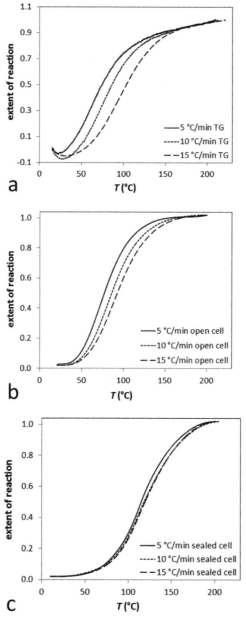

FIGURE 13.11   Effect of heating rate on the extent of reactions for (a) TG, (b) open cell, and (c) sealed cell DSC analyses.

Denaturation (melting) and dehydration of DNA double helix in open cells can be represented by eq 13.11:

$$\text{hydrated DNA double helix} \xrightarrow{\text{heat}} \text{DNA single strand} + H_2O_{(g)}. \quad (13.11)$$

In open cell, the water desorbed from DNA is carried away from the system by the flowing nitrogen gas. On the other hand, during the sealed cell DSC analyses, the water bound to the DNA is released as in the open cell but vapor–liquid equilibrium is established in the cell (Fig. 13.12). The amount of water released above the saturation limit condenses in the cell on the sample. When the vapor pressure of released water reaches 0.3 MPa (at around 135°C), the water vapor starts to escape from the cell. After that, the pressure in the cell remains constant at 0.3 MPa. Thus, the reaction occurring in closed cells can be written as

$$\text{hydrated DNA double helix} \xrightarrow{\text{heat}} \text{DNA single strand} + H_2O_{(g)} + H_2O_{(l)}. \quad (13.12)$$

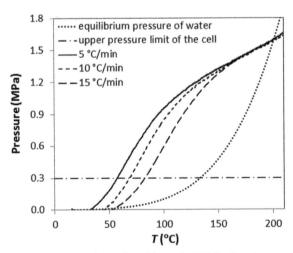

FIGURE 13.12 The hypothetical (predicted from the TG data) and saturation pressures of water vapor in the sealed DSC cells.

Considering the water desorbed from DNA consumed an average heat of vaporization of 2.24 kJ/g water (average value for the temperature range of 10–200°C) and knowing the amount of water desorbed from the TG analyses, the enthalpy change of melting of DNA in the open and sealed DSC cells (ΔH_m) was calculated by subtracting the enthalpy change of vaporization of water (ΔH_{vap}) from the calorimetric enthalpy change (ΔH_{cal}) and is given in Table 13.3. The calorimetric enthalpy corresponds to the area under the first endothermic peaks.

TABLE 13.3 T_m and Enthalpy Changes Calculated from the Open and Hermetically Sealed Cell DSC Analyses.

Heating rate (°C/min)	Open cells					Hermetically sealed cells		
	T_m (°C)	ΔH_{cal} (J/g)	ΔH_m (J/g)	T_d (°C)	ΔH_d (J/g)	T_m (°C)	ΔH_{cal} (J/g)	ΔH_m (J/g)
5	73.3	280.5	34.9	230.0	−171.3	113.6	534.1	297.2
10	82.9	385.8	140.2	238.9	−97.1	112.9	458.1	222.8
15	89.3	418.5	172.9	244.6	-117.9	114.3	538.6	266.5

Higher heat of melting (ΔH_m) values were obtained at higher heating rates in the open cells. The ΔH_m values determined in the present study from the open cell DSC analyses are in good agreement with values given in the literature. Duguid and coworkers[19] reported heat of melting value for calf thymus DNA as 41.0 J/g. Spink and coworkers[49] showed that DNA with A–T base pairs and G–C base pairs has 50.5 J/g and 76.8 J/g heat of melting, respectively, in ethylene glycol.

The calculated ΔH_m values were higher for the hermetically sealed cell than those for the open cell indicating that there were other endothermic reactions besides melting and dehydration of DNA during heating in the sealed cells. Under the 0.3 MPa pressure in the sealed cells, formation of short DNA fragments as a result of incomplete degradation of the DNA is possible.[28] Heating of DNA powder from herring sperm (<50 base pair degraded) in the solid state in hermetic pans at 145°C caused complete denaturation and formation of individual free nucleotides and single-stranded DNA oligomers.[36] Also incomplete degradation of DNA in solution at temperatures above the melting temperature caused formation of DNA fragments.[50] The formation of these fragments also involves enthalpy changes that might lead to the higher heat of melting values found.

13.3.1.3   KISSINGER ANALYSIS

As mentioned above (Section 13.3.1.1), the temperature at which the maximum mass loss occurs (T_p) in the DTG curves increased with the heating rate signifying that the reaction is thermally activated (kinetics limited).[18] The three mass loss peaks in the DTG curves of DNA (see Fig. 13.2) were analyzed by the Kissinger method. The Kissinger plots given by eq 13.10 are shown in Figure 13.13 and the calculated activation energies are given in Table 13.4.

TABLE 13.4 Activation Energies for the DTG Peaks by the Kissinger Method.

Temperature range	Correlation coefficient (R^2)	T_p (°C)			E_a (kJ/mol)
		5°C/min	10°C/min	15°C/min	
15–200°C	0.862	63.4	73.1	92.6	30.55
200–270°C	1.000	219.6	229.3	235.5	135.69
250–400°C	0.986	273.3	296.4	317.3	57.92

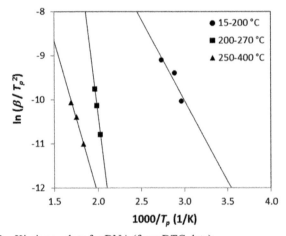

FIGURE 13.13 Kissinger plots for DNA (from DTG data).

The activation energy of the first mass loss step which is attributed to dehydration is lower as compared with the pyrolysis steps at higher temperatures. Furthermore, the activation energy of the first step is lower than those previously calculated by the isoconversional methods for degree of conversion below 0.70. The correlation coefficient of the first step is low as compared with those at higher temperatures. This is resulted from the broadness of the peak and consideration of only the peak maxima by the Kissinger method.

As given in Table 13.3, the maxima of the endothermic and exothermic peaks shifted to higher temperatures with the increasing heating rates for the open cell DSC analyses. On the other hand, shift in the peak maximum temperature was less prominent for the sealed cells. The Kissinger plots for the peaks detected in the DTG, open cell, and sealed cell DSC thermograms in the 15–200°C range are shown in Figure 13.14 and the activation energies calculated are given in Table 13.5.

TABLE 13.5 Activation Energies for the First Peaks in the DTG, Open Cell, and Sealed Cell DSC Thermograms by the Kissinger Method.

Thermal analysis method	Correlation coefficient (R^2)	E_a (kJ/mol)
DTG	0.862	30.55
Open cell DSC	0.999	66.06
Sealed cell DSC	0.117	342.24

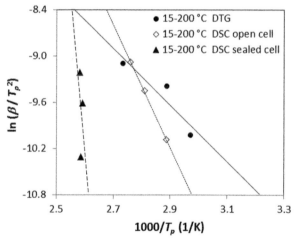

FIGURE 13.14 Kissinger plots for the peaks detected in the DTG, open cell, and sealed cell DSC thermograms in the 15–200°C range.

The activation energy determined from the DTG data by the Kissinger method is the average value for the dehydration in the 15–200°C temperature range. However, we detected two regions in this temperature range by the isoconversional methods as given in Table 13.2. The activation energy determined from the open cell DSC data by the Kissinger method is in good agreement with that determined by the isoconversional methods for the conversions above 0.7. This value also agrees well with the values reported for the desorption of primary hydration water from hydrated wet-spun films of calf thymus CsDNA (58 and 61 kJ/mol).[18,26]

The linearity of the Kissinger plot for the sealed cell DSC data was extremely low indicating that the reactions taking place in these cells are not kinetics limited.[18] Furthermore, if multiple reactions having different activation energies occur simultaneously, the temperature dependence of the overall rate (observed as a single transition in the DSC curve) cannot be fit by a single kinetic equation. Also if the transition has several steps comprising the chemical reaction step, heat, and mass transfer steps, and so on, depending on the experimental conditions, one of these steps can be the rate limiting step. Moreover, for our case, the static atmosphere and high water vapor pressure in the sealed cells might lead to diffusion hindrance and back reactions further complicating the problem. Besides, the Kissinger method assumes that the reaction rate is maximum at the peak maximum temperature (T_p). Thus, the Kissinger plot is linear only if the reacted fraction at the maximum reaction rate is independent of the heating rate.[51] However, there is no evidence for the T_p values in the sealed cell DSC thermograms to be the temperatures where the reaction rate is maximum. Also from the sealed cell DSC data, the reacted fractions at the peak maxima are 0.42–0.46, that is, not independent of the heating rate.

The Kissinger plots for the second peaks in the DTG and open cell DSC thermograms (200–270°C) are shown in Figure 13.15 and the activation energies calculated from these plots are given in Table 13.6.

TABLE 13.6 Activation Energies for the Second Peaks in the DTG and Open Cell DSC Thermograms by the Kissinger Method (200–270°C).

Thermal analysis method	Correlation coefficient (R^2)	E_a (kJ/mol)
DTG	0.999	135.96
Open cell DSC	0.999	154.45

In this temperature range, the DNA undergoes pyrolysis. The correlation coefficients are high suggesting that the complexities are not valid for this temperature range. The values found here are close to that reported for wet-spun films of calf thymus CsDNA (T_p of around 240°C) as around 183 kJ/mol.[18]

The discrepancies in the activation energies calculated using the data from the different thermal methods imply occurrence of different reactions under the experimental conditions and that the thermal analyses are performed in the temperature range of interest.

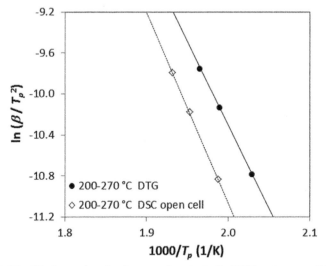

FIGURE 13.15 Kissinger plots for the second peaks in the DTG and open cell DSC.

13.3.2 *MORPHOLOGICAL CHARACTERIZATION OF UNHEATED AND HEATED CALF THYMUS DNA*

Morphological changes occurred during the sealed cell DSC analyses (repeatedly heated/cooled between 10°C and 175°C at different heating/cooling rates) were deduced based on the SEM micrographs of the residues. SEM micrographs for the native and heated/cooled DNA samples are shown in Figure 13.16. The native form of the calf thymus DNA is in the form of a bundle of parallel fibers, each fiber having 0.05–0.2 μm diameter (Fig. 13.16a).

The heating/cooling at 5°C/min caused the fibrous structure to be transformed into a congealed block of melted polymer. Thick fibers and DNA fragments were also observed. No indication for the intumescence action has been observed for this residue (Fig. 13.16b). Formation of the fragments as a result of incomplete degradation of the DNA under the high pressure in the cell was discussed above (Section 13.3.1.2) and suggested as a possible reason for the higher melting enthalpies from the sealed cell DSC data. The DNA residue formed after heating/cooling cycles at 10°C/min rates in the sealed DSC cell exhibited the typical behavior of intumescent fire retardant materials: microbubbles were formed on the fibers (Fig. 13.16c). In the micrograph of the DNA residue formed after heating/cooling cycles at 15°C/min, the bundles of parallel fibers reappeared. Densely packed microbubbles with diameter less than 0.1 µm can be seen on a bundle of about 1 µm diameter (Fig. 13.16d). The samples have become blackened after heating at 175°C, indicating degradation accompanying the morphological changes.

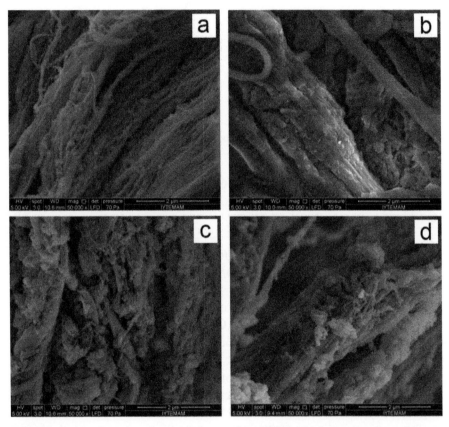

FIGURE 13.16 Scanning electron micrographs of DNA: (a) before heating, after two heating/cooling cycles between 10°C and 175°C in hermetically sealed DSC cells at a rate of (b) 5°C/min, (c) 10°C/min, and (d) 15C/min (scale bar 2 µm for all images).

The morphological changes observed by SEM can be explained by entrapment of the gases released from DNA in the molten DNA.[30,31] This behavior involves transition of DNA from a solid to a viscous liquid on heating, with simultaneous thermal decomposition of weak chemical structures that produce the blowing gases. The thermally induced bond scission is however limited, leading to a stable macromolecular structure, which is essential to the viscoelasticity required by the blowing action.[30] DNA is transformed from a solid to a viscous liquid with simultaneous thermal decomposition of weak chemical structures that produce the blowing gases. In 50–180°C range, Alongi and coworkers[30] detected CO_2 and isocyanic acid besides water as volatiles by TG-FTIR upon degradation of herring sperm DNA in air. In the present study, during heating in the sealed cell under high pressure of water vapor, formation of volatiles as a result of degradation of DNA besides water is possible. These volatiles (blowing gases) have to be trapped and to be diffused slowly in the viscous, melt, degraded DNA in order to exhibit the intumescent behavior. The viscoelastic melt, degraded DNA expands as volatiles are produced and at the same time, cross-linking reactions and charring cause the matrix to harden and to produce highly porous char.[52]

DNA has a significant weight loss between 10°C and 175°C (around 11 wt%) due to desorption of the primary shell hydration water under nitrogen flow. The change of the reaction kinetics for the extent of reaction above 0.7 (above 94°C for 5°C/min) was resolved by the isoconversional analysis of the TG data as discussed before (Section 13.3.1.1).

13.3.3 TRANSMISSION FTIR AND DRIFT SPECTROSCOPY

13.3.3.1 TRANSMISSION FTIR

Transmittance IR spectra of the native DNA and residue after the DNA was heated to 600°C under nitrogen flow are shown in Figure 13.17. The IR vibrations for the native DNA were detected at 960, 1018 (shoulder), 1063, 1089, 1240, 1375, 1425, 1490, 1535, 1653, 1693, 2191, 2930 cm⁻¹ (shoulder) and a broad band in the 2400–3700 cm⁻¹ region with maxima at 3440 cm⁻¹.

The IR spectrum of the residue is very different from that of the native DNA. The bands in the spectrum of native DNA were replaced by 520, 756, 775, 900, 995, 1107, 1170, 1240, 1315, 1620, and 1700 cm⁻¹ bands in the spectrum of the residue. These bands cannot be associated to an organic structure but rather to an inorganic ceramic-like material rich in phosphate and nitrogen as in the case of melamine phosphate thermal degradation.[53] A broad band in the 1800–3700 cm⁻¹ region detected in the spectrum of the residue can be assigned to N–H and C–H stretching vibrations. The broad band at 3440 cm⁻¹ can be assigned to OH groups with a possible contribution from residual water not removed.

Upon heating to 175°C, DNA produced an expanded cellular carbon structure derived from the dehydration of sugar units (probably induced by reacting with phosphate groups) and swollen by the released gases formed from the nitrogen bases. Following breaking of the weak C–N sugar-base bonds at the beginning of the degradation, unsaturated sugar–phosphate chains are formed. Above 400°C, cross-linking reactions take place between the phosphorous and nitrogen-based condensates involving acidic phosphorous and basic nitrogen groups. As a result thermally stable P–N-based structure is formed.[30]

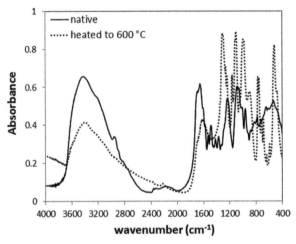

FIGURE 13.17 Transmittance IR spectra of the native DNA and residue after the DNA heated to 600°C under nitrogen flow.

The thermal resistance of the intumescent char is important in terms of its fire retardant performance. A higher amount of char formed may indicate a lower amount of volatile species released during the thermal degradation of a polymer. On heating to 600°C, calf thymus NaDNA produced a residue of about 50 wt% according to the TG analysis. If only Na^+ and phosphate ions remained after the pyrolysis, the residue would be 35 wt%. Thus, it can be concluded that there are other carbonized species in the residue besides phosphates. During cross-linking reactions between phosphorous and nitrogen-based condensates above 400°C, carbon may be trapped and thermally stabilized in the P–N-based structure.[30]

From SEM micrographs (Fig. 13.16), it was seen that DNA exhibited intumescent behavior upon heating to 175°C in the sealed cells as a result of melting and degradation, and release of water was assumed to be the major volatile product. The chemical transformations occurred in the DNA during the sealed cell DSC analysis were followed by the transmittance IR spectroscopy.

Transmittance IR spectra of the native DNA and of the DNA that was repeatedly heated/cooled in the sealed DSC cells at different rates are shown in Figure 13.18. Specifically, the bands in the 1500–1800 cm^{-1} region are assigned to base stacking and pairing and those at 1060 cm^{-1} and 1089 cm^{-1} to backbone conformation.[22] In preparing KBr pellets for the samples, 2 mg DNA was used. Considering errors in weighing of the samples, the band intensities were normalized by the intensity of 1234 cm^{-1} band. This band is due to the PO$_2^-$ asymmetric stretching vibrations and is largely invariant to temperature, thus it is used as a standard for intensity normalization.[22]

Heating of DNA in the sealed cell led to slight increase in the intensities of 1089 cm^{-1} (PO$_2^-$ symmetric stretching vibration) and 1063 cm^{-1} (C–O stretching vibration of the phosphodiester) bands (Fig. 13.18a). These changes imply conformational changes in the deoxyribose–phosphate backbone. The normalized absorbance of 1089 cm^{-1} band is a good measure of thermal denaturation of DNA. These bands are primarily assigned to vibrational modes of the ordered deoxyribose–phosphate backbone and are highly sensitive in frequency and intensity to DNA melting.

The position of 1234 cm^{-1} band did not change after heating to 175°C (Fig. 13.18b). This band is due to the asymmetric PO$_2^-$ stretching vibrations. Not the intensity but the frequency of this band is known to be strongly dependent on the water content of the DNA: at high water content, it appears at 1220 cm^{-1} and linearly shifts to a higher frequency with the decreasing water.[10]

Intensity of the bands in the range of 1400–1500 cm^{-1} (vibrations for coupling between base–sugar entities) increased in the spectra after the DNA was heated/cooled at 175°C at 15°C/min rate (Fig. 13.18c). The bands in this range became less prominent and band at 1480 cm^{-1}, which is assigned mainly to purine imidazole ring vibrations,[54,55] disappeared in the spectra of the sample heated/cooled at 5°C/min. Alongi and coworkers detected a small intensity increase at 1430–1475 cm^{-1} in the spectra of herring sperm DNA heated in air at 180°C and interpreted this as an indication of chemical modifications involved in the degradation step concerning the intumescent behavior. The bands at 1260, 1060, 1006, and 960 cm^{-1} also lost intensity after the heating.[30]

For 5°C/min rate, the bands in the 1550–1750 cm^{-1} region, which are assigned to double-bond in-plane vibrations of the bases (e.g., C=O, C=N, and C=C), were replaced by a single band (Fig. 13.18d) probably due to condensation of bases after breaking of the weak C–N sugar–phosphate bonds at low heating rate. In the literature, shifting of the 1653 cm^{-1} band to a lower frequency was attributed to conversion of carbonyl groups (C–O) into C=N bonds upon the condensation.[30] This band is sensitive to base stacking and pairing.[22] Thus, the observed changes in the intensity can be interpreted

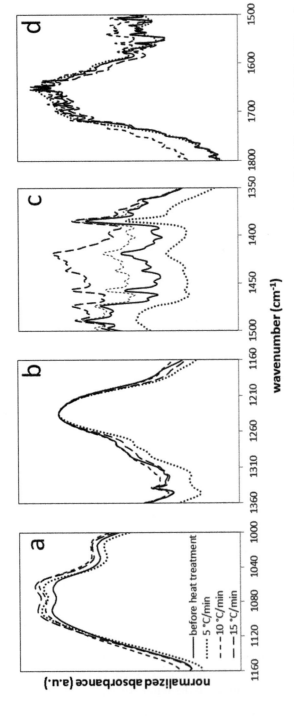

FIGURE 13.18 Transmission FTIR spectra of the native and heated DNA in the spectral regions of (a) 1000–1160 cm⁻¹, (b) 1150–1350 cm⁻¹, (c) 1300–1500 cm⁻¹, and (d) 1500–1800 cm⁻¹. The legend given in (a) is valid for the others.

as disruption of stacking and pairing of bases. The decrease in the intensity of this band was attributed to rupture of hydrogen bonds between the base pairs (as during melting of DNA).[19] No such effects were observed for the residues obtained at 10°C and 15°C/min.

The band at 877 cm^{-1}, which is assigned to sugar–phosphodiester vibrations, has a higher intensity in the spectrum of the residue obtained at heating rate of 15°C/min (Fig. 13.19a). This is due to conversion of carbonyl groups (C–O) into C=N bonds during the bases condensation reactions. It was also reported that disruption of hydrogen bonding between water molecules and polar groups of the sugar–phosphate backbone upon heating the DNA solution might produce the band at 872 cm^{-1}.[19]

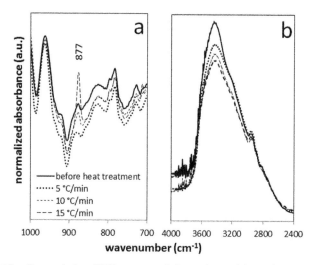

FIGURE 13.19 Transmission FTIR spectra of the native and heated DNA in the spectral regions of (a) 700–1000 cm^{-1} and (b) 2400–4000 cm^{-1}. The legend given in (a) is valid for (b).

The intensity of 3440 cm^{-1} band decreased gradually with the increasing heating rate (Fig. 13.19b). The intensity of this band is strongly dependent on the water content and has been used as a measure of water content of DNA.[11] The lower intensity of this band in the spectra of residues obtained at higher rates might suggest evolution of higher amounts of water. This is in good agreement with the more prominent intumescence characteristic of the residue obtained at 15°C/min observed in the SEM micrograph (Fig. 13.16d). However, the higher intumescent action for this residue cannot be explained only by the amount of water released. Since the overall intumescent process of DNA depends on the competing reactions through which its thermal degradation takes place, the chemical structure of the char may have an important effect on its foaming behavior.[56]

As a consequence, the intumescent behavior of the DNA observed by the SEM analysis is accompanied by the structural modifications (upon dehydration, melting and degradation) as deduced from the transmittance IR spectra of the residues. The effect of heating rate on these structural changes was also interpreted from these spectra.

13.3.3.2   DRIFT SPECTROSCOPY ANALYSIS

Effect of evacuation and heating under vacuum on the structural vibrations of the DNA was examined through the in situ DRIFT spectroscopy analysis. Variations in the normalized intensities of 1089, 1653, and 3440 cm^{-1} bands in the DRIFT spectra with temperature under vacuum are shown in Figure 13.20. Also variations in the normalized band intensities with evacuation and heating during the in situ DRIFT spectroscopy analysis are shown in Figure 13.21.

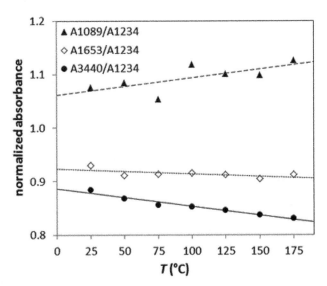

FIGURE 13.20   Variation of the normalized intensities of 1089 cm^{-1}, 1653 cm^{-1}, and 3440 cm^{-1} bands during the in situ DRIFT spectroscopy analysis under vacuum.

The intensity of 1089 cm^{-1} increased upon evacuation at room temperature and continued to increase gradually with the temperature; although small fluctuations are apparent. Upon cooling and exposure to the room atmosphere (60% relative humidity) after the heating, the intensity continued to increase (irreversible changes in deoxyribose–phosphate backbone). Semenov and coworkers reported that the intensity of this band is strongly dependent on the water content and decreases under nitrogen flow at room temperature.[57]

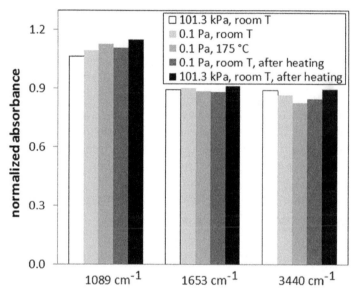

FIGURE 13.21   Variation of normalized intensities of 1089 cm^{-1}, 1653 cm^{-1}, and 3440 cm^{-1} bands with evacuation and heating during the in situ DRIFT spectroscopy analysis.

Slight loss of intensity was observed for 1653 cm^{-1} during heating under vacuum which recovered its initial value upon exposure to the room atmosphere (60% relative humidity), indicating reversibility of the changes in stacking and pairing of the bases upon increase in the relative humidity. As stated before, the loss in the intensity can also be attributed to rupture of hydrogen bonds between base pairs (as during melting of DNA).[19]

The 3440 cm^{-1} band also lost intensity upon evacuation and heating as a function of temperature continuously due to removal of the water. Upon exposure to the room atmosphere (60% relative humidity), the band gained its initial intensity as a result of adsorption of water. The normalized intensity of 3440 cm^{-1} band is positively correlated with that of 1653 cm^{-1} band and negatively with that of 1089 cm^{-1} (not shown).

The recovery of the initial intensities of the 3440 and 1653 cm^{-1} bands upon exposure to the atmosphere and irreversible intensity loss for 1089 cm^{-1} band indicates that the bases were restacked and repaired upon hydration of the DNA but the backbone conformation was irreversibly destroyed. Thus, the endothermic transition observed in the DSC thermograms recorded during the first heating in the sealed cell is not due to denaturation upon loss of the hydration water, but due to the thermal degradation of the DNA.

The intensities of the 1089, 1653, and 3440 cm^{-1} bands are more temperature sensitive under vacuum than under water vapor of 0.3 MPa pressure. Thus, higher enthalpies are required in order to modify the sugar–phosphate backbone conformation, base stacking, and pairing as well as water content of the DNA under the pressure as reflected by the higher melting enthalpies for the sealed cells (Table 13.3). Also during the heating under dynamic evacuation, any gases evolved are continuously removed from the sample. This may affect the intumescence process and thus fire-retardant performance of the DNA. Loss of structural water is strongly affected by the external pressure. Presence of water vapor on the sample leads to a decrease in the dehydration rate as compared to that in vacuum, that is, an increase in the activation energy for dehydration. The presence of water vapor affects not just the dehydration but the subsequent thermal reactions, and hence the characteristics of the residue produced.[58]

13.4 CONCLUSIONS

The intumescent behavior of Na salt of calf thymus DNA fibers was investigated for the first time. The thermal denaturation and dehydration kinetics of sodium salt of calf thymus DNA was investigated using the nonisothermal thermogravimetric data obtained at three different heating rates. The mass loss of DNA due to dehydration was 11% and the remaining mass at 600°C was 50% at 10°C/min heating rate. By the application of the model-free methods, the average apparent activation energy in the temperature range of 15–200°C was found to be dependent on the conversion, indicating the complexity of the dehydration reaction and reflects the changing mechanism (multistep mechanism) during the course of the reaction. On the observed E_a versus α curves, two different regions exist: for α ≤ 0.70, where E_a approximately does not change with conversion and α > 0.70, where E_a increases with the conversion. For α ≤ 0.70 (corresponding to the final temperatures of 93.5°C, 104.4°C, and 121.8°C for 5°C, 10°C, and 15°C/min heating rates, respectively) with fairly constant activation energy with conversion, the volatilization of the structural water in DNA double helix can be assumed to follow a single-step reaction. The averaged activation energies over this conversion range were determined as 39.70 ± 2.29, 35.96 ± 2.41, and 41.67 3.05 kJ/mol using FWO, KAS, and Freidman methods, respectively. These activation energy values are slightly lower than the activation energy of bulk water vaporization implying diffusion-controlled vaporization. For $\alpha > 0.70$, the corresponding average activation energies were found to be 66.06 29.53, 62.85 ± 30.74, and 70.58 ± 29.20 kJ/mol, respectively. The Kissinger method indicated 30.55, 135.69, and 57.92 kJ/mol activation energy for the 15–200°C, 200–270C, and 250–400°C ranges, respectively. The approach described in the present study allowed the determination of the activation energy values for the dehydration and degradation steps for DNA. The DSC analysis of DNA in open cell and hermetically sealed closed cells indicated higher melting peak maximum (T_m), 110°C, than that in open cell, 80°C. The heat of melting value of DNA in the sealed cell is higher than heat of melting value of open cell for 5°C/min heating rate which are 297.2 and 37.9 J/g, respectively. For 10°C/min and 15°C/min heating rate, the heat of melting value of DNA sealed cell experiments was also higher than that measured in open cell. Since it was thought that DNA degradation will occur at higher temperatures, heating was carried out only up to 175°C in closed cells. However, the first peak in sealed cell experiments was not completed during heating up to 175°C. By the application of Kissinger method to the DSC curves, the R^2 values of 0.99 and 0.12 were obtained for the endothermic first peak in open and sealed cells, respectively. When DNA was heated, morphology of the fibers was changed and thicker and irregular fiber structures with swollen bulbs were formed. However, when DNA was cooled, the exothermic crystallization peak could not be seen. Both the transmission FTIR and DRIFT spectroscopy analysis showed that normalized 1653 and 3440 cm^{-1} band intensities decreased as DNA lost water besides melting up to 175°C. The water loss (denaturation) caused the intensity of 1089 cm^{-1} band to increase.

KEYWORDS

- calf thymus DNA
- dehydration
- degradation
- activation energy
- in situ DRIFT spectroscopy

REFERENCES

1. Light, R. J. *A Brief Introduction to Biochemistry*; Benjamin: New York, 1968.
2. Patil, S. D.; Rhodes, D.G.; Burgess, D. J. DNA-Based Therapeutics and DNA Delivery Systems: A Comprehensive Review. *AAPS J.* **2005**, *7*, E61–E77.
3. Hofemeister, J. Achievements in Application of Recombinant DNA Technology for Creation of Industrial Microorganisms. *Zentralbl. Mikrobiol.* **1988**, *143*, 551–560.
4. Teles, F. R. R.; Fonseca, L. P. Trends in DNA Biosensors. *Talanta* **2008**, *77*, 606–623.
5. Thomsen, P. F.; Willerslev, E. Environmental DNA—An Emerging Tool in Conservation for Monitoring Past and Present Biodiversity. *Biol. Conserv.* **2015**, *183*, 4–18.
6. Texter, J. Nucleic Acid–Water Interactions. *Prog. Biophys. Mol. Biol.* **1978**, *33*, 83–97.
7. Westhof, E. Water: An Integral Part of Nucleic Acid Structure. *Annu. Rev. Biophys. Bio.* **1988**, *17*, 125–144.
8. Saenger, W. Structure and Dynamics of Water Surrounding Biomolecules. *Annu. Rev. Biophys. Biophys. Chem.* **1987**, *16*, 93–114.
9. Yakovchuk, P.; Protozanova, E.; Frank-Kamenetskii, M. D. Base-Stacking and Base-Pairing Contributions into Thermal Stability of the DNA Double Helix. *Nucleic Acids Res.* **2006**, *34*(2), 564–574.
10. Lavalle, N.; Lee, S. A.; Rupprecht, A. Counterion Effects on the Physical Properties and the A to B Transition of Calf-Thymus DNA Films. *Biopolymers* **1990**, *30*, 877–887.
11. Falk, M.; Hartman, K. A.; Lord, R. C. Hydration of Deoxyribonucleic Acid II. An Infrared Study. *J. Am. Chem. Soc.* **1963**, *85*, 387–391.
12. Mrevlishvili, G. M.; Carvalho, A. P. S. M. C.; Ribeiro da Silva, M. A.V.; Mdzinarashvili, T. D.; Razmadze, G. Z.; Tarielashvili, T. Z. The Role of Bound Water on the Energetics of DNA Duplex Melting. *J. Therm. Anal. Calorim.* **2001**, *66*, 133–144.
13. Falk, M.; Hartman, K. A. Jr.; Lord, R. C. Hydration of Deoxyribonucleic Acid. I. A Gravimetric Study. *J. Am. Chem. Soc.* **1962**, *84*, 3843–3846.
14. Falk, M.; Hartman, K. A. Jr.; Lord, R. C. Hydration of Deoxyribonucleic Acid. III. A Spectroscopic Study of the Effect of Hydration on the Structure of Deoxyribonucleic Acid. *J. Am. Chem. Soc.* **1963**, *85*, 391–394.
15. Zehfus, M. H.; Johnson, W. C. Conformation of P-Form DNA. *Biopolymers* **1984**, *23*, 1269–1281.
16. Lindsay, S. M.; Lee, S. A.; Powell, J. W.; Weidlich, T.; Demarco, C.; Lewen, G. D.; Tao, N. J. The Origin of the A to B Transition in DNA Fibers and Films. *Biopolymers* **1988**, *27*, 1015–1043.
17. Sclavi, B.; Peticolas, W. L.; Powell, J. W. Fractal-Like Patterns in DNA Films, B Form at 0% Relative Humidity, and Antiheteronomous DNA: An IR Study. *Biopolymers* **1994**, *34*, 1105–1113.
18. Marlowe, R. L.; Lukan, A. M.; Lee, S. A.; Anthony, L.; Chandrasekaran, R.; Rupprecht, A. J. Differential Scanning Calorimetric and X-Ray Study of the Binding of the Water of Primary Hydration to Calf-Thymus DNA. *J. Biomol. Struct. Dyn.* **1996**, *14*, 373–379.
19. Duguid, J.; Bloomfield, V.; Benevides, J.; Thomas, J. DNA Melting Investigated by Differential Scanning Calorimetry and Raman Spectroscopy. *Biol. J.* **1996**, *71*, 3350–3360.
20. Liu, Y.; Tan, F. Calorimetric Studies of Thermal Denaturation of DNA and tRNAs. *J. Therm. Anal.* **1995**, *45*, 35–38.
21. Tsereteli, G. I.; Belopolskaya, T. V.; Grunina, N. A.; Vaveliouk, O. L. Calorimetric Study of the Glass Transition Process in Humid Proteins and DNA. *J. Therm. Anal. Calorim.* **2000**, *62*, 89–99.

22. Lee, S.; Debenedetti, P.; Erringston, J.; Pethica, B.; Moore, D. A Calorimetric and Spectroscopic Study of DNA at Low Hydration. *J. Phys. Chem. B* **2004**, *108*, 3098–3106.

23. Falk, M.; Hartman, K. A.; Lord, R. C. Hydration of Deoxyribonucleic Acid II. An Infrared Study. *J. Am. Chem. Soc.* **1963**, *85*, 387–391.

24. Falk, M.; Poole, A. G.; Goymour, C. G. Infrared Study of the State of Water in the Hydration Shell of DNA. *Can. J. Chem.* **1970**, *48*, 1536–1542.

25. Abdurakhman, H.; Tajiri, K.; Yokoi, H.; Kuroda, N.; Matsui, H.; Yanagimachi, T.; Taniguchi, M.; Kawai, T.; Toyota, N. Infrared Spectroscopy on Poly(dG)-poly(dC) DNA at Low Hydration. *J. Phys. Soc. Jpn.* **2007**, *76*, 4009–4014.

26. Whitson, K. B.; Lukan, A. M.; Marlowe, R. L.; Lee, S. A.; Anthony, L.; Rupprecht, A. Binding of the Water of Primary Hydration to the Sodium and Cesium Salts of Deoxyribonucleic Acid and Potassium Hyaluronate. *Phys. Rev. E* **1998**, *58*, 2370–2377.

27. Hoyer, H. W.; Chow, M.; Gary, V. Differential Scanning Calorimetry Studies on the Dehydration of Deoxyribonucleic Acid and Polyadenylic Acid. *J. Colloid Interface Sci.* **1981**, *80*, 132–135.

28. Karni, M.; Zidon, D.; Polak, P.; Zalevsky, Z.; Shefi, O. Thermal Degradation of DNA. *DNA Cell Biol.* **2013**, *32*, 298–301.

29. Glavin, D.; Schubert, M.; Bada, J. L. Direct Isolation of Purines and Pyrimidines from Nucleic Acids Using Sublimation. *Anal. Chem.* **2002**, *74*, 6408–6412.

30. Alongi, J.; Di Blasio, A.; Milnes, J.; Malucelli, S.; Kandola, B.; Camino, G. Thermal Degradation of DNA, an All-in-One Natural Intumescent Flame Retardant. *Polym. Degrad. Stab.* **2015**, *113*, 110–118.

31. Alongi, J.; Milnes, J.; Malucelli, G.; Bourbigot, S.; Kandola, B. Thermal Degradation of DNA-Treated Cotton Fabrics Under Different Heating Conditions. *J. Anal. Appl. Pyrol.* **2014**, *108*, 212–221.

32. Alongi, J.; Carletto, R. A.; Di Blasio, A.; Carosio, F.; Bosco, F.; Malucelli, G. DNA: A Novel, Green, Natural Flame Retardant and Suppressant for Cotton. *J. Mater. Chem. A* **2013**, *1*, 4779–4785.

33. Alongi, J.; Carletto, R. A.; Di Blasio, A.; Cuttica, F.; Carosio, F.; Bosco, F.; Malucelli, G. Intrinsic Intumescent-Like Flame Retardant Properties of DNA-Treated Cotton Fabrics. *Carbohydr. Polym.* **2013**, *96*, 296304.

34. Carosio, F.; Di Blasio, A.; Alongi, J.; Malucelli, G. Green DNA-Based Flame Retardant Coatings Assembled Through Layer by Layer. *Polymer* **2013**, *54*, 5148–5153.

35. Alongi, J.; Di Blasio, A.; Cuttica, F.; Carosio, F.; Malucelli, G. Bulk or Surface Treatments of Ethylene Vinyl Acetate Copolymers with DNA: Investigation on the Flame Retardant Properties. *Eur. Polym. J.* **2014**, *51*, 112–119.

36. Isarov, S. A.; Lee, P. W.; Towslee, J. H.; Hoffman, K. M.; Davis, R. D.; Maia, J. M.; Pokorski, J. K. DNA as a Flame Retardant Additive for Low-Density Polyethylene. *Polymer* **2016**, *97*, 504–514.

37. Flynn, J. H.; Wall, L. A. A Quick, Direct Method for the Determination of Activation Energy from Thermogravimetric Data. *J. Polym. Sci. B Polym. Lett.* **1966**, *4*, 323328.

38. Ozawa, T. A New Method of Analyzing Thermogravimetric Data. *Bull. Chem. Soc. Jpn.* **1965**, *38*(11), 1881–1886.

39. Doyle, C. D. Estimating Isothermal Life from Thermogravimetric Data. *J. Appl. Polym. Sci.* **1962**, *6*, 639–642.

40. Kissinger, H. E. Variation of Peak Temperature with Heating Rate in Differential Thermal Analysis. *J. Res. Natl. Bur. Stand.* **1956**, *57*, 217–221.

41. Akahira, T.; Sunose, T. Method of Determining Activation Deterioration Constant of Electrical Insulating Materials. *Res. Report. Chiba Inst. Technol.* **1971**, *16*, 22–31.
42. Coats, A. W.; Redfern, J. P. Kinetic Parameters from Thermogravimetric Data. *Nature* **1964**, *201*, 68–69.
43. Friedman, H. L. Kinetics of Thermal Degradation of Char-Forming Plastics from Thermogravimetry: Application to Phenolic Plastic. *J. Polym. Sci. Pol. Sym.* **1964**, *6*, 183–195.
44. Brown, M. E.; Maciejewski, M.; Vyazovkin, S.; Nomen, R.; Sempere, J.; Burnham, A.; Opfermann, J.; Strey, R.; Anderson, H.; Kemmler, A.; Keuleers, R.; Janssens, J.; Desseyn, H. O.; Li, C. R.; Tang, T. B.; Roduit, B.; Malek, J.; Mitsuhashi, T. Computational Aspects of Kinetic Analysis. Part A: The ICTAC Kinetics Project-Data, Methods and Results. *Thermochim. Acta* **2000**, *355*, 125–143.
45. Steinfeld, J. I.; Francisco, J. S.; Hase, W. L. *Chemical Kinetics and Dynamics*; Prentice-Hall: New York, 1999.
46. Galwey, A. K. What Is Meant by the Term 'Variable Activation Energy' When Applied in the Kinetic Analyses of Solid State Decompositions (Crystolysis Reactions)? *Thermochim. Acta* **2003**, *397*, 249–268.
47. Prado, J. R.; Vyazovkin, S. Activation Energies of Water Vaporization from the Bulk and from Laponite, Montmorillonite, and Chitosan Powders. *Thermochim. Acta* **2011**, *524*, 197–201.
48. Vyazovkin, S. V.; Lesnikovich, A. I. An Approach to the Solution of the Inverse Kinetic Problem in the Case of Complex Processes: Part 1. Methods Employing a Series of Thermoanalytical Curves. *Thermochim. Acta* **1990**, *165*, 273–280.
49. Spink, C. H.; Garbett, N.; Chaires, J. B. Enthalpies of DNA Melting in the Presence of Osmolytes. *Biophys. Chem.* **2007**, *126*, 176–185.
50. Yan, L.; Iwasaki, H. Fractal Aggregation of DNA After Thermal Denaturation. *Chaos Soliton Fract.* **2004**, *20*, 877–881.
51. Kissinger, H. E. Reaction Kinetics in Differential Thermal Analysis. *Anal. Chem.* **1957**, *29*, 1702–1706.
52. Alongi, J.; Han, Z.; Bourbigot, S. Intumescence: Tradition Versus Novelty. A Comprehensive Review. *Prog. Polym. Sci.* **2015**, *51*, 28–73.
53. Costa, L.; Camino, G.; Luda, M. P. Mechanism of Thermal Degradation of Fire Retardant Melamine Salts. *ACS Symp. Ser.* **1990**, *425*, 211238.
54. Le-Tien, C.; Lafortune, R.; Shareck, F.; Lacroix, M. DNA Analysis of a Radiotolerant Bacterium *Pantoea agglomerans* by FT-IR Spectroscopy. *Talanta* **2007**, *71*, 1969–1975.
55. Banyay, M.; Sarkar, M.; Gräslund, A. A Library of IR Bands of Nucleic Acids in Solution. *Biophys. Chem.* **2003**, *104*, 477488.
56. Camino, G.; Martinasso, G.; Costa, L.; Gobetto, R. Thermal Degradation of Pentaerythritol Diphosphate, Model Compound for Fire Retardant Intumescent Systems: Part II Intumescence Step. *Polym. Degrad. Stab.* **1990**, *28*, 17–38.
57. Semenov, M.; Bolbukh, T.; Maleev, V. Infrared Study of the Influence of Water on DNA Stability in the Dependence on AT/GC Composition. *J. Mol. Struct.* **1997**, *408/409*, 213–217.
58. Dollimore, D. The Techniques and Theory of Thermal Analysis Applied to Studies on Inorganic Materials with Particular Reference to Dehydration and Single Oxide Systems. *Sel. Annu. Rev. Anal.* **1972**, *2*, 1–81.

PART IV
Special Topics

CHAPTER 14

ENVIRONMENTAL ENGINEERING, MEMBRANE TECHNOLOGY AND NOVEL SEPARATION PROCESSES: A BROAD SCIENTIFIC PERSPECTIVE AND A CRITICAL OVERVIEW

SUKANCHAN PALIT*

Department of Chemical Engineering, University of Petroleum and Energy Studies, Energy Acres, Bidholi via Prem Nagar, Dehradun 248007, Uttarakhand, India

Corresponding author. E-mail: sukanchan68@gmail.com; sukanchan92@gmail.com.

CONTENTS

ABSTRACT

The world of science and technology is moving toward a newer and visionary realm at a rapid pace. Vision of science is entering a new era. Environmental engineering science in a similar vein is undergoing paradigmatic change. Environmental regulations, stringent environmental restrictions, and the futuristic vision has urged scientific domain to devise new and innovative technologies. Membrane science is an innovative technology which needs to be reorganized and redefined. Grave concerns, inimitable disaster and environmental sustainability has plunged human civilization to an unwanted catastrophe. Environmental disaster is bane to human progress. Mankind's progress, civilization's prowess, and the path toward future progress will pave the way toward newer technologies and will surpass visionary frontiers. Membrane science in today's world has an unsevered umbilical cord with industrial wastewater treatment and drinking water treatment. Global water technology scenario is at a deep distress. Developed as well as developing nations are witness to the crisis of groundwater contamination. At this critical juncture of history and time, scientific advancement and application of membrane science are the vicious and vehement parameters toward the immense and innovative growth of human civilization. The author, with a deep and cogent insight, delineates the utmost importance of membrane science in the progress of environmental engineering science. The treatise also delineates the hidden scientific truths of the immense potential of membrane science, fouling and concentration boundary layer.

14.1 INTRODUCTION

The world of environmental engineering and chemical engineering is moving drastically toward a newer and innovative era. Grave concerns and effective environmental restrictions are the torchbearers of the future of industrial wastewater treatment and drinking water treatment. Advancement of science and technology and the vision to excel has urged the human civilization and the scientific domain to aim for new innovations. History of science and engineering are boon to human civilization and human progress. Challenge, vision, and the scientific endurance are opening up new chapters in the application of membrane science in environmental engineering science. The challenge of science needs to be revisited and reenvisioned. Membrane science and its classification are surely and inevitably changing the face of scientific endeavor. Since the scientific endeavor of

Loeb–Sourirajan model in membrane technology, the world of science has revolutionized and reenvisioned. Concentration boundary layer is the next domain of scientific endeavor which needs to be reshaped and restructured. Ground water remediation is the next primordial issue which needs to be targeted vehemently. Arsenic groundwater contamination is a bane to human civilization and human progress. The author cautiously addresses with deep comprehension to the challenges behind the application of membrane technology in environmental engineering science. The challenges and future trends need to be evaluated by the scientific community vehemently and intensely. History of science is ushering in a new and insightful beginning. Challenges, vision, and barriers in the application of membrane science will surely bring forward in the scientific forefront a new generation of scientific vision and scientific truth.[21]

Today, membrane separation processes stands in the midst of deep comprehension and wide scientific vision. The environmental engineering paradigm in today's world is undergoing drastic changes and the future perspectives are surely bright and far-reaching. Human scientific progress, the wide vistas of challenges and the wide future of research pursuit will go a long way in the true evolution of a newer scientific era. This present century is the century of innovative science and technology. The challenge of science and technology and its vision are gearing human mankind toward a positive horizon.[21,22]

14.1.1 VISION OF THE PRESENT TREATISE

The present treatise explores and delineates with cogent and effective insight the scientific truth and scientific vision behind membrane science and technology. Scientific cognizance and vision of science are the torchbearers to a new realm and new era. The treatise redefines the hidden truth behind the application domain of membranes. Loeb–Sourirajan model defined the application domain of membrane science and thus ushered in a new innovative area of science.[1] The years after its discovery were truly far-reaching and scientifically ground breaking. Purpose and aim of this treatise is visionary and the challenge veritably groundbreaking. Groundwater treatment, industrial wastewater treatment, and global drinking water scenario is undergoing a drastic and vicious challenge. Global water initiative is a primordial issue which needs to be targeted with utmost importance. The author with lucid details delineates the future trends in the application of membrane science in alleviating global water crisis.[1–4]

The author with great scientific vision and cogent scientific insight unfolds the intricacies of membrane separation processes and the future of global water shortage. In today's world, global water shortage and environmental sustainability have an unsevered umbilical cord. More devastating crisis is looming large over the scientific horizon. The vision of this treatise unfolds the immense potential of membrane separation processes and novel separation processes. Its application to environmental engineering science is delineated and discussed in details. The author rigorously elucidates the scientific vision and the scientific greatness behind membrane science and novel separation processes with an extended view toward global water crisis. Technological initiative and technological vision in today's human civilization is surpassing wide and versatile frontiers. The author repeatedly focuses on environmental engineering techniques with a positive vision. The future of membrane technology will veritably open up newer innovative areas.

14.2 THE GOAL OF THE STUDY

The mission, goal, and aim of the study is to address the grave cause of industrial water pollution. Challenge needs to be redefined and reenvisioned. Membrane science is the next generation technology. Man's vision, mankind's prowess, and civilization's progress are opening windows of innovation in environmental engineering science. The treatise analyses the immense scope and potential of membrane science in the future of environmental science. Goals of human mankind with respect to environmental engineering science and application of membrane science needs to be reshaped. The world of environmental engineering is ushering in a new and vibrant era. History of human mankind is also moving toward a visionary era. Loeb–Sourirajan model defined the entire path of human scientific endeavor in membrane science. The author skillfully rewrites vision and history of science in the path toward progress in membrane. Scientific challenge, scientific truth, and scientific vision will all go a long way in the emancipation of membrane science and technology. The aim and objective of this scientific endeavor targets is the application of membrane science in solving global water crisis. Science, engineering, and technology are ushering in a new era of scientific truth and vision. Global water crisis is moving from one definite and visionary perspective to another.

14.3 ENVIRONMENTAL ENGINEERING SCIENCE, MEMBRANE SCIENCE, AND GLOBAL WATER CRISIS

Global water crisis and global water shortage are the primordial issues which need to be addressed at this critical juncture of history and time. Environmental engineering science is veritably witnessing dramatic and drastic challenges. Challenges in the path toward progress, scientific vision, and scientific validation are the order of today's technological era. Scientific validation from the laboratory to the common mass and human society needs to be vehemently and intensely addressed and targeted. Global water crisis is a bane and a catastrophe to the human civilization. Mankind's history, civilization's prowess and the road toward progress will all lead a long way in the emancipation of engineering and science of membrane separation processes. Scientific cognizance, scientific doctrine, and the immensely vital domain of scientific validation are the veritable pallbearers of a new dawn of human civilization. Vision of technology needs to be revalidated at every step of human progress.

14.3.1 VISIONARY SCIENTIFIC ENDEAVOUR IN THE FIELD OF MEMBRANE SEPARATION PROCESSES

The scientific world of today is witnessing vicious challenges. Membrane separation processes are one of the primordial challenges in the alleviation of global water crisis. The author with deep and cogent comprehension elucidates the present trends in research in the field of membrane science and the future directions and success. Technology initiative and scientific endeavor are the pillars of modern science. Water technology needs to be revamped with the course of human history. Provision of basic human needs, need to be the prime focus of future scientific research pursuit. Loeb–Sourirajan model[11] revolutionized the science and technology of membrane separation processes. Cheryan (1998)[11] elucidated in a visionary treatise the past, present, and future trends of the application of ultrafiltration and microfiltration. The phenomenon of osmosis, which is the transport of water or solvent through a semi-permeable membrane (defined as a membrane that is permeable to solvent and impermeable to solutes), has been known since 1748, when Abbe Nollet observed that water diffuses from a dilute solution to a more concentrated one when separated by a semi-permeable membrane.[11] Cheryan[11] touches upon performance and engineering models of membrane separation processes and the crucial domain of fouling and cleaning. Process design of diafiltration–continuous, discontinuous and dialysis stands as a major and critical point of his research. The research also encompassed upon applications of ultrafiltration and microfiltration.

Chen et al. (2013)[12] discussed in details in a study on dynamics of ultrafiltration membrane materials preparing and its uses in drinking water treatment. The ultrafiltration membrane technology as a new water treatment technology can be used in a large scale depending largely on membrane material itself and also depending on the membrane production conditions and process. This treatise reviews the development and current and future trends of preparing ultrafiltration membrane material, and analyzes deeply its application prospects in water treatment of drinking water. This paper veritably surpassed visionary boundaries. Commercialization process of preparing ultrafiltration membrane materials and polymer membrane has got unprecedented research and development. The membrane material such as bisphenol A type of cellulose derivatives (PSF), polysulfone polyether sulfone (PES), polysulfone amide (PSA), ployvinylidene fluoride (PVDF) and polyacryonitrile (PAN), have extensively promoted the development of ultrafiltration membranes.

Ahmed et al. (2010)[13] dealt lucidly in a review fabrication of a polymer-based mixed matrix membrane. This short review delineates the main and promising areas of research in gas separation, by considering the materials for membranes, the industrial applications of membrane separations, and finally the opportunities for integration of gas separation units in hybrid systems for the intensifications of the process.

Cath et al. (2006)[14] reviewed forward osmosis, its principles, applications, and recent trends of its development. This paper is a visionary treatise which deals extensively and lucidly the wide world of a modern domain of membrane science-forward osmosis. Osmosis is a physical phenomenon known for long decades. After 1960s, attention was on osmosis through synthetic materials. Osmosis as it is known now as a forward osmosis which has new applications in separation processes for waste water treatment, food processing, and seawater desalination. The success of membrane science and desalination are extremely relevant to the present and future of human scientific pursuit. A state of art has been attempted considering physical principles and applications of forward osmosis.

14.4   GLOBAL WATER CRISIS AND ARSENIC GROUNDWATER CONTAMINATION

Arsenic groundwater contamination is creating unimaginable environmental havoc in developing and developed world. Human scientific endeavor and the futuristic scientific vision is at a deep distress. Global water crisis and scientific challenges are the pallbearers of future technological initiatives and scientific vision.[11] Arsenic and heavy-metal groundwater contamination are ravaging the human mankind and in a similar vein the world of challenges needs to be revamped and veritably readdressed. Water crisis needs to be alleviated. In today's human civilization, arsenic groundwater contamination and global water crisis have an unsevered umbilical cord. The success and efficacy of scientific endeavor wholly and veritably lies on effective and determined scientific validation and scientific vision. Global water challenges are in deep distress and unending whirlpool. In such a crucial juncture, technological and scientific visions are the pallbearers of the future. Both developed and developing nations are ensconced with this ravaging environmental disaster. Science and technology of water initiatives needs to be effectively restructured with the active participation of the civil society. With this aim and objective in mind, the scientific domain will surely move toward true emancipation of water technology and science.[11]

14.4.1   ARSENIC GROUNDWATER CONTAMINATION AND THE FUTURE OF GLOBAL WATER SHORTAGE

Arsenic groundwater contamination and its effective remediation are the primordial issues facing progress of human mankind today.[1-5] Developed and developing countries are in a deep and questionable distress with every step of human progress. Bangladesh and West Bengal state of India are the major regions of the developing economies in such a highly distressful conditions. Heavy metal contamination is a vexing question which looms large in the midst of scientific horizon. The world of challenges, the immense difficulties and the inimitable disaster questions the pursuit of science and technology. History of science is in the midst of immense catastrophe. Civil society decisions, the burgeoning urge to excel and the question of scientific validation will go a long way in alleviating the concern and the inimitable arsenic contamination crisis. Membrane science and its applications are the inevitable pallbearers to a new dimension of scientific hope and scientific cognizance.[1-5]

14.5 TECHNOLOGY INITIATIVES AND TECHNOLOGICAL VISION IN THE FIELD OF WATER TECHNOLOGY

The world of water technology is slowly gearing toward newer visionary initiatives. Developed and developing nations are in the whirlpool of environmental devastations and inimitable crisis. Arsenic groundwater contamination is a burning example. Vision of science has no answers. Bangladesh and the state of West Bengal in India, for centuries are bearing the immense burden of heavy metal groundwater contamination.[11] Technology initiatives need to be reenvisioned with each step of human life. As human mankind moves toward the third decade of twenty first century, the need for scientific and technological validation are enhanced. Science in today's world is a colossus without a will of its own. In such a crucial juxtaposition, the efforts and the scientific urge to excel should move toward effective validation of science. Water technology and water science in a similar vein need to totally focused toward human needs and the effective validation of science.[11]

14.6 SCIENTIFIC VISION, SCIENTIFIC COGNIZANCE, AND THE PATH TOWARD PROGRESS

Scientific vision in the domain of membrane science and technology is inspiring and far-reaching. Innovations, scientific instinct and scientific validation needs to be reenvisioned. Mankind's prowess, civilization's progress, and the vistas of science are the torchbearers to a new scientific generation and a newer scientific realm.[6-9] History of science and technology is in the midst of immense disaster and pessimism with the burgeoning environmental crisis. Environmental disasters and its causes and concerns will inevitably pave the way toward a new dimension of scientific pursuit. Scientific validation is the answer to the grave concerns and the impending crisis.[10] Advancements in science and technology in today's world of scientific vision depends wholly on scientific validation. In such a crucial juxtaposition, membrane science is the only answer to the wide gamut of environmental engineering problems.[11,21,22]

14.7 VISION OF THE APPLICATION OF MEMBRANE SCIENCE

The vision of the application of membrane science in environmental engineering science is absolutely wide and far-reaching.[11] Environmental regulations and stringent restrictions has urged the future scientific genera-tions and scientific domain to scale visionary heights. Membrane science and its varied technologies are the plausible technologies for tomorrow. In today's scientific endeavor, technology is diverse. Advancement of science and technology relies intensely and vehemently on scientific vision and strong scientific understanding. History of human civilization is taking a definitive turn. The challenge for the future is groundbreaking and replete with scientific fortitude. Membrane and water treatment is evolving into new future directions and future dimensions. History of human scientific endeavor and the progress of science is moving toward a definitive direc-tion along with emancipation of engineering. Chemical engineering, process engineering and environmental engineering is in a world of immense and intense rationalization and optimization. History of human civilization and human scientific endeavor is in today's scientific world is taking a defini-tive turn and a visionary turnaround. The world of scientific fortitude and scientific understanding is witnessing a devastating as well as an effec-tive environmental engineering paradigm. Environmental catastrophes are rebuilding environmental engineering techniques and paradigm. Man's vision is emboldened at every step of human history and human scientific endeavor. Mankind is steadily moving toward a new era of effective envi-ronmental engineering frontier. The frontiers of science and engineering of membrane separation processes are surpassed and the world of challenges are overcome at its utmost. Water technology and environmental engineering technologies in today's world have an umbilical cord. The success of envi-ronmental sustainability will move toward alleviating global water crisis. Effective water science, the vision to excel and the visionary frontiers of membrane science will evolve into new future dimensions of holistic science and technology.[21,22]

14.8 SCIENTIFIC DOCTRINE OF MEMBRANE SCIENCE

The doctrine of membrane science is undergoing drastic challenges and vehe-ment changes.[11] History of environmental engineering science is witnessing visionary changes and environmental sustainability is the ultimate goal of human civilization.[11] At such a crucial juxtaposition, a scientist's vision as well as the intense scientific understanding behind environmental engineering techniques are vehemently emboldened. Membrane science is ushering in a new dawn of scientific era. Separation phenomenon in chemical engineering frontiers is a hallmark toward the future of scientific endeavor. Desalination, industrial wastewater treatment, and provision for clean drinking water are the pallbearers to a new scientific understanding and vision.[11,21,22]

14.9 MEMBRANE SCIENCE: CURRENT TRENDS AND FUTURE CHALLENGES

Current trends and future challenges in the field of membrane science are of utmost importance to the growth of science and engineering. History of human mankind is witnessing a glorious scientific rebirth and scientific rejuvenation with the progress of scientific endeavor in membrane science. Membrane fouling, phenomenon of concentration boundary layer and the effectivity of separation phenomenon are the veritable hallmarks of future scientific research pursuit. Reverse osmosis and desalination are the hallmarks of reenvisioning of membrane technology. Future of global water crisis is grave and vehemently disastrous.[11] Current trends of effectivity of membrane separation phenomenon are evolving into new dimensions of research pursuit. Membrane fouling is a vexing and challenging issue. Future challenges and future direction of membrane science should be effectively toward alleviating and wiping global water crisis. Advancement in science and technology and challenges of the future will lead a long way in the ultimate emancipation of membrane separation phenomenon.[11] Membrane science in today's scientific world has an umbilical cord with alleviation of global water crisis. Man's vision, the visionary path toward scientific progress and the civilization's advancement are the hallmarks toward a new era of environmental engineering paradigm. Current trends in the advancement of membrane science are targeted toward greater emancipation toward global water crisis which veritably includes desalination and industrial wastewater treatment. Challenges are surmounting at every step of human scientific endeavor. Yet the urge and vision to excel has propelled the human scientific pursuit to scale one visionary height over another.

14.10 VISION OF SCIENCE, THE WORLD OF CHALLENGES AND THE RESEARCH PURSUIT IN NOVEL SEPARATION PROCESSES

Vision of science in today's scientific world is undergoing remarkable challenges. Human scientific vision, scientific rigor and ardor are drastically changing the scientific landscape. Man's prowess and civilization's immense and wide vision needs to be reshaped with respect to scientific research pursuit. Novel separation processes and membrane science are transforming the scientific horizon. Scientific endeavor, the vision of technology initiatives and the world of challenges are the pallbearers of energy and environmental sustainability. Scientific research pursuit and the world of chemical process engineering are revolutionizing the scientific domain.

14.11   TECHNOLOGY INITIATIVE IN DEVELOPED AND DEVELOPING COUNTRIES AND FUTURE OF GLOBAL WATER SCENARIO

Technology initiative in developed and developing countries in the domain of water technology are changing the face of global science. True emancipation and true realization of environmental sustainability are the need of the hour. The greatness and the glory of membrane science needs to be reenvisioned. Global water scenario is witnessing a disastrous crisis with the burning issue of arsenic groundwater pollution. Global water technology is in the midst of a deepening crisis in the developed as well as developing world. Science and technology has no constructive answers. Human science and vision is endangered. Global arsenic groundwater remediation needs to be effectively addressed. Human scientific vision stands as a primordial issue in the future path of human civilization.

14.12   GLOBAL WATER CRISIS AND ARSENIC GROUNDWATER CONTAMINATION

Arsenic groundwater contamination is a veritable bane to human civilization.[5–8,11] Progress of science and environmental engineering in today's world in a state of distress as well as in a state of reenvisioning. Groundwater remediation needs to be reshaped and revitalized with every step of human progress. Developing and developed economies are entangled in water shortage crisis and ground water contamination. Newer innovative technologies needs to be envisioned with progress of scientific endeavour.[11] Provision of clean drinking water and industrial wastewater treatment stands tall in the midst of immense optimism as well as deep introspection.[10,11] In Bangladesh and the state of West Bengal, India, groundwater contamination is playing havoc to the human civilization. Concern, challenges and barriers has opened up new vistas of scientific hardships in years to come in the field of arsenic groundwater remediation. History of human challenges is having a new beginning with the grave concern of the world's largest unsolvable environmental disaster.[10,11] Bangladesh and West Bengal still today remains in a murky state of affairs with the drinking water crisis. The challenge is absolutely insurmountable and disastrous. Science has met with a disastrous failure at every step of human progress in the path of arsenic groundwater remediation. Human mankind's future targets, the world of immense challenges and the march of science are the primordial issues which needs to be tackled with utmost importance in future.[11,21,22]

14.13 VISION AND CHALLENGES TO MOVE FORWARD IN ENVIRONMENTAL ENGINEERING SCIENCE

Industrial wastewater treatment is witnessing a new dawn in the present century. Environmental engineering science in a similar vein is evolving toward a newer future dimension at every step of life and human civilization.[11] Membrane science stands as the utmost answer and the plausible solution toward global water crisis and controlling industrial pollution. Vision and challenges to move forward are immense and retrogressive. At such a critical juncture of science, history and time, the evolution and emancipation of new and innovative technologies are of utmost importance to the progress of membrane separation phenomenon. The wide world of membrane science needs to be reenvisioned and restructured carrying with the legacy of Loeb–Sourirajan model.[11] Loeb–Sourirajan model has left an indelible impact to the future of environmental engineering science. Drinking water treatment, industrial pollution control and the future of environmental engineering science are in questionable times and in a state of immense distress. History of human civilization needs to be restructured and reshaped at every step of human progress.[11]

14.14 THE WORLD OF VISIONARY FRONTIERS AND THE PROGRESS OF MEMBRANE SCIENCE

Science and engineering in today's world are surpassing visionary frontiers. Scientific vision and scientific candor are in the path of new rejuvenation. Membrane science is a veritable pillar of novel separation process. Loeb–Sourirajan model had revolutionized and rebuilt the membrane science scenario. Scientific progress in today's world is at a deep distress due to ecological imbalance and frequent environmental catastrophes. The vision of science is undergoing tremendous metamorphosis. Water science and technology is advancing toward a newer regeneration. Technological vision needs to be reshaped with the passage of history due to the serious environmental issues. Thus novel separation processes and membrane science are primordial issues of this century[11] Science and engineering of membrane separation processes are surpassing visionary frontiers. Human scientific endeavor in today's world is at its zenith.

14.15 BARRIERS, DIFFICULTIES, AND THE WORLD OF CHALLENGES

Scientific challenges and scientific vision are vast and versatile. Science is witnessing vicious challenges. The effectivity of membrane separation process stands as a primordial issue. Fouling stands as other major impediment. The world of scientific truth is beckoning toward a newer vision and newer goal. Barriers and difficulties are running riot in the scientific research pursuit domain. Yet the goal and vision needs to be reshaped and reenvisioned with each step of human history and time.[11]

Barriers and difficulties in the progress of membrane science are vast and versatile. Scientific understanding and scientific rigor are at its helm with the passage of human history and time. Fouling and concentration polarization stands as a major impediment to the success of membrane separation phenomenon. Research endeavor and scientific validation are the torchbearers of tomorrow's success of membrane science. The author in this treatise stresses study on the immense importance of novel separation processes such as membrane science in the path of environmental engineering science validation. Success of scientific rigor, the immense progress of engineering in this century and the wide world scientific understanding all will lead a long way in the successful validation of membrane separation processes.[15–20]

14.16 VISION OF SCIENCE, SCIENTIFIC IMAGINATION, AND THE AVENUES OF SCIENCE FORWARD

Vision of science and scientific imagination are in the road toward newer vision and newer hope. Scientific fortitude and scientific forbearance needs to be reenvisioned in such difficult human era. Man's visions as well as civilization's prowess are moving toward newer future direction and a visionary eon. This is an age of scientific sagacity and immense scientific cognizance.

Global water crisis is at a state of immense disaster. The road forward and the progress of human civilization is in the midst of unimaginable crisis with the growing concern of ecological imbalance and at the same time the apprehension of environmental catastrophes. Membrane science is changing the scientific horizon. Man's immense instinctive vision, civilization's move forward and the world of challenges has redefined the paradigm of environmental engineering science. The target of human civilization, the progress of scientific endeavor and the urge of the civil society should be geared toward a newer environmental engineering paradigm.[11]

14.17 NOVEL SEPARATION PROCESSES AND THE FUTURE OF SCIENCE AND ENGINEERING

Novel separation processes needs to be redefined with each foray into civilization's history and time. Membrane science and membrane separation processes are in the process of new scientific regeneration and a new scientific imagination. The vision of science in today's world is veritably awe-inspiring. Technological vision is challenged. Novel separation processes today stands as a major pillar of environmental engineering paradigm. Human scientific endeavor is in the midst of deepening crisis. Yet the future of science and engineering needs to be reenvisioned. Loeb–Sourirajan model revolutionized the concept of diffusion process in membranes. The discovered theory and the subsequent research pursuit are the torchbearers of a new visionary scientific age.[11,15-17]

14.18 FUTURE DIRECTIONS AND FUTURE CHALLENGES IN THE EMANCIPATION OF SCIENCE AND TECHNOLOGY

Future directions and future dimensions of research in the field of membrane science are surpassing wide and visionary frontiers. The main crux of the research endeavor should be greater application toward alleviation of global water crisis. Membrane separation phenomenon should be of utmost importance to the avenues of progress in water technology and environmental engineering science. Science and its application are entering into an absolutely visionary era. History of science and technology and its legacy are in a revolutionary phase in the present century. The world of challenges has shrunken with new innovations and newer scientific instincts. The emancipation of science is visible and revealed at every step of human scientific endeavor. Progress of civilization with the advances of environmental engineering are ushering in a new dawn of scientific growth and a new direction toward emancipation of science.[11,18–20]

14.19 FUTURE VISION AND FUTURE FLOW OF THOUGHT

Future vision of membrane science and future flow of thoughts in environmental engineering science needs to have a definite direction.[11] Global water crisis, groundwater remediation, and the future of membrane science are the hallmarks and definite yardsticks to the scientific progress.[11] Arsenic groundwater remediation is the hallmark toward a greater emancipation of science and engineering. Man's scientific progress, civilization's prowess, and the world of challenges have given way to newer windows of innovation and the effective avenues of environmental engineering science. Vistas of science are beyond imagination and beyond comprehension. Groundwater remediation remains as a single yardstick to the progress of science and technology. The world of water technology is in a grave crisis. In such a visionary juncture, application of science and engineering such as membrane science remains entrenched in scientific frontier. The history of science is ushering in a new beginning with the evolution of new technologies in membrane separation processes.[11]

14.20 CONCLUSION

History of human scientific endeavor is moving through a difficult and distressful era.[11] A scientist's vision is emboldened at every step of human progress. Membrane science and the paradigm of environmental engineering science is surpassing wide and visionary frontiers. The question of environmental sustainability is a vexing issue in the future of human mankind. Environmental devastation along with stringent restrictions is the hallmarks of the present environmental engineering paradigm. Membrane separation phenomenon is moving toward a newer direction and a visionary realm.[11] The success of science is befitting to the progress of human civilization. Mankind's history, civilization's legacy and the path of environmental regulations have urged the scientific community to yearn for innovation and scientific instinct. Challenges for tomorrow and the vision for the future have propelled human society to look for green chemistry. In such a juxta position, green environmental engineering techniques will surely enhance the scientific vision and scientific truth. Scientific forbearance and the world of challenges needs to be restructured with the evolution of new environmental engineering paradigm.[11] The future vision of membrane science will inevitably enhance the civilization's growth and mankind's prowess.[11] Membrane science in today's world will open up new doors of scientific innovation in years to come.

Science and engineering are moving fast toward a new visionary frontier. Environmental engineering science is witnessing drastic changes in its path toward true realization of environmental sustainability. This treatise will be a pallbearer toward the future world of membrane science and environmental engineering science. Progress of engineering, the grave concerns of human civilization and the scientific forbearance to excel will all lead a long and definite way in the true emancipation of membrane science.

KEYWORDS

- **membrane**
- **environment**
- **water**
- **arsenic**
- **global**
- **shortage**

REFERENCES

1. Bruggen, V.; Manttari, M.; Nystrom, M. Drawbacks of Applying Nanofiltration and How to Avoid Them: A Review. *Sep. Purif. Technol.* **2008**, *63*, 251–263.
2. Matsuura, T. Progress in Membrane Science and Technology for Seawater Desalination—A Review. *Desalination* **2001**, *134*, 47–54.
3. Sidek, N. M.; Ali, N.; Fauzi, S. A. A. The Governing Factors of Nanofiltration Membrane Separation Process Performance—A Review. In *UMTAS, Empowering Science Technology and Innovation Towards a Better Tomorrow*; 2011; pp 241–248.
4. Hong, S.; Elimelech, M. Chemical and Physical Aspects of Natural Organic Matter (NOM) Fouling of Nanofiltration Membranes. *J. Membr. Sci.* **1997**, *132*, 159–181.
5. Hilal, N.; Mohammad, A. W.; Atkin, B.; Darwish, N. A. Using Atomic Force Microscopy Towards Improvement in Nanofiltration Membranes Properties for Desalination Pretreatment—A Review. *Desalination* **2003**, *157*, 137–144.

6. Hilal, N.; Zoubi, A.; Darwish, N. A.; Mohammad, A. W.; Abu Arabi, M. A Comprehensive Review of Nanofiltration Membranes: Treatment, Pretreatment, Modeling and Atomic Force Microscopy. *Desalination* **2004**, *170*, 281–308.
7. Ashaghi, K. S.; Ebrahimi, M.; Czermak, P. Ceramic Ultrafiltration and Nanofiltration Membranes for Oilfield Produced Water Treatment: A Mini Review. *Open Environ. Sci. J.* **2007**, *1*, 1–8.
8. Wijmanns, J. G.; Baker, R. W. The Solution-Diffusion Model—A Review. *J. Membr. Sci.* **1995**, *107*, 1–21.
9. Nghiem, L. D.; Schafer, A. I.; Elimelech, M. Removal of Natural Hormones by Nanofiltration Membranes: Measurement, Modeling and Mechanisms. *Environ. Sci. Technol.* **2004**, *38*, 1888–1896.
10. Renou, S.; Givaudan, J. G.; Poulain, S.; Dirassouyan, F.; Moulin, P. Landfill Leachate Treatment: Review and Opportunity. *J. Hazard. Mater.* **2008**, *150*, 468–493.
11. Cheryan, M. *Ultrafiltration and Microfiltration Handbook*; Technomic Publishing Company Inc.: USA, 1998.
12. Chen, A.; Fan, Q.; Tian, Q. Study on Dynamic of Ultrafiltration Membrane Materials Preparing and Used in Drinking Water Treatment. *Int. J. Environ. Sci. Dev.* **2013**, *49*(4), 343–385.
13. Ahmed, I.; Yusof, Z. A. M.; Beg, M. D. H. Fabrication of Polymer Based Mix Matrix Membrane—A Short Review. *Int. J. Basic Appl. Sci.* **2010**, *10*(2), 17–27.
14. Cath, T. Y.; Childress, A. E.; Elimelech, M. Forward Osmosis: Principles, Applications and Recent Developments. *J. Membr. Sci.* **2006**, *281*, 70–87.
15. Palit, S. *Filtration: Frontiers of the Engineering and Science of Nanofiltration—A Far-reaching Review, CRC Concise Encyclopedia of Nanotechnology*; Ortiz-Mendez, U., Kharissova, O. V., Kharisov. B. I., Eds.; Taylor and Francis: USA, 2016; pp 205–214.
16. Palit, S. *Advanced Oxidation Processes, Nanofiltration, and Application of Bubble Column Reactor, Nanomaterials for Environmental Protection*; Boris, I. K., Oxana, V. K., Rasika Dias, H. V., Eds.; Wiley: USA, 2015; pp 207–215.
17. Palit, S. Microfiltration, Groundwater Remediation and Environmental Engineering Science—A Scientific Perspective and a Far-reaching Review. *Nat. Environ. Pollut. Technol.* **2015**, *14*(4), 817–825.
18. Palit, S. Nanofiltration and Ultrafiltration—The Next Generation Environmental Engineering Tool and a Vision for the Future. *Int. J. Chem. Technol. Res.* **2016**, *9*(5), 848–856.
19. Palit, S. Frontiers of Nano-electrochemistry and Application of Nanotechnology—A Vision for the Future. *In Handbook of Nanoelectrochemistry*; Springer International Publishing: Switzerland, 2015, pp 1–15.
20. Palit, S. Dependence of Order of Reaction on pH and Oxidation–reduction Potential in the Ozone-oxidation of Textile Dyes in a Bubble Column Reactor. *Int. J. Environ. Pollut. Control Manag.* **2011**, *3*(4), 69–78.
21. www.wikipedia.com (accessed Jan 2017).
22. www.google.com (accessed Jan 2017).

CHAPTER 15

A COMPARATIVE STUDY ON ELECTROCENTRIFUGE SPINNING AND ELECTROSPINNING PROCESS AS TWO DIFFERENT NANOFIBER CREATION TECHNIQUES

S. RAFIEI*

Department of Textile Engineering, University of Guilan, Rasht, Iran

Corresponding author. E-mail: saeedeh.rafieii@gmail.com

CONTENTS

ABSTRACT

In this study, centrifugal and electrical forces have been employed simultaneously for nanofiber production, and the effect of adding the centrifugal force to the electrical forces on the nanofiber diameter and production rate has been investigated in detail.

15.1   INTRODUCTION

When the diameters of polymer fibers are shrunk from micrometers to submicro or nanometers, several new characteristics emerge, including a very high ratio of surface area to volume, flexibility in surface functionalities, and superior mechanical performance (e.g., stiffness and tensile strength). These unique properties make these polymeric fibers ideal candidates for many important applications.[1–3]

Although several methods have been proposed for nanofiber manufacturing so far, an efficient and cost-effective procedure of production is still a challenge as debated by many experts. Electrospinning has been the most successful method to produce nanofibers so far. It is a well-known and prominent procedure which is based on electrostatic force that provides the possibility for spinning nanofibers from many kinds of polymers using melt or solution spinning.[4–6]

The use of mechanical methods to apply high rates of tension to a polymer solution or melt is only possible for polymers with high extension ability, to prevent uneven transmission of tension and stress concentration. In order to apply high rates of tension to a polymer solution or melt, methods which can apply even distribution of stress during the tension process are required. When the centrifugal force acts upon a substance, the particles of that matter will experience a force proportional to their distance from the center of rotation. Thus, this force can be used to apply high rates of tension on a polymer solution. During the tension process, if the polymer solution has sufficient viscosity, it will be stretched as a string, transforming to a polymeric fiber after drying.[7,8]

Electrocentrifuge spinning or has been recently introduced as a beneficial method for nanofiber production. It is necessary to specify the effective parameters on nanofiber production and diameters. In this research, the effective parameters on the process and the influence of each parameter on fibers diameter are addressed. Also the fiber production capability of electrocentrifuge technique is compared with that of the conventional electrospinning method.[9–11]

15.1.1 NANOFIBERS

Nanofibers are defined as fibers with diameters less than 100 nm which can be produced by different methods such as melt processing, interfacial polymerization, electrospinning, electrocentrifuge spinning, and antisolvent-induced polymer precipitation.[5] The fibers which are produced by these methods have a very high specific surface area, unusual strength, high surface energy, surface reactivity, and high thermal and electric conductivity that could be used in products by special application. These unique properties make the polymeric nanofibers ideal candidates for many important applications, such as, nanofiltration,[12] nanocatalysis, tissue scaffolds,[13] protective clothing, filtration, nanoelectronics,[14] high-performance nanofibers,[15,16] drug delivery systems, wound dressings, and composites. Although several methods have been proposed for nanofiber manufacturing so far, an efficient and cost-effective procedure of production is still a challenge and is debated by many experts.[17–20]

15.1.2 ELECTROSPINNING METHOD

Electrospinning is an efficient and simple mean of producing nanofibers by solidification of a polymer solution stretched by an electric field.[21] It is a well-known and prominent procedure based on electrostatic forces that provides the possibility for spinning nanofibers from many kinds of polymers using melt or solution spinning.[22,23]

In typical electrospinning process, an electrical potential is applied between droplets of polymer solution or melt, held through a syringe needle and a grounded target; afterwards the electrostatic field stretches the polymer solution into fibers as the solvent is evaporating. During the process the polymer jet undergoes instabilities, which together with the solution properties, determines the morphology of the forming nanosized structures obtained onto the collector. The electrical forces elongate the jet thousands of times, causing the jet become very thin. Ultimately, the solvent evaporates or the melt solidifies, and a web containing very long nanofibers is collected on the target. The fiber morphology is controlled by the experimental design and is dependent on solution properties and system conditions such as conductivity, solvent polarity, solution concentration, polymer molecular weight, viscosity, and applied voltage.[4,24]

To overcome various limitations of the typical electrospinning method, researchers have applied some modifications to the original setup.[24,25] Some of these reformed techniques which have been done by now include: electrospinning using a collector consists of two pieces of electrical conductive plates separated by a gap, collecting spun nanofibers on a rotating thin wheel with sharp edge, fabricating aligned yarns of nylon 6 nanofibers by rapidly oscillating a grounded frame within the jet[26] and finally using a metal frame as a collector to generate parallel arrays of nanofibers.[27] Matthews et al.[26] have applied a rotating drum to collect aligned nanofibers at a very high speed up to thousands of revolution per minute (RPM). In another approach, Deitzel et al.[28] used a multiple field technique which can straighten the polymer jet to some extent. Luming et al.[29] used this technique to collect aligned nanofibers by increasing the surface velocity of the drum.

Increasing the production rate of nanofibers has always been one of the most important challenges of researchers. Most efforts in this point have been focused on increasing the number of nozzles.[30] Yarin et al. introduced the concept of "upward needleless electrospinning of multiple nanofibers".[31] A similar idea was discussed by Liu and He[32,33] by the name of "Bubble electrospinning" method.

15.1.3 CENTRIFUGAL FORCE

In Newtonian mechanics, the term centrifugal force is used to refer to an inertial force (also called a 'fictitious' force) directed away from the axis of rotation that appears to act on all objects when viewed in a rotating reference frame. The concept of centrifugal force can be applied in rotating devices such as centrifuges, centrifugal pumps, centrifugal governors, centrifugal clutches, etc., as well as in centrifugal railways, planetary orbits, banked curves, etc., when they are analyzed in a rotating coordinate system.[34]

Centrifugal force is an outward force apparent in a rotating reference frame; it does not exist when measurements are made in an inertial frame of reference. All measurements of position and velocity must be made relative to some frame of reference. For example, if we are studying the motion of an object in an airliner traveling at great speed, we could calculate the motion of the object with respect to the interior of the airliner, or to the surface of the earth. An inertial frame of reference is one that is not accelerating (including rotation). The use of an inertial frame of reference, which will be the case for all elementary calculations, is often not explicitly stated but may generally be assumed unless stated otherwise.[35,36]

In terms of an inertial frame of reference, centrifugal force does not exist. All calculations can be performed using only Newton's laws of motion and the real forces. In its current usage, the term "centrifugal force" has no meaning in an inertial frame. In an inertial frame, an object that has no forces acting on it travels in a straight line, according to Newton's first law. When measurements are made with respect to a rotating reference frame, however, the same object would have a curved path, because the frame of reference is rotating. If it is desired to apply Newton's laws in the rotating frame, it is necessary to introduce new, fictitious, forces to account for this curved motion. In the rotating reference frame, all objects, regardless of their state of motion, appear to be under the influence of a radially (from the axis of rotation) outward force that is proportional to their mass, the distance from the axis of rotation of the frame, and to the square of the angular velocity of the frame. This is the centrifugal force.[37,38]

In order to apply high rates of tension to a polymer solution or melt, methods which can apply even distribution of stress during the tension process are required. When the centrifugal force acts upon a substance, the particles of that matter will experience a force proportional to their distance from the center of rotation. Thus, this force can be used to apply high rates of tension on a polymer solution. During the tension process, if the polymer solution has sufficient viscosity, it will be stretched as a string and transformed to a polymeric fiber after drying.

15.1.4 CENTRIFUGAL SPINNING METHODS

Hooper patented the centrifugal spinning in 1924. Classical centrifugal spinning involves supplying centripetal force using a rotary distribution disc with side nozzle holes. The rotation shears the spinning dope, classically thermoplastic materials, for instance, molten mineral or glass to form fibers such as mineral wool. Recent advances in centrifugal nanospinning have demonstrated the potential of this technique for nanofibers production. The initial work done merging electrospinning with centrifugal concepts indicated a competitive edge compared with the technique alone in terms of its ability to produce homogeneous nanofibers of below 100 nm in size and a simultaneous one-step post-spin draw possibility to enhance molecular alignment and fiber crystallinity for improved tensile strength of the resultant fiber.[8,9]

Nanofibers which are produced by traditional electrospinning usually have a wide diameter distribution, whereas the fiber diameter has an important effect on the performance of nanofiber mats in many important applications. Several of variables can influence the electrospun nanofiber diameter, diameter uniformity, and nanofiber quality such as volumetric charge density, distance from nozzle to collector, initial jet/orifice radius, relaxation time, viscosity, concentration, solution density, electric potential, perturbation frequency, and solvent vapor pressure. Furthermore, fiber diameter is increased with increasing the surface tension and flow rate, but is decreased with increasing the electric current. Solvent–polymer interactions critically influence the diameter and morphology of the electrospun fibers. One method for obtaining finer nanofiber is to reduce the concentration of the polymer solution. Also, it has been found that with the increase in electrical conductivity of the polymer solution, there is a significant decrease in the electrospun nanofiber diameter. Also, the control of the process parameters such as applied voltage and temperature can influence electrospun nanofiber diameter. The diameter of the nanofibers is increased by increasing the electrospinning voltage.[2,39]

Although all these approaches can influence nanofiber diameter, there are some limitations due to the addition of unwanted components, insufficient and difficult controllability, and strict thinning effect. Therefore, it is necessary to provide a method that can fabricate nanofibers with high uniformity and fineness without any additional procedures.

15.1.5 CENTRIFUGAL ELECTROSPINNING

Centrifugal electrospinning has been attempted and the setup typically comprises of rotary spinnerets similar to centrifugal spinning. In 2006, centrifugal electrospinning using a rotary spinneret was reported by Dosunmu et al. using a porous ceramic tube spinneret and by Andrady et al. using a rotatable spray head with four individual extrusion elements, which in turn can be made of bundles of multiple nozzles.

Reiter Oberflachen technik GmbH developed the Hyper Bell centrifugal electrospinning technology, which was subsequently acquired by Dienes Apparatebau GmbH. A centrifugal electrospinning unit with three spin heads currently supplied by Dienes Apparatebau is able to increase the throughput of conventional nozzle electrospinning by a thousand fold with a minimum achievable nanofibers diameter of 80 nm. Coupling centripetal force with electrostatic force, highly aligned poly(lactic acid) (PLA) electrospun fibers with improved modulus of 3.3 GPa (PLA in chloroform and tetrahydrofuran) were produced.

15.1.6 CENTRIFUGAL NANOSPINNING

A spinning method using only centripetal force to produce nanofibers was recently developed. Force spinning by FibeRio Technology Corporation achieved a minimum as-spun fibers diameter of 45 nm. The first polyethylene oxide (PEO) nanofibers obtained by force spinning demonstrated homogeneity with an average diameter of 105–300 nm. Furthermore, aligned PLA nanofibrous scaffolds were prepared using a similar method by Badrossamay et al.[35] under the name "rotary jet-spinning," and were used to seed cardiomyocytes in mice. The result showed good tolerance of the nanofibers, and the cell seedings successfully developed into pulsating multicellular tissues. The temperature, rotational speed of the spinneret, and collection distance are the parameters influencing the geometry and morphology of centrifugally spun nanofibers.

Centrifugal nanospinning is a versatile method that overcomes many of the limitations. The process offers higher productivity and simplicity in an equipment setup without the complication of high voltage in up scaled processing and the material constraint on electrical conductivity or relative permittivity compared to electrospinning. Nevertheless, centrifugal nanospinning is an extrusion process limited by challenges related to the material properties and the designs of the spinneret, which can lead to large differences in fiber quality and productivity. The fiber diameter in a single PLA sample can vary from 50 to 3.5 mm. The degree of complexity in the spinneret design is also proportional to the cost incurred. Although centrifugal nanospinning is a facile technique to generate three-dimensional scaffolds with a moderately high degree of uniaxial alignment, formation of more complex three-dimensional nanostructures with functionalized nano features or alignment of fibers in more than one direction may be difficult and has yet been demonstrated with this technique to date.

There has been much research and progress in the development of various designs and modification to the electrospinning process over the last century. Nevertheless, there are still many areas where further refinement of the process will be welcomed. To begin with, although there are several setups designed to achieve fiber alignment, there is still a serious shortcoming in getting highly aligned nanofibers over a large area of substantial thickness. Generally, a drum collector is not able to get highly aligned fibers even though it is able to get a larger area of fibrous mesh. Aligned nanofibers have been shown to induce cell elongation and proliferation in the direction of the fiber alignment. The ability to fabricate highly aligned fibers in large quantity over a large area will allow more investigation in cellular response to the fiber alignment in terms of gene expression and cell interaction. Typically, electrospun assemblies are in a two-dimensional form and in the case of yarn a one-dimensional form. The only three-dimensional electrospun structure with significant length, width, and height is a fibrous tube. However, researchers have yet found a way to consistently fabricate a solid three-dimensional structure through electrospinning. With the ability to fabricate three-dimensional structure, other applications such as bone replacement scaffold can be considered. Recently, Smit et al. and Khil et al. demonstrated the fabrication of continuous yarn made out of purely nanofibers. However, the spinning speed of 30 m min^{-1} is still much slower than that of the industrial fiber spinning process, which runs from 200 to 1500 m min^{-1} for dry spinning (Gupta & Kothari, 1997). Yarn made out of electrospun fibers has many applications, especially when fabricated into textiles. However, for electrospun yarn to be adopted by the textile industry, its yarn spinning speed has to be improved significantly.

Although Sun et al. made a breakthrough in the electrospinning process by creating controlled pattern using electrospun nanofibers, the small volume of solution that can be spun at one time significantly reduced the practicality of the process. An advantage of electrospinning is its ability to spin long continuous fibers at high speed. However, it is still not possible to form highly ordered structures rapidly. Arrayed nanofiber assemblies created thus far are based on arranging aligned nanofibers with very little control on the distance between each fiber.

In the following experimental part of this study, besides exploring the effects of centrifugal force on the nanofibers diameter, the fiber production capability of electrocentrifuge technique is compared with that of the conventional electrospinning method.

15.2 EXPERIMENTAL STUDY BY FOCUSING ON FLOW RATE FOR BOTH DEVICES

15.2.1 MATERIALS

Commercial polyacrylonitrile (PAN) polymer powder with $\overline{Mw}$ = 100,000 g/mol and $\overline{Mn}$ = 70,000 g/mol was supplied by Polyacryl, Iran. The used solvent was Dimethyl Formamide (DMF) from Merck Company. The polymer solutions of PAN in DMF with the concentration of 13–16 wt% were prepared using a digital scale (Libror AEU-210, Shimadzu) with accuracy to gage of 0.0001 g. Dissolving and stirring of mixture was performed with constant speed at room temperature after that, the solution was kept on 70°C for 2 h to complete the dissolution.

Scanning electron microscopy (SEM) images were used to measure the nanofibers, diameter of different samples using a field-emission scanning electron microscope (Philips SEM, XL-30). At least 100 fibers were chosen to compare each sample layer and diameter with the image scale. The average of the results was applied as the diameter of fibers produced within the process.

In order to measure the flow rate, $0.3 + X$ mg of solution with four decimal places accuracy was poured into the needle container. Then the needle was located into the centrifuge apparatus and centrifugal action was continued until the solution weight run to $0.3 - X$ mg. The difference between the two weights indicates the amount of pumped solution in grams. By measuring the density of solvent and polymer into account, the volume of pumped solution can be calculated.

15.2.2 ELECTRO CENTRIFUGE METHOD AND SETUP

Figure 15.1 represents a schematic picture of the centrifuge apparatus. In order to exert centrifugal force, a rotating disk with capability of controlling speed in the range of 0–10,000 rounds per min has been used. A tube with an inner diameter of around 4 mm is eccentrically placed within the cylinder body as polymer container which is connected to a needle with geometric characteristics of 0.165 mm inner diameter, 0.3 mm outer diameter, and 17 mm of length as nozzle. The assembly of the polymer solution container and the nozzle inside the disk reduces the effects of air stream on the nozzle tip during the rotation. The intense impact between the air and the needle causes the polymer solution to dry at the needle tip and block the flow of the polymer solution through the nozzle.

The centrifuge method for PAN solution with the concentration of 12–16 wt% at a rotational speed in the range of 0–9540 rpm can produce defected nanofiber with drop sprinkling and bead. To improve the centrifuge method in order to obtain intact nanofibers, electrical force was used simultaneously.

A high voltage power supply was used to apply an electrostatic force which is able to generate DC voltages up to 22 kV. A metallic cylinder of 26.6 cm diameter and 10 cm height which is connected to a negative electrode of high voltage supply was used as the fiber collector and the nozzle was connected to the positive electrode as shown in Figure 15.2. The distances between the disk center and the needle tip and the interior surface of the collector cylinder and the needle tip are 5.3 and 8 cm, respectively. Only 0.30 mL of the volume of the container is filled up by the solution in all experiments.[37,38]

Applying centrifugal force causes the polymer solution to flow out of the nozzle. A jet is formed at the nozzle exit, if the viscosity of the solution will be high enough; then the electrical and centrifugal forces elongate the jet thousands of times and it becomes very thin. Ultimately, by the evaporating the solvent, very long nanofibers are collected on the interior surface of the collector

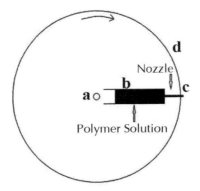

FIGURE 15.1   Schematic of centrifuge spinning setup: (a) axle of rotation, (b) polymer solution container, (c) nozzle tip, and (d) cylinder.

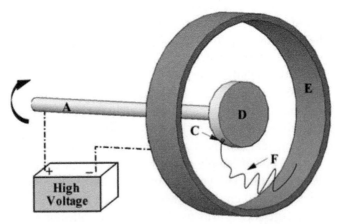

FIGURE 15.2   Schematic of electrocentrifuge setup: (a) axle of rotation, (c) nozzle tip, (d) cylinder, (e) collector, and (f) polymeric jet.

15.2.3 THEORY OF FLOW RATE

It should be noted that the flow rate is controlled only by the centrifugal force in the present experimental setup. When the free surface of the polymer solution in the container and the nozzle tip are at the same atmospheric pressure, the pressure within the container increases as the centrifugal force is exactly equal to the pressure loss across the nozzle due to the viscosity effect, that is, $\Delta P_{1,2} = \Delta P_{2,3}$. On the other hand, the pressure rise in the container can be estimated as:

$$\Delta P_{1,2} = \int_{x_0}^{L_0} \rho\, \omega^2\, dl \text{ or } \Delta P_{1,2} = \frac{1}{2} \rho \left(L_0^2 - x_0^2 \right) \omega^2, \tag{15.1}$$

where ρ is the liquid density, L_0 is the distance between the center of rotation and the nozzle entrance, x_0 is the distance between the center of rotation and the liquid free surface, and ω is the angular speed.

Equation 15.1 shows that the pressure loss across the nozzle depends on the length of the liquid column $(L_0 - x_0)$ within the container. Since the flow rate is also proportional to the same pressure difference, we can control the flow rate by changing x_0 and ω.

15.3 RESULTS AND DISCUSSION

Electrical forces which are known as Columbus and external field forces play the drawing role in electrospinning method. The electrocentrifuge method has been established with the intention of adding centrifugal force to the above forces. Therefore, the effect of centrifugal force on the production rate and fiber diameter is a matter of importance and has been investigated in this study.

15.3.1 EFFECT OF ROTATIONAL SPEED AT CONSTANT FLOW RATE ON THE NANOFIBER DIAMETER

In order to investigate the effect of rotational speed at the constant flow rate on the fibers' diameter, fibers were produced from a 15 wt% solution at five different quantities of the solution. For each run, the rotational speed was adjusted by trial and error in a way that the flow rate in all rotational speeds remains unchanged (1.2140 mL/h). As reported in Table 15.1 the obtained results reveal that the fibers, diameter increases by increasing the rotational speed at the constant flow rate. This can be interpreted in this way that by increasing the rotational speed, the aerodynamic effects of the air stream formed in between of rotating cylinders become more prominent around the fiber. As the velocity of the air flow increases, the polymer jet gets dry before it can be elongated further by the centrifugal force. Therefore, it seems that at every flow rate there is a rotational speed in which the minimum fiber diameter can be spun.

TABLE 15.1 Nanofiber Diameters Using Different Rotational Speeds at a Constant Flow Rate 1.2140 mL/h.

Nanofiber diameter (nm)	Rotational speed (rpm)	Nanofiber production method
344	6360	Electrocentrifuge
396	7950	Electrocentrifuge
470	9540	Electrocentrifuge
352	0	Electrospinning

15.3.2 COMPARISON BETWEEN THE NANOFIBER PRODUCTION RATES IN ELECTROSPINNING AND ELECTROCENTRIFUGE SPINNING SYSTEMS

Before comparing the two systems of electrocentrifuge and electrospinning, it is essential to know that all the effective variables (excluding the centrifugal force) of the two systems have the same conditions.

The variation of surface tension cannot be significant when the polymer concentration is in a range of 13–16%, and, therefore, we assume that this parameter is constant for all experiments. Comparisons between the two systems are made at two different applied voltages of 10 and 15 kV, and for 13, 14, 15, and 16 wt% solutions of PAN at the same conditions. The most effective voltage which is possible to be applied to an 8 cm gap is equal to 15 kV. At higher voltage, the surrounding air will be ionized, which results the generation of electric current between the positive and negative poles. The setup parameters are exactly the same for both systems, except for the flow rate. In the electrospinning system, a syringe pump provides the necessary flow rates, while in the electrocentrifuge system the flow rate is adjusted by the centrifugal force. For the specified parameters, the mean diameter of the nanofibers obtained from the two systems was in a range of 200–600 nm. Figure 15.3 shows typical SEM images of nanofibers produced from a 15 wt% polymer solution with a rotational speed of 6360 rpm and an average diameter of 410 nm.

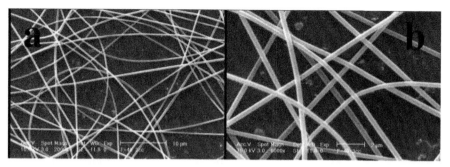

FIGURE 15.3   SEM images of nanofibers produced by electrocentrifuge method with 6360 rpm and 0.3 cc polymer solution in container: (a) low magnification (2000×) and (b) high magnification (6000×).

Keeping other parameters constant, by increasing the flow rate to a specific limit, bead formation and drop sprinkling occur in electrospinning. The flow rate of the solution at the beginning of the bead formation is defined as the maximum rate of solution pumping for electrospinning. However, in the electrocentrifuge case, by increasing the rotational speed the flow rate of the solution will increase when other parameters are constant. Therefore, the production rate of fibers by the electrocentrifuge system is calculated by measuring the solution flow rate at the maximum possible rotational speed that leads to the production of fibers which are free of defects and beads. The increase in the production rate of fibers by the electrocentrifuge method is defined as the ratio of flow rates of the two systems when the produced fibers have the same diameter.

As it can be seen, the electrocentrifuge method can increase the rate of fiber production to a large extent. For the concentration of 16 wt%, the rpm of 9540 is not the maximum possible rotational speed and it is feasible to produce fibers with this concentration at higher velocities. This case was not investigated here due to the apparatus limitations.

Table 15.2 shows a decline in the production rate index as the concentration of the polymer solution are increased in constant voltage. The observations show that at a constant rotational velocity and applied voltage, upon increasing the concentration, the flow rate decreases and the production rate of the electrocentrifuge system rapidly approaches that of the electrospinning method. Interestingly, increasing the applied voltage reduces the production rate when other parameters are constant. The effect of the centrifugal force on the increase in production rate is more prominent for lower voltages due to the lower electrospinning production rate in that range (Table 15.3).

TABLE 15.2 Effect of Polymer Concentration on the Production Rate at Constant Voltage of 15 kV.

Production rate percent	Polymer solution concentration (wt%)	Rotational speed (rpm)	Nanofiber production method
1200	13	6360	Electrocentrifuge
723	14	7950	Electrocentrifuge
450	15	9540	Electrocentrifuge
195	16	–	Electrocentrifuge
197	13	0	Electrospinning

It can be concluded from the experiment that at lower polymer concentrations, the centrifugal forces are more effective. For example, at the concentration of 13 wt% and voltage of 10 kV, the production rate is 12 times larger than that of the conventional electrospinning approach.

TABLE 15.3 Effect of Applied Voltage on the Production Rate in Constant Concentration of 13%.

Production rate percent	Applied voltage (kV)	Rotational speed (rpm)	Nanofiber production method
187	10	6360	Electrocentrifuge
444	12	7950	Electrocentrifuge
678	14	9540	Electrocentrifuge
1200	15	–	Electrocentrifuge
177	10	0	Electrospinning

15.4 CONCLUSION

Centrifugal and electrical forces have been employed simultaneously for nanofiber production, and the effect of adding the centrifugal force to the electrical forces on the nanofiber diameter and production rate (which are the most important factors for nanofiber production) has been investigated in this study.

At first, the effect of increasing the rotational speed at a constant flow rate on the fiber diameter was investigated, which causes an increase in the diameter of the produced fibers upon increasing the rotational speed. As a result of an increase in the impact intensity of the air surrounding the nozzle with the exiting jet, the jet dries quickly.

In addition, the nanofiber production capability of the novel method was investigated and compared with conventional electrospinning. Results demonstrate a significant increase in the production rate of the electrocentrifuge method compared with electrospinning which is depended on the concentration of the polymer solution and the applied voltage. Employing the electrocentrifuge technique can overcome the low production rate of electrospinning method. Therefore, this approach can be a satisfactory alternative to electrospinning.

KEYWORDS

- **tensile strength**
- **nanofibers**
- **electrospinning**
- **polymers**
- **electrocentrifuge**

REFERENCES

1. Ramakrishna, S., et al. *An Introduction to Electrospinning and Nanofibers;* World Scientific, 2005; Vol. 90.
2. Frenot, A.; Chronakis, I. S. Polymer Nanofibers Assembled by Electrospinning. *Curr. Opin. Colloid Interface Sci.* **2003**, *8*(1), 64–75.
3. Ding, B.; Yu, J. *Electrospun Nanofibers for Energy and Environmental Applications*; Springer, 2014.
4. Doshi, J.; Reneker, D. H. In *Electrospinning Process and Applications of Electrospun Fibers*, Industry Applications Society Annual Meeting, 1993, Conference Record of the 1993 IEEE.
5. Reneker, D. H.; Chun, I. Nanometre Diameter Fibres of Polymer, Produced by Electrospinning. *Nanotechnology* **1996**, *7*(3), 216–223.
6. Greiner, A.; Wendorff, J. H. Electrospinning: A Fascinating Method for the Preparation of Ultrathin Fibers. *Angew. Chem. Int. Ed. Engl.* **2007**, *46*(30), 5670–5703.
7. Rist, R. C. Influencing the Policy Process with Qualitative Research. In *Handbook of Qualitative Research;* Denzin, N. K., Lincoln, Y. S., Eds.; Sage Publications: USA, 1994, pp 547–557.
8. Voelker, H., et al. Production of Fibers by Centrifugal Spinning. Google Patents, U.S. Patent 5494616 A, Feb 27, 1996.
9. Edmondson, D., et al. Centrifugal Electrospinning of Highly Aligned Polymer Nanofibers Over a Large Area. *J. Mater. Chem.* **2012**, *22*(35), 18646–18652.

10. Liu, S.-L., et al. Assembly of Oriented Ultrafine Polymer Fibers by Centrifugal Electrospinning. *J. Nanomater.* **2013**, *2013*(2514103), 8. http://dx.doi.org/10.1155/2013/713275.

11. Mary, L. A., et al. Centrifugal Spun Ultrafine Fibrous Web as a Potential Drug Delivery Vehicle. *eXPRESS Polym. Lett.* **2013**, *7*(3), 238–248.

12. Subramanian, S.; Seeram, R. New Directions in Nanofiltration Applications—Are Nanofibers the Right Materials as Membranes in Desalination? *Desalination* **2013**, *308*, 198–208.

13. Pham, Q. P.; Sharma, U.; Mikos, A. G. Electrospinning of Polymeric Nanofibers for Tissue Engineering Applications: A Review. *Tissue Eng.* **2006**, *12*(5), 1197–1211.

14. Lee, J.; Feng, P. X.-L.; Kaul, A. B. In *Characterization of Plasma Synthesized Vertical Carbon Nanofibers for Nanoelectronics Applications*, MRS Proceedings; Cambridge University Press: Cambridge UK, 2012.

15. Zhang, W.; Pintauro, P. N. High-Performance Nanofiber Fuel Cell Electrodes. *ChemSusChem* **2011**, *4*(12), 1753–1757.

16. Tan, K.; Obendorf, S. K. Fabrication and Evaluation of Electrospun Nanofibrous Antimicrobial Nylon 6 Membranes. *J. Membr. Sci.* **2007**, *305*(1), 287–298.

17. Kenawy, E.-R., et al. Processing of Polymer Nanofibers Through Electrospinning as Drug Delivery Systems. *Mater. Chem. Phys.* **2009**, *113*(1), 296–302.

18. Chen, J.-P.; Chang, G.-Y.; Chen, J.-K. Electrospun Collagen/chitosan Nanofibrous Membrane as Wound Dressing. *Colloids Surf. A Physicochem. Eng. Asp.* **2008**, *313*, 183–188.

19. Huang, Z.-M., et al. A Review on Polymer Nanofibers by Electrospinning and Their Applications in Nanocomposites. *Compos. Sci. Technol.* **2003**, *63*(15), 2223–2253.

20. Jayaraman, K., et al. Recent Advances in Polymer Nanofibers. *J. Nanosci. Nanotechnol.* **2004**, *4*(1–2), 52–65.

21. Frenot, A.; Chronakis, I. S. Polymer Nanofibers Assembled by Electrospinning. *Curr. Opin. Colloid Interface Sci.* **2003**, *8*, 64–75.

22. Persano, L., et al. Industrial Upscaling of Electrospinning and Applications of Polymer Nanofibers: A Review. *Macromol. Mater. Eng.* **2013**, *298*(5), 504–520.

23. Rafiei, S., et al. Mathematical Modeling in Electrospinning Process of Nanofibers: A Detailed Review. *Cellul. Chem. Technol.* **2013**, *47*, 323–338.

24. Luo, C., et al. Electrospinning Versus Fibre Production Methods: From Specifics to Technological Convergence. *Chem. Soc. Rev.* **2012**, *41*(13), 4708–4735.

25. Sahay, R.; Thavasi, V.; Ramakrishna, S. Design Modifications in Electrospinning Setup for Advanced Applications. *J. Nanomater.* **2011**, *2011*, 17. http://dx.doi.org/10.1155/2011/317673.

26. Fong, H., et al. Generation of Electrospun Fibers of Nylon 6 and Nylon 6-Montmorillonite Nanocomposite. *Polymer* **2002**, *43*(3), 775–780.

27. Dersch, R., et al. Electrospun Nanofibers: Internal Structure and Intrinsic Orientation. *J. Polym. Sci. Part A Polym. Chem.* **2003**, *41*(4), 545–553.

28. Deitzel, J., et al. Controlled Deposition of Electrospun Poly (ethylene oxide) Fibers. *Polymer* **2001**, *42*(19), 8163–8170.

29. Pan, H., et al. Continuous Aligned Polymer Fibers Produced by a Modified Electrospinning Method. *Polymer* **2006**, *47*(14), 4901–4904.

30. Fang, D.; Hsiao, B.; Chu, B. Multiple-jet Electrospinning of Non-woven Nanofiber Articles. *Polym. Prepr.* **2003**, *44*(2), 59–60.

31. Yarin, A.; Zussman, E. Upward Needleless Electrospinning of Multiple Nanofibers. *Polymer* **2004**, *45*(9), 2977–2980.

32. Liu, Y.; He, J.-H. Bubble Electrospinning for Mass Production of Nanofibers. *Int. J. Nonlinear Sci. Numer. Simul.* **2007,** *8*(3), 393–396.
33. Dosunmu, O., et al. Electrospinning of Polymer Nanofibres from Multiple Jets on a Porous Tubular Surface. *Nanotechnology* **2006,** *17*(4), 1123–1127.
34. Arya, A. P. *Introduction to Classical Mechanics*; Allyn and Bacon: Boston, Massachusetts, 1990.
35. Badrossamay, M. R., et al. Nanofiber Assembly by Rotary Jet-spinning. *Nano Lett.* **2010,** *10*(6), 2257–2261.
36. Truesdell, C. *Rational Mechanics*; Academic Press: New York, 1983.
37. Synge, J. L. *Principles of Mechanics*; Read Books Ltd.: UK, 2013.
38. Gantmakher, F. R. *Lectures in Analytical Mechanics*; Mir Publishers: Russia, 1970.
39. Sandou, T.; Oya, A. Preparation of Carbon Nanotubes by Centrifugal Spinning of Coreshell Polymer Particles. *Carbon* **2005,** *43*(9), 2015–2017.

CHAPTER 16

RESULTS OF TESTING OF HELIODRYING APPARATUS WITH POLYCARBONATE COVERING

K. T. ARCHVADZE[1,2,*], T. I. MEGRELIDZE[1,2], L. V. TABATADZE[1,2], and I. R. CHACHAVA[1,2]

[1]Food Industry Department, Georgian Technical University, 77 Kostava 0175, Tbilisi, Georgia

[2]Department of Technology, Sukhumi State University, Ana Politkobskaia 9, 0186, Tbilisi, Georgia

*Corresponding author. E-mail: kiti987a@gmail.com

CONTENTS

ABSTRACT

For drying of agricultural products, there are offered three heliodrying apparatus—convective, large-scaled, and sheet with polycarbonate coverings, developed and probated at Georgian Technical University. These apparatus are used for getting of dried mushrooms and other agricultural products. Driers can be used in private subsidiary, farm economies, and agricultural companies. They do not need big capital investment, they are simple in production, and keep qualitative characteristics of dried products.

16.1　INTRODUCTION

Use of solar energy in rational combination with other sources of energy in many cases allows to economize significant amount of fuel-energetic resources. Effect from the use of solar energy is specially felt during realization of most power-consuming thermotechnological processes in helioapparatus. Nowadays, problems of searching new alternative sources of energy have become especially actual. Today, the different forms of renewable energy resources include hydropower, solar, wind, and animal and vegetative biomass.

In recent years, the extended assortment of products with new species of nutritional concentrates more often included cultivated mushrooms. Provision of safety and quality of food products shall be one of the key directions of the state policy of Georgia in the sphere of healthy nutrition. Edible mushrooms do not take last place in the food ration of human. Mushrooms are used in fresh, frozen, conserved, and dried forms. Although included last in listed productions, mushrooms are included in recipes of other food products in form of aroma and/or enriching supplements. Besides, mushrooms are used in products of nonfood and medical purposes. Mushrooms have high nutritional value, and they are under solidly high demand at the markets of European countries and in other countries worldwide. Edible mushrooms most often are used as additional source of vegetable protein.

16.2　EXPERIMENTAL

16.2.1　MATERIALS

Sheet heliodrying apparatus, a large heliodrying apparatus, the convective heliodrying apparatus, dog rose, honey agarics, food products, and polycarbonate.

16.2.2 OBJECTS OF STUDY

Heliodrying apparatus, polycarbonate, dog rose, and honey agarics.

Drying in heliodrying apparatus with polycarbonate (Fig. 16.1) covering is recommended for effectively drying agricultural products.

FIGURE 16.1 Polycarbonate.

There are inventions of different complicated collectors of solar energy. We offer three heliodrying apparatus with polycarbonate covering, constructed and developed at Georgian Technical University—convective, large-scaled, and sheet heliodrying apparatus, and also 11 variants of mentioned construc- tions. Use of such heliodrying apparatus, according to the data of studies, decreases time of drying, improves keeping of aroma, useful substances, and taste, provides sterility of products, and also simplifies storage of dried product (the product is not spoiled and is kept for longer time than usual). As experiments have shown, use of noncomplicated heliodrying apparatus with polycarbonate covering in small peasant farms, as well as in city habitants allow to economize means and physical labor, and the method is ecologically friendly (there is no heat and carbonic acid release in environment). Use of high temperatures (in electric drying machine and oven) for preparation of dried fruits often leads to destruction of vitamins. Drying of raw materials by natural solar–air method usually takes a long time, but this does not affect the quality of product and its vitamin composition. But drying in heliodrying apparatus with polycarbonate coverings, according to the analysis of data of dried products on vitamins, is the most optimal because vitamins composi- tion is maintained at maximally high level, as well as consumption and taste of products, and also in dried products ferments and microorganisms are inactivating. We shall note that during the storage process, the products dried in heliodrying apparatus (Fig. 16.2) were more resistant to effect of mold, and they were stored for longer time than products dried by natural drying in open air.

FIGURE 16.2   One of heliodrying apparatus: sheet heliodrying apparatus.

16.3   RESULTS AND DISCUSSION

16.3.1   *DRYING OF MUSHROOMS "HONEY AGARICS"*

16.3.1.1   *NUTRITIONAL VALUE AND CHEMICAL COMPOSITION "HONEY AGARICS".*

Table 16.1 shows contents of nutritional substances (calorie, proteins, fats, carbohydrates, vitamins, and minerals) per 100 g of edible part. The energetic balance of "honey agarics" is presented in Figure 16.3.

TABLE 16.1 Nutritional Value and Chemical Composition "Honey Agarics."

Nutritional value		Vitamins	
Calorie	22 kcal	Vitamin PP	10.3 mg
Proteins	2.2 g	Vitamin B1 (thiamine)	0.02 mg
Fats	1.2 g	Vitamin B2 (riboflavin)	0.38 mg
Carbohydrates	0.5 g	Vitamin C	11 mg
Dietary fibers	5.1 g	Vitamin E (TE)	0.1 mg
Water 90 g		Vitamin PP	10.7 mg
Unsaturated fatty acids	0.2 g		
Mono and disaccharides	0.5 g		
Ash 1 g			
Saturated fatty acids	0.2 g		

Macroelements		Microelements	
Calcium	5 mg	Iron	0.8 mg
Magnesium	20 mg		
Sodium	5 mg		
Potassium	400 mg		
Phosphor	45 mg		

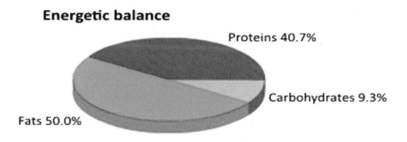

FIGURE 16.3 Energetic balance of "honey agarics."

The results of experiments are given in Figures 16.4–16.6 and Tables 16.2–16.6.

1 Day of Drying

TABLE 16.2 Data Sheet Table for Drying of Product "Honey Agarics."

Time (h)	Relative humidity of air (%)	Atmospheric pressure (kPa)	Wind velocity (km/h)	Tempera-ture of air in shadow (°C)	Temperature of air in sun (°C)	Notes
10.00	56	102.2	12	14	17	
12.00	41	102.2	12	16	20	
14.00	24	102.0	14	19	26	
16.00	21	101.9	16	20	26	
18.00	22	102.0	16	18	21	
20.00	30	101.8	16	14	14	

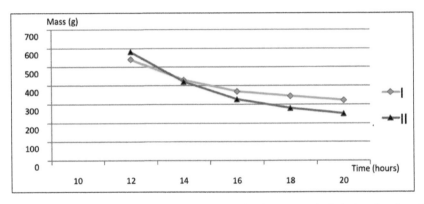

FIGURE 16.4 Changing of mass of raw materials "honey agarics" during the day: (I) changing of the mass of raw material, dried by natural drying (n/d) in the open air and (II) changing of the mass of raw material, dried in heliodrying (h/d).

2 Days of Drying

TABLE 16.3 Data Sheet Table for Drying of Product "Honey Agarics."

Time (h)	Relative humidity of air (%)	Atmospheric pressure (kPa)	Wind velocity (km/h)	Temperature of air in shadow (°C)	Temperature of air in sun (°C)	Notes
10.00	48	101.7	12	18	20	
12.00	38	101.6	10	20	24	
14.00	34	102.4	8	22	28.5	
16.00	32	102.2	8	23	26	
18.00	28	102.1	8	20	21	
20.00	34	101.4	8	17	17	

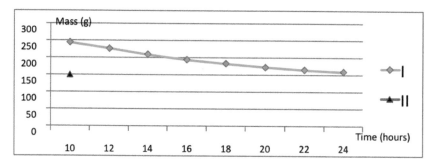

FIGURE 16.5 Changing of mass of raw materials "honey agarics" during the day: (I) changing of the mass of raw material, dried by natural drying on the open air and (II) changing of the mass of raw material, dried in heliodrying.

TABLE 16.4   Analysis of Experienced Data.

Name of product		
Mushrooms "honey agarics"		
	h/d	n/d
Initial mass (g)	580	540
Final mass (g)	150	160
Decrease of mass (%)	74.1	70.4
Duration of drying (h)	22	36

Duration of drying in heliodrying apparatus was 1.7 times shorter than duration of drying in open air.

TABLE 16.5   Nutritional Value and Chemical Composition of Product "Dog Rose."

Nutritional value		Vitamins	
Calorie	109 kcal	Vitamin PP	0.6 mg
Proteins	1.6 g	Beta-carotin	2.6 mg
Fats	0.7 g	Vitamin A (RE)	434 mcg
Carbohydrates	22.4 g	Vitamin B1 (thiamine)	0.05 mg
Dietary fibers	10.8 g	Vitamin B2 (riboflavin)	0.13 mg
Organic acids	2.3 g	Vitamin C	650 mg
Water	60 g	Vitamin E (TE)	1.7 mg
Mono and disaccharides	19.4 g	Vitamin PP	0.7 mg
Starch	3 g		
Ash	2.2 g		
Saturated fatty acids	0.1 g		

TABLE 16.5 *(Continued)*

Macroelements		Microelements	
Calcium	28 mg	Iron	1.3 mg
Magnesium	8 mg	Zinc	1.1 mg
Sodium	5 mg	Copper	37,000 mcg
Potassium	23 mg	Manganese	19 mg
Phosphor	8 mg	Molybdenum	4330 mcg

Energetic balance

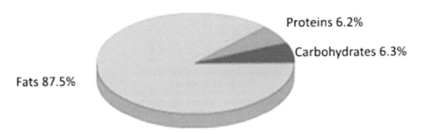

Proteins 6.2%

Carbohydrates 6.3%

Fats 87.5%

FIGURE 16.6 Energetic balance of product "dog rose."

TABLE 16.6 Analysis of Experienced Data.

	Name of product	
	Dog rose	
	h/d	n/d
Initial mass (g)	700	690
Final mass (g)	350	350
Decrease of mass (%)	50	49.3
Duration of drying (24 h)	5	20

16.3.2 DRYING OF DOG ROSE

Drying of dog rose by natural drying in shadow took 20 day–nights, and in heliodrying apparatus with polycarbonate covering 5 day–nights. Therefore, drying in the heliodrying apparatus took four times shorter period.

Laboratory analysis of vitamin C showed that 23% of mentioned vitamin is destructed during natural drying and 15% was destructed during the use of heliodrying apparatus.

Solar drying of fruits and vegetables is one of the most acceptable and cheap methods of their conservation, which makes possible the preservation of maximum amount of vitamins and minerals. The dried vegetables, fruits, and mushrooms differ in the high content of sugar, organic acids, mineral salts, and vitamins and, also, the fibers and carbohydrates riches which not were destroyed in the drying process. They can keep the useful substances for a long time and indemnify the deficiency of the vitamins. As a result of the drying, we receive the stable food products, which do not keep the conservants, dyestuffs, aromatizators, and other strange substances. The dried fruits well substitute the refined sugar, actively promote the radionuclides and slags help out from organism, and operate as antioxidant.

For the effective drying, it is recommended to use the heliodrying apparatus with polycarbonate covering, so far as it shortens the drying time, better preserves aroma, useful matter, and taste of the product, and provides sterility of the product; the dried products are easier to keep, also for longer time. As experiments on drying of various kinds of agricultural products show, drying rate increases—three to five times in comparison with traditional open-air drying. Products dried in the heliodrying device have better consumer properties than products prepared by means of natural drying.

KEYWORDS

- polycarbonate
- heliodrying apparatus
- drying
- product
- mushrooms

REFERENCES

1. Zuev, V. V.; Uspenskaia, M. V.; Olehnovich, A. O. *Physics and Chemistry of Polymers*; Proc. in Russian. SPb.: ITMO, 2010; pp 45–85.
2. Magazine Fast canning; Publisher: OOO "News Express." Special Issue. No. 06, 2011.
3. Kondrashova, E. A.; Konik, N. V.; Peshkova, T. A. *Commodity Food Products. M. "Alpha-book"*; 2009, 338 p.

4. Antipov, S. T.; Cretov, I. T.; Ostrikov, A. N. *Machinery and Equipment for Food Production, a Textbook for High Schools* (In two volumes); M: High School; Acad RAAS Panfilov, V. A., Ed.; Russian Academy of Sciences: Moscow, 2007; 1379 p.
5. Semchikov, Y. D. *Macromolecular Compounds* (in Russian); Russian Academy of Sciences: Moscow, 2003, 368 p.
6. Jenning, D. H. *The Physiology of Fungal Nutrition;* Cambridge University Press: Cambridge, UK, 2007; 622 p.
7. Plaksin, Y. M; Malakhov, H. N.; Larin, B. A. *Processes and Devices for Foodstuffs Processing,* 2nd Edition; Rev. and add. M.: Colossus: Tbilisi, Georgia, 2007; 760 p.
8. Chagin, O. V.; Kokin, N. R; Pastine, V. V. *Equipment for Drying Foodstuffs*; Ivan. Chemical—Primary Process: Univ.: Ivanovo, 2007; 138 p.
9. Morozov, A. I. *Large Mushroom Encyclopedia;* M.: AST; Stalker: Donetsk, 2005; 479 p.
10. Kiseleva, T. F. Drying Technology. In *Training and Methodology Complex*; KemTIPP: Kemerovo, 2007; 117 p.

CHAPTER 17

DEPENDENCE OF MORPHOLOGY ON OPTICAL AND ELECTRICAL PROPERTIES OF METAL OXIDE NANOSTRUCTURES

PAULOSE THOMAS[1] and AJITH JAMES JOSE[2,*]

[1]*Optoelectronic Lab, St. Berchmans College, Changanassery, Kerala, India*

[2]*Research Department of Chemistry, St. Berchmans College, Changanassery, Kerala, India*

Corresponding author. E-mail: ajithjamesjose@gmail.com

CONTENTS

ABSTRACT

Synthesis and study of nanomaterials are challenging area of materials research. Researchers have focused on developing diverse properties of nanoamaterials through different methods. There are large factors affecting the changes of physics properties of nanomaterials as compared with its bulk form. One of the important property controlling parameter is its surface morphology. Nanoparticles possess variety of shapes and their names are characterized by their different shapes. For examples nanowires, nanorods, nanotubes, nanorings, nanospheres, nanoflakes, nanoflowers, nanobelts, etc. These shapes or morphologies sometimes arise spontaneously as an effect of a templating or directing agent during synthesis. Morphology of different nanostructures may vary significantly depending on their material composition, crystal structure and manufacturing method. Variation of different synthesis parameters such as temperature, pressure, reagent concentration, treatment time and pH results in different morphologies. Controlling the morphology of nanoparticles is of key importance for exploiting their properties for their use in several emerging technologies. Also morphology variation is an effective way of controlling functionality of nanomaterials because the variation of large number of surface atoms with respect to their surface morphology that determine their physical and chemical properties. In the present chapter, the tuning of optical and electrical properties of metal oxide nanostructures explained on account of various surface morphologies.

17.1   INTRODUCTION

Nanomaterial research is an application oriented study conducted in science and engineering. It has been done on atomic or molecular level to build up discrete number of assembled atomic/molecular units for wide variety of applications. Nanotechnology and nanoscience are like two sides of a coin and both are complement to each other. Basically, both concentrate on the study and application of small matters. It can be employed in all science fields, such as chemistry, physics, materials science, biology, and engineering.[1-5] For the last 50 years, huge developments have been created in all areas of science,[6-9] particularly in miniaturization of materials and appliances with the effective use of nanoscience and nanotechnology. The size reduction in electronic equipments, advanced medical diagnostic apparatus, fast and precise communication systems, developments in industrial and agricultural field, etc. are the examples of miniaturization followed by the developments of nanoscience and nanotechnology.

However, nanoscience need not be considered as a new science. In early 15th century, many physicist and chemists had studied properties of matter down to the atomic and molecular level. The novelty of present status of nanoscience lies on the exceptional properties of matter at the nanoscale level and the methods used to build new materials, systems, and devices. There are multiple factors that control the properties of nanomaterials. The crystallite structure, reduced particle size dimensions, multiple morphology, etc. are the prime important parameter for changing the property of nano-materials from its bulk form.[10] The present chapter explains the various morphologies of nanomaterials and how they can be dependent on optical and electrical property of metal oxide nanostructures.

17.1.1 SURFACE MORPHOLOGY OF NANOMATERIALS

Morphology of different nanostructures may vary significantly depending on their material composition, crystal structure, and manufacturing method. At the earlier stages of nanoresearch, the researchers suggested that particle morphology is a strong dependent parameter for all physical and chemical properties of nanomaterials. On account of this assumption, over the last decades researchers in all over the world fabricated various morphological nanostructures by numerous physical and chemical synthesis methods. The nanotube, nanowires, nanorods, nanoflowers, nanoflakes, nanospheres, nanocubes, etc. are the important attractive nanomorphologies. The surface-to-volume ratio and quantum confinement effect of nanostructures vary with respect to different morphologies.[5-8] These shapes or morphologies sometimes arise spontaneously as an effect of reactants or directing/capping agents during synthesis. The morphologies of nanoparticles help serve their various purposes such as long carbon nanotubes being used to bridge an elec-trical junction, pores shaped nanostructures helps for construction of energy storage devices such as supercapacitors, etc. Existing synthesis methods allow the production of nanoparticles with a variety of shapes and sizes. The nanolithographic method, chemical vapor deposition method, laser ablation method, sputtering method, hydrothermal method, etc. are the most important synthesis methods for fabricating verity of nanomorphologies. Over these methods, hydrothermal method has got more attention in nanore-search as its synthesis parameters such as temperature, pressure, reagent concentration, treatment time, and pH results in different morphologies, compositions, and better crystallinity of the products. Hence, hydrothermal method is considered as an easy low-cost method for fabricating different nanomorphologies. Sometimes a single synthesis method is not suitable for fabricating the same morphology of different material. For example, hydrothermal method is much useful for fabricating ZnO nanotubes, whereas it is inefficient for CdO nanotube synthesis. The surface effect, particle size, and size-induced quantum confinement property are closely dependent on morphology of nanostructures. Controlling the morphology of nanoparticles is of key importance for exploiting their properties for their use in several emerging technologies. Optical filters and biosensors are among the many applications that use optical properties of metal or metal oxide nanoparticles. Nanostructures aspect ratio is one of the important parameters for morphological characterization. The aspect ratio is the average ratio of the highest to the lowest dimension over a number of similar particles. Recent studies indicated that the aspect ratio of nanoparticles varies with different morphologies. Nanoparticles with high and low aspect ratio are classified separately. High aspect ratio nanoparticles include nanotubes and nanowires, with various shapes, such as helices, zigzags, belts, or perhaps nanowires with diameter that varies with length. Small aspect ratio morphologies include spherical, oval, cubic, prism, helical, or pillar. A nanomaterial has high aspect ratio that indicates the high surface energy or high activity of those nanoparticles.

17.1.2 DIMENSIONALITY AND CLASSIFICATION OF NANOMATERIALS

The capability of nanostructure materials can produce improved properties such as increased hardness, ductility, selective absorption showing more efficient optical and electronic behavior, etc. There are several types of nanostructured materials that are categorized according to their dimensions such as zero-dimensional, one-dimensional, two-dimensional, and three-dimensional nanostructures.[11-16]

The quantum dot is an example of zero-dimensional nanostructure. One-dimensional nanostructures include whiskers, fibers, nanowires, and nanorods. Two-dimensional nanostructures include thin films which have thickness of the order of few nanometers. Three-dimensional nanomaterials or bulk nanomaterials are materials that are not confined to the nanoscale in any dimension. Generally, nanomaterials are synthesized either by top-down or bottom-up approaches[12-15] In top-down approach, a bulk material is crushed into fine particles using the processes as mechanical and chemical fabrication techniques such as laser ablation, ball mill, sputtering, etc. Other examples of top-down approaches are: nanolithography, plasma arc deposition, electron beam evaporation, etc. In ancient days, the mankind used stone ball miller to grind rice in to powder form. This is the better example for understanding a layman to top-down approach in nanofabrication. Bottom-up fabrication is similar to building a brick house. The desired nanomaterials are prepared by arranging atom by atom or molecule by molecule. Researchers prefer bottom-up approach to the other for nanoparticles synthesis. There are several reasons behind which the bottom-up approach plays an important role in the fabrication of nanostructures. In bottom-up approach, the prepared nanostructures has better chemical composition, phase purity, perfect morphology, less defect, etc. The examples of bottom-up processes are: chemical vapor deposition, coprecipitation, electro deposition, sol–gel method, hydro-thermal method, etc.

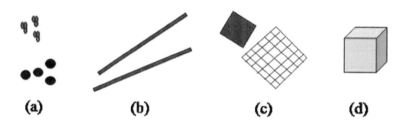

FIGURE 17.1   Classification of nanomaterials (a) 0D spheres and clusters, (b) 1D nanofibers, wires, and rods, (c) 2D films or flakes, and (d) 3D nanomaterials.

Basically, two principal factors affect the property of nanomaterials namely, increased relative surface or larger surface to volume ratio area and quantum confinement effect.[16–19] Nanomaterials have a larger surface to volume ratio compared to same volume or mass of its corresponding bulk material. For example, consider a sphere of radius "r." Its surface area and volume are $4\pi r^2$ and $4/3\pi r^3$, respectively. Its surface area to volume ratio is equal to $3/r$. Hence, as the radius of the sphere decreases its surface to volume ratio increases. Accordingly, nanoparticles of the same material but different size (nanoscale) may have different properties. This surface to volume ratio and its effects on nanomaterials properties is a key feature of nanoscience and nanotechnology. For these reasons, a nanomaterial's surface morphology is of great interest because various morphologies will produce distinct surface to volume ratio and therefore different properties.

Size induced restriction to the free movement of charge carriers is an example of quantum confinement. When the radius of the nanoparticles is smaller than or equal to the size of its exciton Bohr radius, excitons are pinched in to the discrete energy levels created in between the valance and conduction band. As a result, its effective bang gap is increased as compared to bulk. Hence, the band gap of nanomaterials gets widened as the size of nanoparticle is reduced. This phenomenon is a quantum confinement effect. In addition, quantum confinement leads to a collapse of the continuous energy bands of a bulk material into discrete, atomic-like energy levels.[20–22] There are two types of quantum confinement effect: weak confinement and strong confinement. The nanoparticles are said to be in the weak confinement regime, if their radii are of the order of the exciton Bohr radius. The strong confinement effect is mainly originated when the nanoparticles radii is smaller than the exciton Bohr radius. If the size of the nanoparticles is less than 10 nm, the quantum confinement effects dominate and the electronic and optical properties becomes highly tunable. Based on the confinement direction, quantum confined structure will be classified into three categories as quantum well, quantum wire, and quantum dots. For zero-dimensional nanomaterials, the electrons are confined in all the three directions, that is, no electron is free to move or delocalize. Quantum wires are formed when two dimensions of the system are confined. In quantum well, charge carriers are confined to move in a plane and are free to move in two dimensions.[23,24]

17.1.3 MORPHOLOGICAL CHARACTERIZATION OF NANOSTRUCTURES

Electron microscopy is a good tool for studying the surface and morphology of nanostructures. The maximum magnification of optical microscope is 1000× which is insufficient for the study of surface morphology of nanoparticles. In an electron microscope, a magnification of about 10,000,000× can be smoothly achieved with a better resolution. There are different types of electron microscope used in nanoparticles research, that is, scanning electron microscopy (SEM), transmission electron microscopy (TEM), field emission scanning electron microscopy (FE-SEM), focused ion beam—scanning electron microscopy (FIB-SEM), scanning transmission electron microscopy (STEM), and ultrahigh-resolution SEM (UHR-SEM).

An electron microscope (SEM) can produces images of a sample by scanning it with a focused beam of electrons. These high-energy electrons interact with electrons in the sample and reveal information about external morphology (texture), chemical composition, crystalline structure, and orientation of materials of the sample. The electron beam is generally scanned in a raster scan pattern, and the beam position combined with the detected signal to produce a two-dimensional image that displays spatial variations in the scanned objects. An accelerated electron in electron microscopy carries significant amounts of kinetic energy which is dissipated as signals through electron–sample interactions. Essential components of all electron microscopy instruments include the electron source, electron lenses, sample stage, detectors, and display devices. For the surface analysis of thin films, mainly atomic force microscopy (AFM) is used. AFM uses a cantilever with a very sharp tip to scan over a sample surface. As the tip approaches the surface, the close-range, attractive force between the surface and the tip cause the cantilever to deflect toward the surface. A laser beam is used to detect cantilever deflections toward or away from the surface. The raised and lowered features on the sample surface influence the deflection of the cantilever, which is monitored by the position-sensitive photo diode (PSPD).

The main disadvantages of electron microscopes are that it is much expensive to build and maintain. Microscopes designed to achieve high resolutions must be housed in stable buildings with special services such as magnetic field cancelling systems, ultrahigh vacuum for perfect working condition, etc.

17.1.4 MORPHOLOGY-DEPENDENT OPTICAL AND ELECTRICAL ANALYSES OF METAL OXIDE NANOSTRUCTURES

The synthesis and characterization of semiconductor metal oxide nanostructure is a major research topic in materials research.[25–29] Semiconductor metal oxide nanomaterials have attracted wide attention due to their striking optical and electrical properties, which make these materials potentially suitable for applications in electronics, optics, and photonics. For examples, high-sensitivity sensors, high-performance electronic devices, nonlinear applications, storage devices, solar cell fabrications, fuel cells, optical limiters, antibacterial activity, etc. A few metal oxides have been synthesized in wide variety of morphologies such as nanowires, nanorods, nanotubes, nanorings, nanosphere, nanoflakes, nanoflowers, and nanobelts. Synthesis and investigation of these metal oxide nanostructures are beneficial not only for the understanding of the fundamental phenomena in nanoscale but also for development of new generation nanodevices with high performances. Researchers are still continuing their research in metal oxide because of many potential applications at nanometer scale of these materials. The structural characters of metal oxide nanostructures such as lattice symmetry, cell parameters, surface morphology, etc. have been changed in nanometer size range. Some of important semiconductor metal oxide nanostructures are ZnO, CuO, TiO_2, MnO_2, CdO, Fe_2O_3, MgO, SnO_2, etc.

It is well known that the crystalline phase, particle sizes, and surface morphologies of nanomaterials have immense influence on their optical, magnetic, and electrical properties. Accordingly, the controlled synthesis of nanostructured materials with novel morphologies has recently received much attention[25,26] In the recent years, many researchers have focused on synthesis of various metal oxide semiconductor nanomorphotypes[26-29] However, cost-effective single-step production of various nanomorphotypes with high yield and reproducibility are still remaining an important problem. The electrical and optical properties of metal oxides have been studied on account of their particle size, crystalline phase, chemical composition, synthesis route, etc. Unfortunately, very little work is available on the study of dependency of structural, electrical, and optical property of metal oxides upon their surface morphology. Here, we analyze the morphology-dependent optical and electrical properties of metal oxide nanostructures on surface morphology.

The optical analysis is essential for determining the electronic band structure of a material. The electronic band structure would lighten to the optoelectronic capability of nanostructures. The UV-visible spectrum helps to find out the band edge energy of these nanomaterials. Though band gap is a material property, variation in band gap of nanomaterials can be observed due to change in particles size dimensions. Thomas et al. observed a clear picture of change in band gap of CdO nanostructure owing to particle size and surface morphology.[65] They presented variation of band gap of nanomaterial originated by the size induced and morphology assisted quantum confinement effect on different morphologies of CdO nanostructures. The photoluminescence analysis helps to understand the energy gap absorption and subsequent photon emission of the materials. There are numerous studies on the photoluminescence emission spectra of metal oxides and changes in the band edge emission spectral behavior with excitation wavelength as well as morphology. [34-38] The evolution of the excitation wavelength-dependent PL emission line is observed in crystalline ZnO nanomaterials and suggested that every PL peak corresponds to an individual excitation wavelength.[44]

Ellingson et al. observed evolution of excitation energy-dependent efficiency of charge carrier relaxation and photoluminescence in colloidal InP quantum dots.[39, 44] Micic et al. carried out a study on size-dependent PL spectroscopy of InP quantum dots and suggest the emission and absorption features shift to higher energy with decreasing quantum dots size.[40-42, 45] Similarly in the case of polymer nanocomposite materials, fine excitation energy-dependent photoluminescence property is observed. For example, Hairong Zhang et al. presented excitation-wavelength-dependent photoluminescence of a pyromellitic diimide (PMDI) nanowires network and suggest that the luminescence peaks of PMDI nanowires red-shifted as the excitation wavelength gets increased.[46] Ghoshal et al. presents the synthesis of CdO nanostructures in hexagonal sheet and rods structures.[30] Fabrication of CdO microspheres through Ostwald ripening method was developed by Wang et al.[31] Barakat et al. illustrated the photoluminescence and optical characterization of CdO nanorods.[32] Jia et al. fabricated porous CdO nanostructures such as nanowires, nanobelts, nanorods, and one-dimensional hierarchical structures.[33]

Nonlinear optics is one of the leading areas of optoelectronic research in nanomaterial for its importance in optical switching, optical limiting, photothermal cancer therapy, and sensor and eye protection applications.[47,48] But very rare literatures are available for the nonlinear optical properties of nanostructures compared with surface morphology. Chang et al. elucidates the optical limiting property of CdO nanowires in ethanol and water suspensions and concluded that the optical limiting property is due to nonlinear scattering.[49] Thomas and coworkers presented the morphology-dependent nanosecond and ultrafast optical limiting properties of CdO nanostructures in their recent literature.[66,67] They suggest that the nonlinear optical behavior is closely dependent on variation of surface energy of different nanomorphologies of CdO nanostructures and also shows the difference of surface energy of nanostructure upon different morphologies of nanostructures through BET surface area analysis.

Singh et al.[50] reveal the comparative study of optical limiting properties as a function of size in iron oxide nanoparticles and indicate that the prevailing mechanism for the optical limiting in iron oxide nanoparticles is nonlinear scattering. Many researchers have studied the optical limiting property of nanostructures with various morphotypes. For example, Pan et al. presented the nonlinear analysis of Pd, Cu, Ni, Pt, Ag, and Co nanowires.[51] Chang et al. and Venkatram et al. conducted the study on optical limiting property of CdS nanowires and C_{60} TPY-Pb nanowires.[52,53]

Over the last decades, the dielectric and conductivity properties of metal oxide semiconductor materials have caught additional attention because of its colossal dielectric constant values obtained at nanosize range. Due to the large surface to volume ratio and quantum confinement effect, dielectric and conductivity behavior of nanomaterials have been found altered much compared with its bulk materials. In the past decades, many researchers satisfactorily examined electrical transport and dielectric behavior of semiconductor metal oxide nanomaterials with different morphologies. For example, Zhi-Min Liao et al. presented the surface state effect on electron transport in individual ZnO nanowires and confirm that the surface states have greatly affected the electronic transport in single ZnO nanowires.[54] Zhimin Dang and coworkers examined the dielectric properties and morphologies of composites filled with whisker and nanosized ZnO.[55] They claim that dielectric constant and losses increases with increasing w-ZnO content in composites. Sagadevan Suresh conducted the dielectric relaxation behavior of CdS nanoparticles and found that dielectric properties of CdS nanoparticles are found to be significantly enhanced especially in the low-frequency range due to confinement.[56] Velayutham et al. presented the theoretical and experimental approach on dielectric properties of ZnO nanoparticles and polyurethane/ZnO nanocomposites.[57]

Sayed and coworkers carried out a study on synthesis, characterization, optical and dielectric properties of polyvinyl chloride/cadmium oxide nanocomposite films.[58] They concluded that optical and dielectric properties were reinforced by adding the CdO nanoparticles in PVC matrix. Mallikarjuna and team reported the novel high dielectric constant nanocomposites of polyaniline dispersed with Fe_2O_3 nanoparticles and they suggested that conductivity and dielectric constant values are increased by increasing the amount of Fe_2O_3 in the matrix.[59] Srikrishna Ramya and Mahadevan conducted a work on the effect of calcination on the electrical properties and quantum confinement of Fe_2O_3 nanoparticles and reported that dielectric property is strongly dependent to quantum confinement effect.[60] Sahoo et al. studied the characterization of α and γ -Fe_2O_3 nanopowder synthesized by emulsion precipitation–calcination route and also examine the rheological behavior of γ-Fe_2O_3.[61] Reda tests out the electric and dielectric properties of Fe_2O_3/silica nanocomposites and suggested that AC conductivity and dielectric loss of both composites increased gradually with increasing annealing temperature and particle size.[62] Shinde and coworkers fabricated the hematite Fe_2O_3 thin films and examine its application to photoelectron chemical solar cells.[63] Ambika Prasad et al. studied the electrical and sensing properties of polyaniline/iron oxide nanocomposites and reported that prepared nanocomposite has potential application in sensing devices.[64]

17.2 CONCLUSIONS

Out of these literatures, we robustly observed that nanostructures morphology is a dependable parameter for change in optical and electrical properties of metal oxide nanostructures. The real mechanism for these property variations are created by the change in morphology of nanostructures which generates large variation in both surface to volume ratio and confinement effect on nanostructures.

KEYWORDS

- nanotechnology
- nanomaterials
- surface morphology

REFERENCES

1. Roucoux, A.; Schulz, J.; Patin, H. *J. Chem. Rev.* **2002**, *102*, 3757.
2. Lewis, L. N. *Chem. Rev.* **1993**, *93*, 2693.
3. Niemeyar, C. M. *Angew Chem. Int. Ed.* **2001**, *40*, 4128.
4. Niemeyar, C. M. *Angew. Chem. Int. Ed.* **2003**, *42*, 5734.
5. Parak, W. J.; Pellegrino, T.; Plank, C. *Nanotechnology* **2005**, *16*, 9.
6. Hagfeldt, A.; Graetzel, M. *Acc. Chem. Res.* **2000**, *33*, 269.
7. Fitchner, M. *Adv. Eng. Mater.* **2005**, *7*, 443.
8. Moran, C. E; Steele, J. M.; Halas, N. *J. Nano Lett.* **2004**, *4*, 1497.
9. Simon, U. *Nanoparticles: From Theory to Application*; Schmid, G., Ed.; Wiley-VCH: Weinheim, Germany, 2004.
10. Maier, S. A., et al. *J. Adv. Mater.* **2001**, *13*, 1501.
11. Matejivic, E. *Annu. J. Rev. Mater. Sci.* **1985**, *15*, 483.
12. Shull, R. D.; McMichael, R. D.; Swartzendruber, L. J.; Benett, L. H. *Studies of Magnetic Properties of Fine Particles and Their Relevance to Material Science*; Pormann, J. J., Fiorani, D., Eds.; Elsevier Publishers: Amsterdam, 1992; p 161.
13. Heath, J. R.; Kuekes, P. J.; Snider, G. S.; Williams, R. S. *Science* **1998**, *280*, 1716.
14. Andres, R. P., et al. *J. Mater. Res.* **1989**, *4*, 704.
15. Roco, M. C.; Williams, R. S.; Alivisatos, P. (Eds.) *Nanotechnology Research Directions; Vision for Nanotechnology R and D in the Next Decade*; Interagency Working Group in Nanoscience Engineering and Technology [IWGN] Workshop Report, Int. Tech. Research Institutes: WTEC Division, Loyala College, Maryland, USA, 1999.
16. Koper, O. B.; Lagadic, I.; Volodin, A.; Klabunde, K. *J. Chem. Mater.* **1997**, *9*, 2468.

17. Chung, S. W.; Yu, J. Y.; Heath, J. R. *J. Appl. Phys. Lett.* **2000,** *76,* 2068.
18. Cuscó, R.; Ibáñez, J.; Domenech-Amador, N.; Artús, L.; Zúñiga-Pérez, J.; Muñoz-Sanjosé, V. *J. Appl. Phys.* **2010,** *107,* 063519.
19. Reddy, S.; Kumara Swamy, B. E.; Chandra, U.; Sherigara, B. S.; Jayadevappa, H. *Int. J. Electro Chem. Sci.* **2010,** *5,* 10.
20. Gregory, D. S.; Garry, R. *Nat. Mater. Rev. Artic.* **2006,** 5.
21. Smith, C.; Binks, D. *Nanomaterials* **2014,** *4,* 19.
22. Meulenberg, R. W., et al. *ACS Nano* **2009,** *3,* 325.
23. Norris, D. J.; Bawendi, M. G. *Phys. Rev. B* **1996,** *53*(24), 16338.
24. Brus, L. E. *J. Chem. Phys.* **1983,** *79*(11), 5566.
25. Eskizeybek, V.; Avci, A.; Chhowalla, M. *J. Cryst. Res. Technol.* **2011,** *10,* 1093.
26. Guo, Z.; Li, M.; Liu, J. *J. Nanotechnol.* **2008,** *19,* 245611.
27. Wang, Y. W.; Liang, C. H.; Wang, G. Z.; Gao, T.; Wang, S. X.; Fan, J. C. *J. Mater. Sci. Lett.* **2001,** *20,* 1687.
28. Saghatforoush, L. A.; Sanati, S.; Mehdizadehn, R.; Hasanzadeh, M. *J. Superlattices Microstruct.* **2012,** *52,* 885.
29. Balu, A. R.; Nagarethinam, V. S.; Suganya, M.; Arunkumar, N.; Selvan, G. *J. Electron Dev.* **2012,** *12,* 739.
30. Ghoshal, T.; Kar, S.; Chaudhuri, S. *J. Appl. Surf. Sci.* **2007,** *253,* 7578.
31. Wang, W.-S.; Zhen, L.; Xu, C.-Y.; Shao, W. Z. *J. Phys. Chem. C* **2008,** *112,* 14360.
32. Barakat, A. M. N.; Al-Deyab, S.; Kim, H. Y. *J. Mater. Lett.* **2012,** *66,* 225.
33. Jia, Z.; Tang, Y.; Luo, L.; Li, B. J. *Crys. Growth Des.* **2008,** *8,* 2713.
34. Wang, Y. W.; Liang, C. H.; Wang, G. Z.; Gao, T.; Wang, S. X.; Fan, J. C. *J. Mater. Sci. Lett.* **2001,** *20,* 1687.
35. Al-Kuhaili, M. F.; Saleem, M.; Durrani, S. M. A. *J. Alloys Compd.* **2012,** *52,* 1781821.
36. Ma, M.; Zhang, Y.; Guo, Z.; Gu, N. *Nanoscale Res. Lett.* **2013,** *8,* 16.
37. Jubb, A. M.; Allen, H. C. *Appl. Mater. Interfaces* **2010,** *2,* 10.
38. Nasibulin, A. G.; Rackauskas, S.; Jiang, H.; Tian, Y.; Reddy, P. M.; Shandakov, S. D.; Nasibulina, L.; Sainio, J.; Kauppinen, E. I. *Nano Res.* **2009,** *2,* 373.
39. Liou, H.-W.; Lin, H.-M.; Hwu, Y. K.; Chen, W.-C.; Liou, W.-J.; Lai, L. C.; Lin, W. S.; Chiou, W. A. *J. Biomater. Nanobiotechnol.* **2010,** *50,* 60.
40. Hirano, T.; Oku, T.; Suganuma, K. *Diam. Relat. Mater.* **2000,** *9,* 476479.
41. Chaudhari, N. K.; Kim, H. C.; Son, D.; Yu, J.-S. *Cryst. Eng. Comm.* **2009,** *2,* 264.
42. De Montferrand, C.; Hu, L.; Milosevic, I.; Russier, V.; Bonnin, D.; Motte, L.; Brioude, A.; Lalatonne, Y. *Acta Biomater.* **2013,** 7.
43. Zhang, W. C.; Wu, X. L.; Chen, H. T.; Zhu, J.; Huang, G. S. *J. Appl. Phys.* **2008,** *103,* 093718.
44. Ellingson, R. J.; Blackburn, J. L.; Yu, P.; Rumbles, G.; Micic, O. I.; Nozik, A. J. *J. Phys. Chem. B* **2002,** *106,* 7758.
45. Micic, O. I.; Cheong, H. M.; Fu, H.; Zunger, A.; Sprague, J. R.; Mascarenhas, A.; Nozik, A. J. *J. Phys. Chem. B.* **1997,** *101,* 4904.
46. Zhang, H.; Xu, X.; Ji, H.-F. *J. Chem Comm.* **2010,** *46,* 1917.
47. Khatei, J.; Suchand Sandeep, C. S.; Philip, R.; Koteswara Rao, K. S. R. *J. Appl. Phys. Lett.* **2012,** *100,* 081901.
48. Barik, A. R.; Adarsh, K. V.; Naik, R.; Suchand Sandeep, C. S.; Philip, R.; Zhao, D.; Jain, H. *J. Appl. Phys. Lett.* **2011,** *98,* 201111.
49. Chang, Q., et al. *Int. J. Photo Energy* **2012,** *10,* 1155//857345.

50. Singh, C. P.; Bindra, K. S.; Bhalerao, G. M.; Oak, S. M. *Opt. Express* **2008,** *16*, 12.

51. Pan, H.; Chen, W.; Feng, Y. P.; Ji, W. *J. Appl. Phys. Lett.* **2006,** 88, 223106.

52. Chang, Q.; Chang, C.; Zhang, X.; Ye, H.; Shi, G.; Zhang, W.; Wang, Y.; Xin, X.; Song, Y. *J. Opt. Commun.* **2007,** *274*, 201.

53. Venkatram, N.; Narayana Rao, D.; Akundi, M. A. *J. Opt. Express* **2005,** *13*, 867.

54. Liao, Z.-M., et al. *Phys. Lett. A* **2007,** *376*, 207.

55. Dang, Z., et al. *Mater. Res. Bull.* **2003,** *38*, 499.

56. Sagadevan, S. *Appl. Nanosci.* **2014,** *4*, 325.

57. Velayutham, T. S., et al. *J. Appl. Phys.* **2012,** *112*, 054106.

58. El-Sayed, M. A. *Acc. Chem. Res.* **2001,** *34*, 257.

59. Mallikarjuna, N. N.; Manohar, S. K.; Kulkarni, P. V.; Venkataraman, A.; Aminabhavi, T. M. *J. Appl. Polym. Sci.* **2005,** *97*, 1868.

60. Srikrishna Ramya, S. I.; Mahadevan, C. K. *Int. J. Res. Eng. Technol.* **2014,** *3*, 570.

61. Sahoo, S. K.; Agarwal, K.; Singh, A. K.; Polke, B. G.; Raha, K. C. *Int. J. Eng. Sci. Technol.* **2010,** *2*, 118.

62. Reda, S. M. *Int. J. Nano Sci. Technol.* **2013,** *1*, 17.

63. Shinde, S. S., et al. *J. Semicond.* **2011,** *32*, 013001.

64. Ambika Prasad, M. V. N., et al. *Int. J. Eng. Res. Appl.* **2014,** *4*,198.

65. Thomas, P.; Abraham, K. E. *J. Lumin.* **2015,** *158*, 422–427.

66. Thomas, P.; Sreekanth, P.; Abraham, K. E. *J. Appl. Phys.* **2015,** *117*, 053103.

67. Thomas, P.; Sreekanth, P.; Philip, R.; Abraham, K. E. *J. RSC Adv.* **2015,** *5*, 35017.

CHAPTER 18

MECHANISM OF Ni(Fe)ARD ACTION IN METHIONINE SALVAGE PATHWAY, IN BIOSYNTHESIS OF ETHYLENE, AND ROLE OF TYR-FRAGMENT AS REGULATORY FACTOR

LUDMILA I. MATIENKO*, LARISA A. MOSOLOVA,
VLADIMIR I. BINYUKOV, ELENA M. MIL, and GENNADY E. ZAIKOV

Emanuel Institute of Biochemical Physics, Russian Academy of Sciences, 4, Kosygin Str., Moscow 119334, Russia

Corresponding author. E-mail: matienko@sky.chph.ras.ru

CONTENTS

ABSTRACT

Role of Ni(Fe)-macrostructures due to H-bonds in mechanisms of Ni(Fe) ARD action in methionine salvage pathway, in biosynthesis of ethylene, is discussed. The AFM method was used for research of possibility of the stable supramolecular nanostructures formation based on Ni(Fe)ARD model system {Ni^{II}(acac)$_2$ + MP + Tyr} (Tyr = L-Tyrosine)—with the assistance of intermolecular H-bonds. Using UV-spectroscopy, we received the confirmation of the of triple complexes {Ni(acac)$_2$·MP·Tyr} formation.

18.1 INTRODUCTION

The methionine salvage pathway (MSP) (Scheme 18.1) plays a critical role in regulating a number of important metabolites in prokaryotes and eukaryotes. MSP is a ubiquitous pathway found in plants, animals, and bacteria. Methylthioadenosine (MTA) is the first intermediate in this pathway and is formed from S-adenosyl methionine (SAM) during polyamine synthesis in animals and ethylene synthesis in plants (Scheme 18.2). Polyamine is required for cell growth and proliferation, and ethylene is required for ripening of fruits and vegetables. MTA is an inhibitor of both polyamine synthesis and reactions of transmethylation. Inhibition of polyamine synthesis arrests DNA replication, and elevated polyamine is associated with tumor formation. Hence, the concentration of MTA in cells must be tightly regulated. The MSP controls the concentration of MTA by returning it through a series of reactions to methionine, thereby "salvaging" the thiomethyl group of SAM. Acireductone dioxygenases (ARDs) Ni(Fe)-ARD are enzymes involved in the methionine recycle pathway. Fe-ARD catalyze the penultimate step in the pathway, the oxidative decomposition of substrate acireductone (1,2-dihydoxy-3-keto-5-(thiomethyl)pent-1-ene) to formate and 2-keto-4-(thiomethyl)butyrate (KMTB), the keto acid precursor of methionine. The purpose of the off-pathway reaction catalyzed by Ni-ARD is unknown. These represent the only known pair of naturally occurring metalloenzymes with distinct chemical and physical properties determined solely by the identity of the metal ion in the active site.[1]

Both enzymes Ni(Fe)-ARD are members of the structural super family, known as cupins, which also include Fe-acetyl acetone dioxygenase (Dke1)[2,3] and cysteine dioxygenase. These family of cupins use a triad of histidine-ligands (His), and also one or two oxygens from water and a carboxylate oxygen (Glu), for binding with Fe (Ni)-center.[2] Being the members of structural super family of cupins, the Ni(Fe)-ARDs present the unusual case of catalysis, as differ in the mechanism of action in relation to general substrates (1,2-dihydroxy-3-oxo-5 (methylthio)pent-1-ene (acireductone) and dioxygen).

In recent years, the studies in the field of homogeneous catalytic oxidation of hydrocarbons with molecular oxygen were developed in two directions, namely, the free-radical chain oxidation catalyzed by transition metal complexes and the catalysis by metal complexes that mimic enzymes.[4,5] The findings on the mechanism of action of enzymes, and, in particular, dioxygenases and their models, are very useful in the treatment of the mechanism of catalysis by heteroligand Ni(Fe) complexes in the processes of oxidation of hydrocarbons with molecular oxygen in our works.[4–6] Moreover, as one will see below, the investigation of the mechanism of catalysis by metal complexes that modeled the enzymes actions can give the necessary material for the study of the mechanism of action of enzymes.

We have offered the new approach to research of mechanism of homogenous catalysis, and the mechanism of action of enzymes also.[7,8] The first time we have successfully used the method of atomic force microscopy (AFM) to study the possibility of formation of supramolecular nanostructures, based on heteroligand nickel and iron complexes, which are selective catalysts on the one hand and also are models of dioxygenases on the other hand, due to the intermolecular hydrogen bonds.[7,8] Namely, the complexes $Ni_2(acac)(OAc)_3 \cdot MP \cdot 2H_2O$ ("A"), $Fe^{III}_x(acac)_y 18C6_m(H_2O)_n$ ("B"), $\{Ni(acac)_2 \cdot L^2 \cdot PhOH\}$ ("C") are effective catalysts of selective ethyl benzene oxidation to α-phenyl ethyl hydro peroxide and also are structure and functional models of dioxygenases Ni(Fe)-ARD (A–C) and Fe^{II}-Dke1 (B). We assumed that the stability of the complexes A–B as the alkylarenes oxidation catalysts could be related to formation of the stable supramolecular structures due to the intermolecular hydrogen bonds. And specific activities of Ni(Fe)-ARD toward common substrates (acireductone and dioxygen) in synthesis and reproduction of methionine (and ethylene) as one of the reasons—with self-organization into various macrostructures due to intermolecular hydrogen bonds. These assumptions are supported by our AFM research outlined in this chapter.

In this chapter, we discuss the possible role of Tyr-fragment in mechanism of Ni(ARD) dioxygenase actions, based on experience data that we received at the first time with AFM and UV-spectroscopy on model systems.

18.2   MATERIALS AND METHODOLOGY

AFM SOLVER P47/SMENA/ with Silicon Cantilevers NSG11S (NT MDT) with curvature radius 10 nm, tip height: 10–15 μm, and cone angle ≤22° in taping mode on resonant frequency 150 kHz was used.[7,8]

As substrate, a special chemically modified polished silicone surface was used.

Waterproof modified silicone surface was exploit for the self-assembly-driven growth due to H-bonding of complexes $Fe^{III}_x(acac)_y18C6_m(H_2O)_n$, $Ni_xL^1_y(L^1_{ox})_z(L^2)_n(H_2O)_m$, $\{Ni^{II}(acac)_2 \cdot L^2 \cdot PhOH\}$ (L^2 = MP (N-metylpyrrolidone-2), HMPA, MSt), systems $\{Ni^{II}(acac)_2 + \textbf{MP} + Tyr\}$ and $\{Ni^{II}(acac)_2\}$ + Tyr} (Tyr = L-Tyrosine) with silicone surface. The saturated chloroform $(CHCl_3)$ or water solutions of complexes was put on a surface, maintained for some time, and then solvent was deleted from a surface by means of special method—spin-coating process.

In the course of scanning of investigated samples, it has been found that the structures are fixed on a surface strongly enough due to H-bonding. The self-assembly-driven growth of the supramolecular structures on modified silicone surface on the basis of researched complexes, due to H-bonds and perhaps the other noncovalent interactions, was observed.

Method of UV-spectroscopy we used first to prove the possible regulatory role of Tyr-fragment (the formation of triple complexes $Ni(acac)_2 \cdot MP \cdot Tyr$).

18.3   RESULTS AND DISCUSSION

18.3.1   *THE MECHANISM OF FORMATION OF HIGH EFFECTIVE CATALYSTS, HETEROLIGAND NI (OR FE) COMPLEXES, IN THE HYDROCARBON OXIDATIONS WITH DIOXYGEN: THE ROLE OF H-BONDS*

In our works, we have modeled efficient catalytic systems $\{ML^1_n + L^2\}$ (M = Ni, Fe, L^1 = acac⁻, L^2 are crown ethers or quaternary ammonium salts, different electron-donating modifying extraligands) for selective ethylbenzene oxidation to α-phenyl ethyl hydro peroxide, that was based on the established (for Ni complexes) and hypothetical (for Fe complexes) mechanisms of formation of catalytically active species and their operation.[4,5] The high activity of systems $\{ML^1_n + L^2\}$ is associated with the fact that during the ethylbenzene oxidation, the active primary $(M^{II}L^1_2)_x(L^2)_y$ complexes and heteroligand $M^{II}_xL^1_y(L^1_{ox})_z(L^2)_n(H_2O)_m$ complexes are formed to be involved in the oxidation process.

We established mechanism of formation of high effective catalysts, heteroligand complexes $M^{II}_xL^1_y(L^1_{ox})_z(L^2)_n(H_2O)_m$. The axially coordinated electron-donating ligand L^2 controls the formation of primary active complexes ML^1_2, L^2 and the subsequent reactions of β-diketonate ligands in the outer coordination sphere of these complexes. The coordination of an electron-donating extraligand L^2 with an $M^{II}L^1_2$ complex, favorable for stabilization of the transient zwitter-ion $L^2[L^1M(L^1)^+O_2^-]$, enhances the probability of regioselective O_2 addition to the methine C–H bond of an acetylacetonate ligand activated by its coordination with metal ions. The outer-sphere reaction of O_2 incorporation into the chelate ring depends on the nature of the metal and the modifying ligand L^2.[4,5] Thus, formation of nickel complexes $Ni^{II}_xL^1_y(L^1_{ox})_z(L^2)_n$, as a result of the reaction of oxygenation of ligand $L^1 = acac^-$ in $Ni^{II}(acac)_2$, follows a mechanism analogous to those of Ni^{II}-containing ARD[1] or Cu- and Fe-containing quercetin 2,3-dioxygenases.[9,10] Namely, incorporation of O_2 into the chelate acac ring was accompanied by the proton transfer and the redistribution of bonds in the transition complex leading to the scission of the cyclic system to form a chelate ligand $L^1_{ox} = OAc^-$, acetaldehyde, and CO (in the Criegee rearrangement).

In the effect of iron(II) acetylacetonate complexes $Fe^{II}_xL^1_y(L^1_{ox})_z(L^2)_n$, we have found an analogy with the action of Fe^{II}-ARD[1] or Fe^{II}-acetyl acetone dioxygenase (Dke1).[3] For iron complexes oxygen adds to C–C bond (rather than inserts into the C=C bond as in the case of catalysis with nickel(II) complexes) to afford intermediate, that is, an Fe complex with a chelate ligand containing 1,2-dioxetane fragment. The process is completed with the formation of the $(OAc)^-$ chelate ligand and methylglyoxal as the second decomposition product of a modified acac ring (as it has been shown in [3]).

High effectivity of catalytic complexes $M^{II}_xL^1_y(L^1_{ox})_z(L^2)_n$, (M = Ni, Fe, $L^1 = acac^-$, $L^1_{ox} = OAc^-$, $L^2 = $ crown ethers or quaternary ammonium salts), which are formed in the process of selective oxidation of ethylbenzene to PEH at catalysis with primary complexes $(M^{II,III}L^1_n)_x(L^2)_y$ seems to be associated with the formation of stable supramolecular structures due to intermolecular H-bonds.

One of the most effective catalytic systems of the ethylbenzene oxidation to the α-phenyl ethyl hydroperoxide are the triple systems.[4–6] Namely, the phenomenon of a substantial increase in the selectivity (S) and conversion (C) of the ethylbenzene oxidation to the α-phenyl ethyl hydro peroxide upon addition of PhOH together with ligands N-metylpyrrolidone-2 (MP), hexamethylphosphorotriamide (HMPA) or alkali metal stearate MSt (M = Li, Na) to metal complex $Ni^{II}(acac)_2$ was discovered in works Matienko and Moso-lova.[4-6] The role of intramolecular H-bonds was established by us in mecha-nism of formation of triple catalytic complexes $\{Ni(II)(acac)_2 \cdot L^2 \cdot PhOH\}$ (L^2 = MP) in the process of ethylbenzene oxidation with molecular oxygen.[5,6] The formation of triple complexes $Ni^{II}(acac)_2 \cdot L^2 \cdot PhOH$ from the earliest stages of oxidation was established with *kinetic methods*.[4-6] We assumed that the stability of complexes $Ni(acac)_2 \cdot L^2 \cdot PhOH$ in the process of ethyl benzene oxidation can be associated as one of reasons with the supramo-lecular structures formation due to intermolecular H-bonds (phenol–carboxy-late) (see below) and, possible, the other noncovalent interactions:

$$\{Ni^{II}(acac)_2 + L^2 + PhOH\} \rightarrow Ni(acac)_2 \cdot L^2 \cdot PhOH \rightarrow \{Ni(acac)_2 \cdot L^2 \cdot PhOH\}_n$$

In favor of formation of supramolecular macrostructures based on the triple complexes $\{Ni(acac)_2 \cdot L^2 \cdot PhOH\}$ (and $Ni(acac)_2 \cdot MP \cdot Tyr$ (Tyr = Tyrosine-L)) in the real systems of homogeneous (and enzymatic catalysis (see below)), show data of AFM and UV-spectroscopy.

18.3.2 ROLE OF SUPRAMOLECULAR NANOSTRUCTURES FORMED DUE TO H-BONDING, IN MECHANISMS OF CATALYSIS, MODELS OF NI(FE)ARD DIOXYGENASES: ROLE OF TYR-FRAGMENT

Hydrogen bonds vary enormously in bond energy from ~15–40 kcal/mol for the strongest interactions to less than 4 kcal/mol for the weakest. It is proposed, largely based on calculations, that strong hydrogen bonds have more covalent character, whereas electrostatics are more important for weak hydrogen bonds, but the precise contribution of electrostatics to hydrogen bonding is widely debated.[11] Hydrogen bonds are important in noncovalent aromatic interactions, where π-electrons play the role of the proton acceptor, which are a very common phenomenon in chemistry and biology. They play an important role in the structures of proteins and DNA, as well as in drug–receptor binding and catalysis.[12]

Proton-coupled bicarboxylates top the list as the earliest and still the best studied systems suspected of forming low-barrier hydrogen bonds (LBHBs) in the vicinity of the active sites of enzymes.[12] Proton-coupled bicarboxylates appear in 16% of all protein X-ray structures. There are at least five X-ray structures showing short (and therefore strong) hydrogen bonds between an enzyme carboxylate and a reaction intermediate or transition state analogue bound at the enzyme active site. The authors[12] consider these structures to be the best de facto evidence of the existence of LBHBs stabilizing high-energy reaction intermediates at enzyme active sites. Carboxylates figure prominently in the LBHB enzymatic story in part because all negative charges on proteins are carboxylates.

The porphyrin linkage through H-bonds is the binding type generally observed in nature. One of the simplest artificial self-assembling supramolecular porphyrin systems is the formation of a dimer based on carboxylic acid functionality.[13]

18.3.2.1   THE POSSIBLE ROLE OF THE SELF-ASSEMBLING SUPRAMOLECULAR MACROSTRUCTURES IN MECHANISM OF ACTION OF ARDS NI(FE)-ARD INVOLVED IN THE METHIONINE RECYCLE PATHWAY

Structural and functional differences between the two Ni-ARD and Fe-ARD enzymes are determined by the type of metal ion bound in the active site of the enzyme.[14]

Ni-ARD and Fe-ARD act on the same substrate, the acireductone, 1,2-dihydroxy-3-keto-5-methylthiopentene anion, but they yield different products. Fe-ARD catalyzes a 1,2-oxygenolytic reaction, yielding formate and 2-keto-4-methylthiobutyrate, a precursor of methionine (and ethylene), and thereby part of the MSP, while Ni-ARD catalyzes a 1,3-oxygenolytic reaction, yielding formate, carbon monoxide, and 3-methylthiopropionate, an off-pathway transformation of the acireductone. At that the role of a reaction catalyzed by the enzyme Ni-ARD, still not clear[14,15] (see Schemes 18.1 and 18.2).

Pochapsky et al. showed that this dual chemistry can also occur in mammals (MmARD) (before Ni(Fe)ARD activity was discovered by Pochapsky et al. for bacteria and plants). It has been established, that the Fe-bound protein, which shows about 10-fold higher activity than that of Ni-protein, catalyzes on-pathway chemistry, whereas the Ni-ARD forms exhibit off-pathway chemistry.[15]

Interestingly, that in the case of KoARD (*Klebsiella oxytoca* ARD, bacterial enzyme ARD) unlike MmARD, Ni-bound form has higher activity than Fe-KoARD.[1] All forms remain monomeric regardless of bound metal ion. While both Fe- and Ni-ARD from *Klebsiella oxytoca* are monomers,[15] Fe-ARD from *Oryza sativa L* (OsARD) is a trimer, and Ni-bound OsARD is a polymer consisting of several types of oligomers.[15]

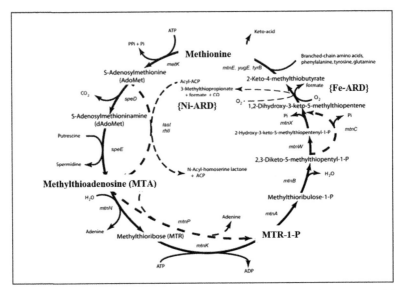

SCHEME 18.1   Acireductone dioxygenases Ni-ARD and Fe-ARD[14] are involved in the methionine recycle pathway.

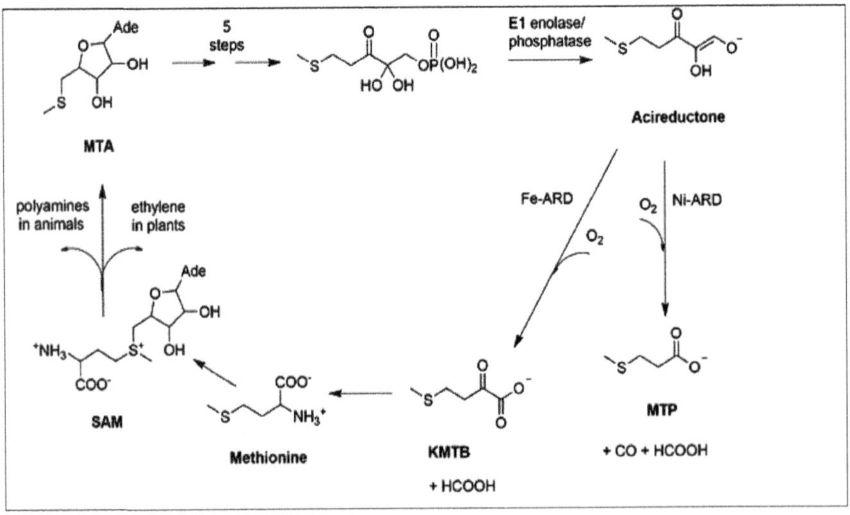

SCHEME 18.2   Acireductone dioxygenases Ni-ARD and Fe-ARD in methionine salvage pathway, in biosynthesis of ethylene.[15]

We assumed that one of the reasons for the different activity of $Ni^{II}(Fe^{II})$-ARD in the functioning of enzymes in relation to the common substrates (acireductone and O_2) can be the association of catalyst in various macrostructure due to intermolecular H-bonds.

We assumed that the $Fe^{II}ARD$ operation comprises the step of oxygen activation.

$(Fe^{II} + O_2 \rightarrow Fe^{III}–O_2^-·)$ (by analogy with Dke1 action).[3] Specific structural organization of iron complexes may facilitate the following regioselective addition of activated oxygen to acireductone ligand and the reactions leading to formation of methionine (and ethylene also). Association of the catalyst in macrostructures with the assistance of the intermolecular H-bonds may be one of reasons of reducing $Ni^{II}ARD$ activity in mechanisms of $Ni^{II}(Fe^{II})ARD$.[14]

Earlier, we demonstrated a specific structural organization of functional models of iron (nickel) enzymes $Ni^{II}(Fe^{II})ARD$. We used an AFM method to research the possibility of the formation of stable supramolecular nanostructures based on iron (nickel) heteroligand complexes due to intermolecular H-bonds.[7,8]

So, in Figures 18.1 and 18.2, three-dimensional and two-dimensional AFM image of the structures on the basis of iron complex with 18C6 $Fe^{III}_x(acac)_y18C6_m(H_2O)_n$, formed at putting a uterine solution on a hydrophobic surface of modified silicone are presented. It is visible that the generated structures are organized in certain way forming structures resembling the shape of tubule microfiber cavity (Fig. 18.2c). The heights of particles are about 3–4 nm. In control experiments, it was shown that for similar complexes of nickel $Ni^{II}(acac)_2·18C6·(H_2O)_n$ (as well as complexes $Ni_2(OAc)_3(acac)·MP·2H_2O$) this structural organization is not observed. It was established that these iron constructions are not formed in the absence of the aqueous environment. Earlier we showed the participation of H_2O molecules in mechanism of $Fe^{III,II}_x(acac)_y18C6_m(H_2O)_n$ transformation by analogy with Dke1 action, and also the increase in catalytic activity of iron complexes $(Fe^{III}_x(acac)_y18C6_m(H_2O)_n$, $Fe^{II}_x(acac)_y18C6_m(H_2O)_n$, and $Fe^{II}_xL^1_y(L^1_{ox})_z(18C6)_n(H_2O)_m)$ in the ethyl benzene oxidation in the presence of small amounts of water.[16]

After our works, it was found that the possibility of decomposition of the β-diketone in iron complex by analogy with Fe-ARD action increases in aquatic environment.[17] That apparently is consistent with data, published by us earlier in our initial work.[16]

Unlike catalysis with Fe-ARD, mechanism of catalysis by Ni-ARD does not include O_2 activation, and oxygenation of acireductone leads to the formation of products not being precursors of methionine.[1,14,15] In our previous works, we have shown that formation of multidimensional forms based on nickel complexes can be one of the ways of regulating the activity of two enzymes.[7]

The association of complexes $Ni_2(AcO)_3(acac) \cdot MP \cdot 2H_2O$, which are functional and structure models of Ni-ARD, to supramolecular nanostructure due to intermolecular H-bonds $[H_2O–MP, H_2O–(OAc^-)]$ [or $(acac^-)$], is demonstrated on the next Figure 18.3. All structures (Fig. 18.3) have various heights from the minimal 3–4 nm to ~20–25 nm for maximal values (in the form reminding three almost merged spheres)[7]

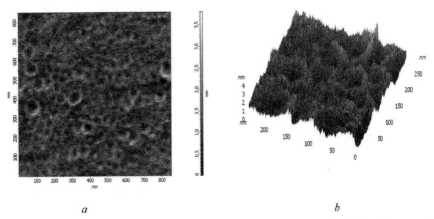

a   *b*

FIGURE 18.1   The AFM two-dimensional (a) and three-dimensional (b) images of nanoparticles on the basis $Fe_x(acac)_y 18C6_m(H_2O)_n$ formed on the surface of modified silicone.

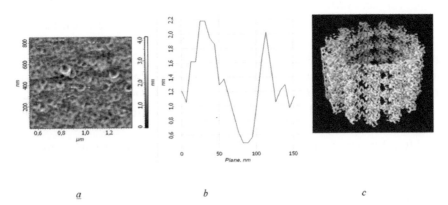

a   *b*   *c*

FIGURE 18.2   The AFM two-dimensional image (a) of nanoparticles on the basis $Fe_x(acac)_y 18C6_m(H_2O)_n$ formed on the hydrophobic surface of modified silicone, (b) the section of a circular shape with fixed length and orientation is about 50–80 nm, and (c) the structure of the cell microtubules.

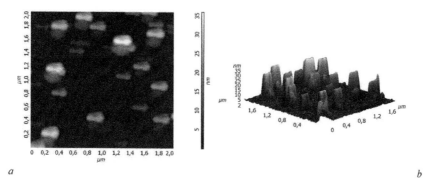

FIGURE 18.3 The AFM two-dimensional (a) and three-dimensional (b) images of nanoparticles on the basis $Ni_2(AcO)_3(acac)\cdot L^2\cdot 2H_2O$ formed on the hydrophobic surface of modified silicone.

In case of binary complexes $\{Ni(acac)_2\cdot MP\}$ (see below Fig. 18.9c), we also observed formation of nanostructures due to H-bonds. But these nanoparticles differ on form and are characterized with less height: $h \sim 8$ nm as compared with nanostructures based on complexes $Ni_2(AcO)_3(acac)\cdot L^2\cdot 2H_2O$ (Fig. 18.3).

18.3.2.2 POSSIBLE EFFECT OF TYR-FRAGMENT, BEING IN THE SECOND COORDINATION SPHERE OF METAL COMPLEX

We assume that it may be necessary to take into account the role of the second coordination sphere, including Tyr-fragment (see Fig. 18.4).[14] We first suggest the participation of Tyrosine moiety in mechanisms of action of Ni(Fe)-ARD enzymes.

It is known that Tyr residues are located in different regions of protein by virtue of the relatively large phenol amphiphilic side chain capable of (a) interacting with water and participating in H-bond formation and (b) undergoing cation-π and nonpolar interactions.[18] The versatile physicochemical properties of Tyrosine allow it to play a central role in conformation and molecular recognition.[19]

Tyrosine can participate in different enzymatic reactions. Recently, it has been researched role of Tyrosine residue in mechanism of heme oxygenase (HO) action. HO is responsible for the degradation of a histidine-ligated ferric protoporphyrin IX (Por) to biliverdin, CO, and the free ferrous ion. The role of reactions of Tyrosyl radical formation which occurs after oxidation of

Fe(III)(Por) to Fe(IV) = O(Por(·+)) in mechanism of human heme oxygenase isoform-1 (hHO-1) and the structurally homologous protein from *Coryne-bacterium diphtheriae* (cdHO) is described.[20]

It assumed that Tyr-fragment may be involved in substrate H-binding in step of O_2 activation by iron catalyst, and this can decrease the oxygenation rate of substrate in the case of homoprotocatechuate 2,3-dioxygenase action.[21]

Tyr-fragment is discussed as important in methyl group transfer from S-adenosylmethionine (AdoMet) to dopamine.[22] The experimental findings with the model of methyltransferase and structure survey imply that methyl CH–O hydrogen bonding (with participation of Tyr-fragment) represents a convergent evolutionary feature of AdoMet-dependent methyltransferases, mediating a universal mechanism for methyl transfer.[23]

Tyrosine residue Tyr149 is found in the Met-turn for astacin endopepti-dases and serralisines. Tyr149 giving a proton, forms a hydrogen bond with zinc and becomes the fifth ligand. This switch plays a specific role, partici-pating in the stabilization of the transition state during the binding of the substrate to the enzyme.[24]

In the case of Ni-ARD, Tyr-fragment, involved in mechanism, can reduce the Ni-ARD activity. The structure of the active center of Ni-ARD with Tyr residue in second coordination sphere is shown in Figure 18.4.

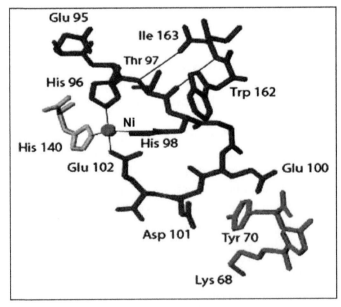

FIGURE 18.4 The structure of Ni[II]ARD with Tyr residue in the second coordination sphere.[14]

Really, as mentioned above we have found[5] that the inclusion of PhOH in complex Ni(acac)$_2$·L^2 (L^2 = N-methylpirrolidone-2), which is the primary model of Ni^{II}ARD, leads to the stabilization of formed triple complex Ni(acac)$_2$•L^2•PhOH. In this case, as we have emphasized above, ligand (acac)$^-$ is not oxygenated with molecular O$_2$. Also the stability of triple complexes Ni(acac)$_2$•L^2•PhOH seems to be due to the formation of supramolecular macrostructures that are stable to oxidation with dioxygen. Formation of supramolecular macrostructures due to intermolecular (phenol–carboxylate) H-bonds and, possible, the other noncovalent interactions[25–27] based on the triple complexes Ni(acac)$_2$·L^2·PhOH, that we have established with the AFM-method[7,8,28] (in the case of L^2 = MP, HMPA, NaSt, LiSt), is in favor of this hypothesis (Fig. 18.5).

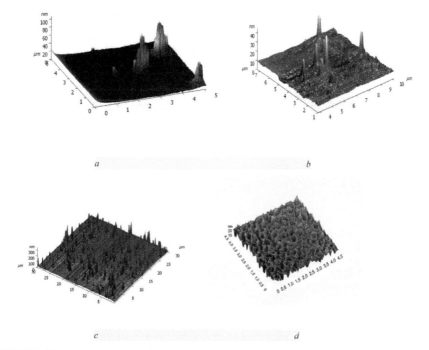

FIGURE 18.5 (a) The AFM three-dimensional image [5.0 × 5.0 (μm)] of the structures (h ~ 80–100 nm) formed on a surface of modified silicone on the basis of triple complexes Ni^{II}(acac)$_2$·MP·PhOH, (b) AFM three-dimensional image [6.0 × 6.0 (μm)] of the structures (h ~ 40 nm) formed on a surface of modified silicone on the basis of triple complexes {Ni^{II}(acac)$_2$·HMPA·PhOH}, (c) AFM three-dimensional image [30 × 30 (μm)] of the structures (h ~ 300 nm) formed on a surface of modified silicone on the basis of triple complexes Ni^{II}(acac)$_2$·NaSt·PhOH, and (d) AFM three-dimensional image [4.5 × 4.5 (μm)] of the structures (h ~ 10 nm) formed on a surface of modified silicone on the basis of triple complexes Ni^{II}(acac)$_2$·LiSt·PhOH.

Conclusive evidence in favor of the participation of tyrosine fragment in stabilizing primary Ni- complexes as one of regulatory factors in mechanism of action of Ni-ARD has been obtained by AFM. We observed first the formation of nanostructures based on Ni-systems using L-Tyrosine (Tyr) as an extraligand. The growth of self-assembly of supramolecular macrostructures due to intermolecular (phenol–carboxylate) H-bonds and, possible, the other noncovalent interactions,[25–27] based on the triple systems {Ni(acac)$_2$ + MP + Tyr}, we observed at the apartment of a uterine H$_2$O solution of triple system {Ni(acac)$_2$ + MP + Tyr} on surfaces of modified silicon (Fig. 18.6). Spontaneous organization process, that is, self-organization, of researched triple complexes (Figs. 18.5 and 18.6) at the apartment of a uterine solution of complexes on surfaces of modified silicon are driven by the balance between intermolecular and molecule-surface interactions, which may be the consequence of hydrogen bonds and the other noncovalent interactions.[29]

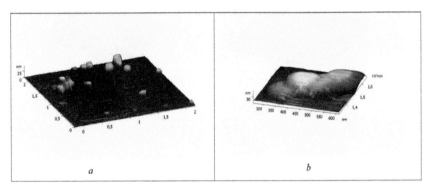

FIGURE 18.6 The AFM three-dimensional image [2.0 × 2.0 (μm)] of the structures ($h \sim$ 25 nm) (a) and three-dimensional image [0.3 × 0.6 (μm)] of the structures ($h \sim$ 50 nm) (b), formed on a surface of modified silicone on the basis of triple systems {Ni^{II}(acac)$_2$ + MP + Tyr}.

Histogram of volumes of the particles based on systems {Ni^{II}(acac)$_2$ + MP + Tyr}, and also the empirical and theoretical cumulative normal probability distribution of volumes, and the empirical and theoretical cumulative log-normal distribution of volumes of the particles based on systems {Ni^{II}(acac)$_2$ + MP + Tyr}, formed on the surfaces of modified silicon, are presented in Figure 18.7. As can be seen, distribution of volumes of the particles in this case is well described by a log-normal law.

But as one can see in Figure 18.8, in case of binary systems {Ni(acac)$_2$ + Tyr}, we also observed formation of nanostructures due to H-bonds. But these nanoparticles as well as particle based on {Ni(acac)$_2$·MP} complexes

(Fig. 18.8c) differ in form and high from the nanostructures on the basis of triple systems {Ni(acac)$_2$ + MP + Tyr} (Fig. 18.6).

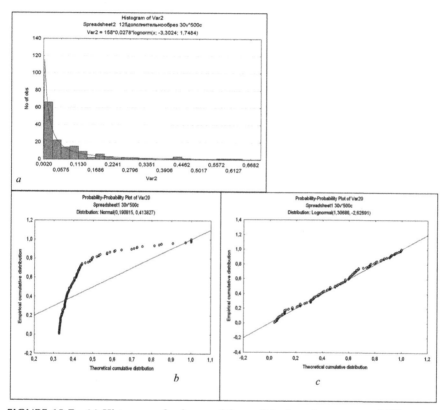

FIGURE 18.7 (a) Histogram of volumes of the particles based on systems {Ni^{II}(acac)$_2$ + MP + Tyr}, (b) the empirical and theoretical cumulative normal distribution of volumes of the particles based on systems {Ni^{II}(acac)$_2$ + MP + Tyr}, and (c) the empirical and theoretical cumulative log-normal distribution of volumes of the particles based on systems {Ni^{II}(acac)$_2$ + MP + Tyr}.

Here, first the formation of complexes {Ni(acac)$_2$·MP Tyr} was confirmed by electron absorption spectra (method of UV-spectroscopy).

As can be seen from Figure 18.9, when an aqueous solution of Tyr is added to the Ni(acac)$_2$ aqueous solution, a decrease in absorption intensity of acetylacetonate ion (acac)$^-$ (λ_{max} = 296 nm) and a small short-wavelength shift of the absorption maximum (to λ_{max} ~ 294 nm) (hypsochromic shift of the absorption maximum) takes place. A similar change in the intensity of the (acac)$^-$ absorption band is observed in the absorption spectra of Ni(acac)$_2$

when it is coordinated with monodentate ligand MP, or crown-ether 18C6, and in the case of the coordination of axial monodentate ligands with the other metal acetylacetonates.[5]

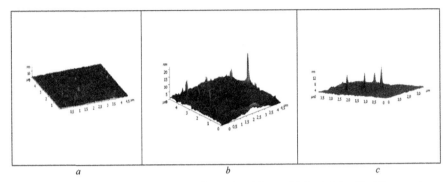

FIGURE 18.8 (a), (b) The AFM three-dimensional image [4.5 × 4.5 (μm)] of the structures [$h \sim$ 10–15 (20) nm] based on binary systems {Ni^{II}(acac)$_2$ + Tyr} formed on a surface of modified silicone. (c) The AFM of three-dimensional image [4.0 × 4.0 (μm)] of nanoparticles based on {Ni(acac)$_2$·MP} formed on the surface of modified silicone. Figure 18.8 shows our original data published for the first time.

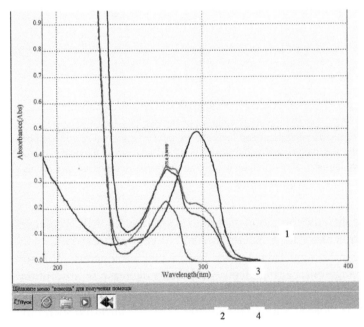

FIGURE 18.9 Electron absorption spectra of aqueous solutions: (1) Ni(acac)$_2$, (2) Tyr, (3) {Ni(acac)$_2$ + Tyr}, and (4) {Ni(acac)$_2$ + Tyr + MP} at 20°C.

With the introduction of the third component MP, there was an additional reduction of the absorption maximum, indicating MP coordination[5] with {Ni(acac)$_2$·Tyr} and triple complex {Ni(acac)$_2$·MP·Tyr} formation.

At that, we observed the growth of max of absorption spectra of Tyr with small bathochromic shift (from λ_{max} ~ 274 nm to λ_{max} ~ 280, 275 nm). It is known, that phenol absorption band (λ_{max} ~ 275 nm), associated with π–π^* transition, can undergo a bathochromic shift, for example, due to the outer sphere interaction of π-system of PhOH with π-inner-ligand systems (at the outer sphere complex OSC formation).[30]

The data shown in Figure 18.10 (sum of the spectra of Tyr and (acac)$^-$ ligand with respect to the spectrum of a mixture {Ni^{II}(acac)$_2$ + MP + Tyr} (Fig. 18.10a), as well as the result of subtraction of Tyr spectrum from spectrum of mixture {Ni^{II}(acac)$_2$ + MP + Tyr} (Fig. 18.10b) also favor the formation of triple complexes {Ni(acac)$_2$·MP·Tyr}.

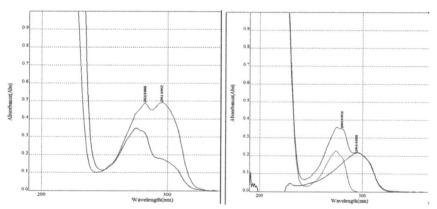

FIGURE 18.10 (a) Sum of the electron absorption spectra of aqueous solutions of Tyr (spectrum 2, Fig. 18.9) and (acac)$^-$ ligand (spectrum 1, Fig. 18.9) with respect to the spectrum of a mixture {Ni^{II}(acac)$_2$ + MP + Tyr} (spectrum 4, Fig. 18.9). (b) The result of subtraction of Tyr spectrum (2, Fig. 18.9) from spectrum of mixture {Ni^{II}(acac)$_2$ + MP + Tyr} (4, Fig. 18.9).

It should be noted that the absorption at λ = 295 nm does not disappear completely, as it takes place in the presence of CH$_3$COOH acid by displacing the acac$^-$ ligand to outer coordination sphere of the Ni complex that be accompanied by appearance of (acac)$^-$ absorption band at λ = 275 nm.[31] So received experimental data cannot be explained by coordination of the carboxyl group of tyrosine with the nickel ion. So apparently, the formation of triple complexes {Ni(acac)$_2$·MP·Tyr} is due to coordination by OH– group of tyrosine.

At the same time, it is necessary to mean that important function of Ni^{II}ARD in cells is established now. Namely, carbon monoxide (CO) is formed as a result of action of nickel-containing dioxygenase Ni^{II}ARD. It was established that CO is a representative of the new class of neural messengers, and seems to be a signal transducer such as nitrogen oxide, NO.[1,14,15]

18.4 CONCLUSION

Usually in the quest for axial modifying ligands that control the activity and selectivity of homogeneous metal complex catalysts, the attention of scientists is focused on their steric and electronic properties. The interactions in the outer coordination sphere, the role of hydrogen bonds and also the other noncovalent interactions are less studied.

We have assumed that the high stability of heteroligand $M^{II}_x L^1_y (L^1_{ox})_z$ $(L^2)_n (H_2O)_m$ (M = Ni, Fe, L^1 = acac⁻, L^1_{ox} = OAc⁻, L^2 = electron-donating mono, or multidentate activating ligands) complexes as selective catalysts of the ethylbenzene oxidation to PEH, formed during the ethylbenzene oxidation in the presence of {$ML^1_n + L^2$} systems as a result of oxygenation of the primary complexes $(M^{II}L^1_2)_x(L^2)_y$, can be associated with the formation of the supramolecular structures due to the intermolecular H-bonds.

The supramolecular nanostructures on the basis of catalytic active iron $Fe^{III}_x(acac)_y 18C6_m(H_2O)_n$ and nickel complexes $Ni^{II}_x L^1_y (L^1_{ox})_z (L^2)_n (H_2O)_m$ (L^1 = acac⁻, L^1_{ox} = OAc⁻, L^2 = N-methylpirrolidone-2, x = 2, y = 1, z = 3, n = 1, m = 2), $Ni_2(AcO)_3(acac)\cdot L^2 \cdot 2H_2O$; {$Ni(acac)_2 \cdot L^2 \cdot PhOH$} ($L^2$ = MP, HMPA, NaSt, LiSt), obtained with AFM method, indicate high probability of supramolecular structures formation due to H-bonds in the real systems, namely, in the processes of alkylarens oxidation.

Since the investigated complexes are structural and functional models of $Ni^{II}(Fe^{II})$ARD dioxygenases, the data could be useful in the interpretation of the action of these enzymes.

Specific structural organization of iron complexes may facilitate the first step in Fe^{II}ARD operation: O_2 activation and following regioselective addition of activated oxygen to acireductone ligand (unlike mechanism of regioselective addition of no activated O_2 to acireductone ligand in the case of Ni^{II}ARD action), and reactions leading to formation of methionine.

The formation of multidimensional forms (in the case of Ni^{II}ARD) may be one way of controlling $Ni^{II}(Fe^{II})$ARD activity.

We assumed the participation of Tyr-fragment which is in the second coordination sphere in mechanism of $Ni^{II}(Fe^{II})$ARD operation, as one of possible mechanisms of reduction in enzyme activity in $Ni^{II}(Fe^{II})$ARD enzymes operation, and we received experimental facts in favor this assumption. So with AFM method, we observed first the formation of supramolecular macrostructures based on triple systems {Ni(acac)$_2$ + MP + Tyr} that included L-Tyrosine as extraligand, formed on a surface of modified silicone due to intermolecular (phenol–carboxylate) H-bonds and, possible, the other noncovalent interactions. Earlier, self-assembly based on the triple complexes Ni(acac)$_2$·L^2·PhOH, we observed with the AFM-method for L^2 = MP, HMPA, NaSt, LiSt).

In this **chapter**, using method of UV-spectroscopy, we received the confirmation of the of triple complexes {Ni(acac)$_2$·MP·Tyr} formation.

KEYWORDS

- **models of Ni(Fe)ARD dioxygenases**
- **dioxygen**
- **AFM method**
- **nanostructures based on model triple systems {Ni^{II}(acac)$_2$ + MP + Tyr}**
- **triple complexes {Ni(acac)$_2$·MP·Tyr} with UV-spectroscopy**

REFERENCES

1. Dai, Y.; Pochapsky, T. C.; Abeles, R. H. Mechanistic Studies of Two Dioxygenases in the Methionine Salvage Pathway of *Klebsiella pneumonia*. *Biochemistry* **2001**, *40*, 6379–6387.

2. Leitgeb, S.; Straganz, G. D.; Nidetzky, B. Functional Characterization of an Orphan Cupin Protein from *Burkholderia xenovorans* Reveals a Mononuclear Non-heme Fe^{2+}-dependent Oxygenase that Cleaves β-diketones. *FEBS J.* **2009**, *276*, 5983–5997.

3. Straganz, G. D.; Nidetzky, B. Reaction Coordinate Analysis for β-diketone Cleavage by the Non-Heme Fe^{2+}-dependent Dioxygenase Dke 1. *J. Am. Chem. Soc.* **2005**, *127*, 12306–12314.

4. Matienko, L. I. Solution of the Problem of Selective Oxidation of Alkylarenes by Molecular Oxygen to Corresponding Hydro Peroxides. Catalysis Initiated by Ni(II), Co(II), and Fe(III) Complexes Activated by Additives of Electron-donor Mono or Multidentate

Extra-ligands. In *Reactions and Properties of Monomers and Polymers*; D'Amore, A., Zaikov, G., Eds.; Nova Science Publ. Inc: New York, 2007; pp 21–41.

5. Matienko, L. I.; Mosolova, L. A.; Zaikov, G. E. *Selective Catalytic Hydrocarbons Oxidation. New Perspectives*; Nova Science Publ. Inc: New York, 2010; p 150.

6. Matienko, L. I.; Binyukov, V. I.; Mosolova, L. A. Mechanism of Selective Catalysis with Triple System {bis(acetylacetonate)Ni(II) + metalloligand + phenol} in Ethylbenzene Oxidation with Dioxygen. Role of H-bonding Interactions. *Oxid. Commun.* **2014**, *37*, 20–31.

7. Matienko, L. I.; Binyukov, V. I.; Mosolova, L. A.; Mil, E. M.; Zaikov, G. E. *The New Approach to Research of Mechanism Catalysis with Nickel Complexes in Alkylarens Oxidation Polymer Yearbook 2011;* Nova Science Publ. Inc: New York, 2012; pp 221–230.

8. Matienko, L. I.; Binyukov, V. I.; Mosolova, L. A.; Mil, E. M.; Zaikov, G. E. Supramolecular Nanostructures on the Basis of Catalytic Active Heteroligand Nickel Complexes and Their Possible Roles in Chemical and Biological Systems. *J. Biol. Res.* **2012**, *1*, 37–44.

9. Gopal, B.; Madan, L. L.; Betz, S. F.; Kossiakoff, A. A. The Crystal Structure of a Quercetin 2,3-Dioxygenase from *Bacillus subtilis* Suggests Modulation of Enzyme Activity by a Change in the Metal Ion at the Active Site(s). *Biochemistry* **2005**, *44*, 193–201.

10. Balogh-Hergovich, E.; Kaizer, J.; Speier, G. Kinetics and Mechanism of the Cu(I) and Cu(II) Flavonolate-catalyzed Oxygenation of Flavonols, Functional Quercetin 2,3-Dioxygenase Models. *J. Mol. Catal. A: Chem.* **2000**, *159*, 215–224.

11. Ma, J. C.; Dougherty, D. A. The Cation–π Interaction. *Chem. Rev.* **1997**, *97*, 1303–1324.

12. Graham, J. D.; Buytendyk, A. M.; Wang, Di.; Bowen, K. H.; Collins, K. D. Strong, Low-Barrier Hydrogen Bonds May Be Available to Enzymes. *Biochemistry* **2014**, *53*, 344–349.

13. Beletskaya, I.; Tyurin, V. S.; Tsivadze, A. Y.; Guilard, R. R.; Stem, C. Supramolecular Chemistry of Metalloporphyrins. *Chem. Rev.* **2009**, *109*, 1659–1713.

14. Chai, S. C.; Ju, T.; Dang, M.; Goldsmith, R. B.; Maroney, M. J.; Pochapsky, T. C. Characterization of Metal Binding in the Active Sites of Acireductone Dioxygenase Isoforms from *Klebsiella* ATCC 8724. *Biochemistry* **2008**, *47*, 2428–2435.

15. Deshpande, A. R.; Wagenpfail, K.; Pochapsky, T. C.; Petsko, G. A.; Ringe, D. Metal-Dependent Function of a Mammalian Acireductone Dioxygenase. *Biochemistry* **2016**, *55*, 1398–1407.

16. Matienko, L. I.; Mosolova, L. A. The Modeling of Catalytic Activity of complexes Fe(II,III)(acac)$_n$ with R$_4$NBr or 18-crown-6 in the Ethylbenzene Oxidation by Dioxygen in the Presence of Small Amounts of H$_2$O. *Oxid. Commun.* **2010**, *33*, 830–844.

17. Allpress, C. J.; Grubel, K.; Szajna-Fuller, E.; Arif, A. M.; Berreau, L. M. Regioselective Aliphatic Carbon–Carbon Bond Cleavage by Model System of Relevance to Iron-containing Acireductone Dioxygenase. *J. Am. Chem. Soc.* **2013**, *135*, 659–668.

18. Radi, R. Protein Tyrosine Nitration: Biochemical Mechanisms and Structure Basis of Functional Effects. *Acc. Chem. Res.* **2013**, *46*, 550–559.

19. Koide, S.; Sidhu, S. S. The Importance of Being Tyrosine: Lessons in Molecular Recognition from Minimalist Synthetic Binding Proteins. *ACS Chem. Biol.* **2009**, *4*, 325–334.

20. Smirnov, V. V.; Roth, J. P. Tyrosine Oxidation in Heme Oxygenase: Examination of Long-range Proton-coupled Electron Transfer. *J. Biol. Inorg. Chem.* **2014**, *19*, 1137–1148.

21. Mbughuni, M. M.; Meier, K. K.; Münck, E.; Lipscomb, J. D. Substrate-mediated Oxygen Activation by Homoprotocatechuate 2,3-Dioxygenase: Intermediates Formed by a Tyrosine 257 Variant. *Biochemistry* **2012**, *51*, 8743–8754.

22. Zhang, J.; Klinman, J. P. Enzymatic Methyl Transfer: Role of an Active Site Residue in Generating Active Site Compactio that Correlates with Catalytic Efficiency. *J. Am. Chem. Soc.* **2011**, *133*, 17134–17137.

23. Horowitz, S.; Dirk, L. M. A.; Yesselman, J. D.; Nimtz, J. S.; Adhikari, U.; Mehl, R. A.; Scheiner S.; Houtz, R. L.; Al-Hashimi, H. M.; Trievel, R. C. Conservation and Functional Importance of Carbon–Oxygen Hydrogen Bonding in AdoMet-Dependent Methyltransferases. *J. Am. Chem. Soc.* **2013**, *135*, 15536–15548.

24. Bond, J. S.; Beynon, R. J. The Astacin Family of Metalloendpeptidases. *Protein Sci.* **1995**, *4*, 1247–1261.

25. Dubey, M.; Koner, R. R.; Ray, M. Sodium and Potassium Ion Directed Self-assembled Multinuclear Assembly of Divalent Nickel or Copper and L-Leucine Derived Ligand. *Inorg. Chem.* **2009**, *48*, 9294–9302.

26. Basiuk, E. V.; Basiuk, V. V.; Gomez-Lara, J.; Toscano, R. A. A Bridged High-spin Complex bis-[Ni(II)(rac-5,5,7,12,12,14-hexamethyl-1,4,8,11-tetraazacyclotetradecane)]-2,5-pyridinedicaboxylate Diperchlorate Monohydrate. *J. Incl. Phenom. Macrocycl. Chem.* **2000**, *38*, 45–56.

27. Mukherjee, P.; Drew, M. G. B.; Gómez-Garcia, C. J.; Ghosh, A. (Ni_2), (Ni_3), and $(Ni_2 + Ni_3)$: A Unique Example of Isolated and Cocrystallized Ni_2 and Ni_3 Complexes. *Inorg. Chem.* **2009**, *48*, 4817–4825.

28. Matienko, L.; Binyukov, V.; Mosolova, L.; Zaikov, G. The Selective Ethylbenzene Oxidation by Dioxygen into α-phenyl Ethyl Hydroperoxide, Catalyzed with Triple Catalytic System {$Ni^{II}(acac)_2$ + NaSt(LiSt) + PhOH}. Formation of Nanostructures {$Ni^{II}(acac)_2 \cdot NaSt \cdot (PhOH)$}$_n$ with Assistance of Intermolecular H-bonds. *Polym. Res. J.* **2011**, *5*, 423–431.

29. Gentili, D.; Valle, F.; Albonetti, C.; Liscio, F.; Cavallini, M. Self-organization of Functional Materials in Confinement. *Acc. Chem. Res.* **2014**, *47*, 2692–2699.

30. Nekipelov, V. M.; Zamaraev, K. I. Outer-sphere Coordination of Organic Molecules to Electric Neutral Metal Complexes. *Coord. Chem. Rev.* **1985**, *61*, 185–240.

31. Matienko, L.I.; Maizus, Z. K. Mechanism of Self-inhibition of Oxidation Processes Catalyzed by Nickel Compounds. *Kinet. Catal.* **1974**, *15*, 317–322 (in Russian).

INDEX